INTRODUCTION TO COMPUTER-ASSISTED EXPERIMENTATION

INTRODUCTION TO COMPUTER-ASSISTED EXPERIMENTATION

Kenneth L. Ratzlaff
Director, Instrumentation Design Laboratory
Department of Chemistry
University of Kansas
Lawrence, Kansas

A WILEY-INTERSCIENCE PUBLICATION

JOHN WILEY & SONS
NEW YORK   CHICHESTER   BRISBANE   TORONTO   SINGAPORE

Library of Congress Cataloging in Publication Data:

Ratzlaff, Kenneth L.
    Introduction to computer-assisted experimentation.

    "A Wiley-Interscience Publication."
    Includes index.
    1. Physical laboratories—Data processing.
2. Microcomputers.   I. Title.

QC51.A1R38 1987   502.8'5   86-19011
ISBN 0-471-86525-7

Printed in the United States of America

10 9 8 7 6 5 4 3 2

To my wife, Virginia, for her love and unfailing support, and to our children, Michael, Jonathan, and Rebekah

Preface

During the past several years, a profound change has taken place in the scientific laboratories of universities, industries, government labs, and other research centers. The personal computer has arrived. This machine has the computing power rivaling that of many mainframes in centralized facilities a little more than a decade ago. However, the cost is much *less* than that of a typical scientific instrument. Even more important, it is becoming increasingly "friendly." No longer does computer operation require specialized training. The mystique of the "computer jock" is finally disappearing.

Computers demonstrate a wide range of capabilities. On the one hand, they can be and have been used for our detriment. Computers are popularly known for their ability to act on behalf of "Big Brother." It is sobering also to note that a primary impetus in their development has been support of the tools of war; indeed, some of today's most challenging computer problems exist in technologies which threaten world peace, such as the "Star Wars" initiative.

On the other hand, computers also are aids to efforts that can help bring about peace and improvements in mankind's lot. They are indispensable aids to modern fundamental research; educators find them to be valuable assistants; and they can take over much of the drudgery associated with repetitive tasks. The computer is, for most scientists, a positive tool for research with positive benefits. It is hoped that this book will assist those studies.

Despite the small computer's relative "friendliness," efficient and efficacious use of the small computer is not effortless. To make the greatest and most effective use of the small computer, some study is required. The objective of this book is to present a fundamental introduction to the use of the small computer in the laboratory.

In preparing this book, a few assumptions are made abut the reader:

- Some familiarity with the scientific laboratory in the form of either professional practice or some study in an experimental science, such as a few lab courses in chemistry, physics, biology, or a related field
- An understanding of electricity such as that taught in high school physics
- Some experience in programming a computer [however, the language of that experience (FORTRAN, BASIC, Pascal, or even Logo) or the computer-type (mainframe or microcomputer) matters little. For the most part, examples are written in BASIC, even though the understanding of BASIC is not assumed.]

This book attempts to work from the top down. That is, the beginning chapters deal with the larger picture, defining the small computer and its place in the lab. Early in the book (Chapter 4), interfacing techniques are introduced at a level sufficient to satisfy many needs.

Ensuing chapters become somewhat more specialized. Chapters 5 and 6 present an operational approach to modern electronics, while Chapters 7 and 8 detail specialized transducers for sensing and controlling real-world phenomena. Communications, graphics, and computational methods, in Chapters 9 through 11, are presented with the laboratory computer in mind even though the subjects are not limited to lab applications, but are encountered by most small computer users. General topics related to the organization of a project round out Chapter 12.

A serious attempt has been made to limit the amount of jargon that must be learned as compared to what is needed for understanding other manuals and guides. A poster behind my word-processing computer shows Albert Einstein with the quotation "Everything should be made as simple as possible, but not simpler," and that has been a guide.

After studying this book, a scientific investigator should understand what types of problems a small lab computer can handle and how they are solved. Readers needing technical details concerning such topics as the advanced techniques of computer interface design, machine language programming, and numerical analysis are referred to advanced references.

Having gained the background provided here, the investigator should be able to use a small computer in the laboratory following one of two pathways. In many cases, simple interfacing tasks can be handled using only the information found in this text. On the other hand (and possibly of wider significance), a better understanding of laboratory computers should enable discussion of research directions involving computers with the electronics or data-processing personnel. The informed scientist will have the ability to communicate with them without being intimidated by strange concepts and unintelligible jargon.

Finally, the most desirable objective is that the investigator will come to treat the small computer as a useful, nonthreatening, and even enjoyable tool. With the barriers to the acceptance of computers dismantled, the computer can become a positive and challenging laboratory aid.

KENNETH L. RATZLAFF

Lawrence, Kansas
January 1987

Acknowledgments

The preparation of this book has been a most interesting challenge. There are many people who, knowingly or unknowingly, have made it possible. Expression of my appreciation is a pleasant task.

Christopher Gunn (Kansas University Graphics Lab) has been an expert editor, and the comments of Dr. Eugene Ratzlaff (IBM Instruments) and Professor John Walters (St. Olaf's College) have been invaluable.

Charlotte Pauls prepared most of the artwork, endeavoring to work within a schedule that was not her own. My colleague Tom Peters has been the source of new ideas and help; the Instrumentation Design Lab's steering committee has encouraged the effort.

Dr. Jean Johnson, Mathematics Department, University of Kansas, developed the tables in Appendix Two.

The most important are those who provided the nontechnical support. Jean, Francis, Audrey, Joe, Bob, Patrice, Jim, Carey, and Jean of the Lawrence Mennonite Fellowship have particularly helped to keep me on track. But most of all, my wife, Virginia, has given unfailing support, encouraging me and freeing me from shared family duties to spend time at the word processor, and Michael, Jonathan, and Rebekah (our children) have adjusted, especially during the final push. I express my sincere appreciation and gratitude to all of you.

K.L.R.

Contents

INTRODUCTION TO
COMPUTER-ASSISTED
EXPERIMENTATION

CHAPTER ONE

Introduction

1.1. INTRODUCTION

The reader of this book is likely to be engaged in science, as student or practitioner, with the knowledge that the small computer is having a profound impact on laboratory-based experimentation. For the scientist who would be inclined to read the book from the beginning, the place of the laboratory computer in the universe of computers is not necessarily obvious. That understanding can be useful in defining the scope of this study.

At the very outset, a *preliminary* definition of what will be considered to be a laboratory computer is necessary. We will consider it to be a machine that resides physically in or adjacent to the laboratory, has a limited number of users at any given time (usually one, but sometimes a few more), is programmable in one or more programming languages, and is capable of passing information to and from an experiment.

Furthermore, we want to learn to view the computer as "friendly"; that is, the computer should relate to its scientist users largely on their terms without the need for the user to undergo training for the status of "computer jockey." Indeed, over the past decade, the manufacturers of computers have changed the way that they view the computer user. People who use small computers are no longer expected to be computer professionals.

A parallel transformation also has occurred in the way society views the computer. This transformation is reflected in two important popular films, *2001, A Space Odyssey* and *Star Wars*. In the former, a single *highly centralized* computer, known as HAL, was capable of controlling many aspects of the lives and existence of the astronaut-protagonists until the computer finally was disabled; in HAL, we saw the artistic extension of the computer that instructed us not to "fold, spindle, or mutilate": that is, to work on the computer's terms. However, *Star Wars* depicted *distributed* computing power in the form of robots ("'droids") that were universally available to serve the user; the inherent capacity for good and evil remained with the humanoid users. (This view of robots in science fiction novels was spelled out in Isaac Asimov's *I, Robot* but did not become part of popular filmmaking until later.)

The equivalent transformation for the working scientist is equally profound. By bringing computers into the laboratory, the scientist in need of computing power is released from the whims and constraints of the "Computer Center"; control of the computer, to the extent that both the computing power and the split-second time response of the computer are directly available to the experiment, is brought into the laboratory.

1.1.1. Historical Development

One perspective on the laboratory computer's role can be gained by examining the historical development of computing—first in general, and then in the laboratory. Much like a tree (Fig. 1.1), this development has roots in several areas; these roots came together to form a single trunk as computers developed with fantastically increasing power. Although the quest for raw power continues, we will be interested in a fairly recent branch: a quest for the low-cost, easy-to-use power that has made routine laboratory operation feasible.

The first root in this historical tree was a calculating machine: specifically, the Analytical Engine of the Englishman Charles Babbage, designed in the 1830s. Babbage began with a small but clever device that he called the Difference Engine (Fig. 1.2), capable of automatically generating successive values of algebraic functions by the method of finite differences. The sequences were developed by Augusta Ada Lovelace Byron, daughter of poet Lord Byron, often called the first programmer. The machine was mechanical and only a small version was ever constructed successfully. However, that Difference Engine was successfully used for generating mathematical tables for several years.

Babbage, however, turned his attention to the more grandiose Analytical Engine which, although never actually constructed, was designed with the concepts of programmed sequences and punched card input and output (I/O). This made possible greatly increased precision through the sequential steps of multiple precision arithmetic and led to his claim, "I have converted the infinity of space, which was required by the conditions of the problem, into the infinity of time." The difficulty of machining the parts to the required precision led to the abandonment of the Analytical Engine.

A second root is often associated with a loom developed by J. M. Jacquard, patented in 1801. Punched cards were used to determine the sequence

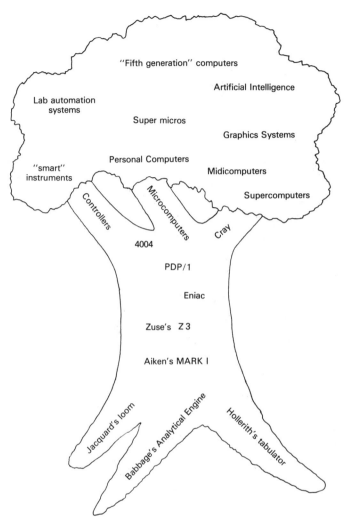

FIGURE 1.1 The historical family tree for computers. The roots lie in the nineteenth century, and the crown includes the present and the near future.

of operations, that is, to program the machine. This required logical decision making and is related to the concept of the servo-mechanism whereby a particular result (that is, the weaving operation) is compared with a control (the card) to determine if a change is necessary. The antitechnology Luddite rebellion developed partially in reaction to the introduction of this machine into England.

The name of Herman Hollerith is associated with a third root, the tabulating machine. The 1880 U.S. census showed that manual tabulation could not be completed within the census period, so Hollerith developed the

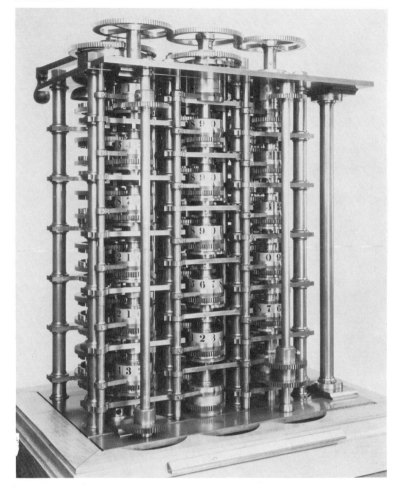

FIGURE 1.2 Part of Charles Babbage's first difference engine, 1832. [Courtesy of "Trustees of the Science Museum" (London).]

punched card machine for the 1890 census as a means of dramatically increasing the efficiency of sorting, tabulating, and storing large amounts of data. Even in the business end of computer development, Hollerith had a major impact since his Tabulating Machine Company became part of a company which led to the formation of International Business Machines (IBM) in 1924. His name is now applied to the field of a line of FORTRAN-language code which contains text.

By the late 1930s, the concept of a general-purpose digital computer was being developed. In Berlin, Konrad Zuse built the Z1 and Z2, storing data in memory constructed from old telephone relays. However, he broke new ground in computer logic with the Z3, which used floating-point representation to store 64 numbers, was program-controlled, and incorporated digital logic; it was the best working computer produced until 1941.

At about this time, Howard Aiken, a physics professor at Harvard, began a collaboration with four IBM engineers that resulted in the creation of the Mark I. It contained more than 3000 relays for data memory and weighed two tons! This machine was immediately put to work to support the U.S. Navy for classified computations.

The next step—to speed up operation limited by mechanical relays—was to introduce electronic computation. Zuse recognized this need but failed to get support from war-torn Germany, and Aiken distrusted electronics. In the absence of progress by these pioneers, the Electronic Numerical Integrator and Calculator (ENIAC) became the next major step. It was completed in 1946 at the University of Pennsylvania's Moore School of Engineering. The ENIAC led the way on a grandiose scale; over 19,000 vacuum tubes were incorporated in the device, consuming 200 KW of power. Although difficult to program, it was more than 3000 times faster in execution than the Mark 1.

The ENIAC, however, was based on the fundamental concepts embodied in the prototype computer of Dr. John V. Atanasoff at Iowa State College (now University), built in 1939. Atanasoff was the first to use the base-2 numbering system, dynamic electronic memory, and calculation by logic rather than enumeration.

For most of the next two decades, the emphasis in large computer development—and there were no other kinds—was on increased speed of operation and increased storage. The driving force for computers capable of handling scientific computations came from the military and aerospace sectors, but the business community also saw the value of computers for data management. Many companies emerged, spurred on by the demand of the consumer.

1.1.2. The Small Computer

In 1960, Digital Equipment Corporation (DEC) released the Programmed Data Processor 1 (PDP-1), the first of what would become known as the minicomputer. In its overall capability, it equalled the best computer of a decade earlier, but at a price of under $100,000.

Then, in 1965, the first computer built on an assembly line was produced, the DEC PDP-8. Tens of thousands of machines in this series were sold. The PDP-8 marked a significant shift in the philosophy of computers. Recall that a decade earlier computational capability was the overriding consideration. The cost and complexity of a computer was much too great to tie it directly to experiments or instruments in the real world, since the computer would then be required to operate in the time frame of actual experimental events (or in real-time, to introduce the jargon term). That would unacceptably require the computer to wait on the experiment or on other real-time events.

The cheaper, more rugged, and smaller minicomputer generated applications in process control, business, communications, and aerospace activities. It also found its way into scientific activities, primarily for the control of sophisticated experiments and instruments. The latter category accounted for little more than 10% of the market, however.

By 1970, there were scores of manufacturers of "minis" selling products in the $15,000 to $200,000 range. Clearly, the market could not support this proliferation of products, and a shakeout occurred, leaving a handful of dominant companies including recent leaders DEC, Data General, and Hewlett–Packard. However, the potential of such a machine for doing more than computing was clear to many scientists; it could take information directly from an experiment and use that information in the control of the experiment.

The development of large scale integration (LSI) of electronic components also was well underway. Powerful calculator circuits were beginning to drive the new and highly competitive hand-held calculator market. In 1968, a Japanese calculator company had approached Intel Corporation, a small electronics company, and asked them to design circuits for a scientific calculator. Engineer Ted Hoff, noting that the company PDP-8 computer's central processing board was less complex than the proposed calculator circuit, suggested that a fully *programmable* unit could do the job. The calculator company was not interested, but Intel management approved the project. The result, in 1971, was the Intel 4004, a 4-bit microprocessor with 2300 transistors, 45 instructions, and the capacity to address 4000 bytes. It quickly found its way into various controllers and timers, and was followed by a variety of new microprocessors with increasing internal complexity and greater capabilities.

By 1978, the Intel 8086 single-integrated circuit microprocessor had 29,000 transistors, 133 instructions, and could address a megabyte. Since then, microprocessor development has substantially surpassed the 8086 in complexity, addressability, and speed; with 32-bit microprocessors coming of age, no end is in sight. Intel, National Semiconductor, and Motorola have emerged as the leaders.

Although possibly not the most sophisticated microprocessor, the Intel 80386 chip is representative of the latest generation. By comparison, it has over 250,000 transistors on a single integrated circuit. The power of the instruction set, the operation speed, and the memory capacity have all been increased, and the need for other circuits to support the operation of the microprocessor has decreased (Fig. 1.3).

Hand-in-hand with microprocessor development came the production of integrated circuits that support the microprocessor: Inexpensive, compact memory circuits enable large capacity in a small unit, while integrated control chips for various peripheral devices (disk drives, graphic output systems, terminals, etc.) provide convenience and efficiency. The fully developed

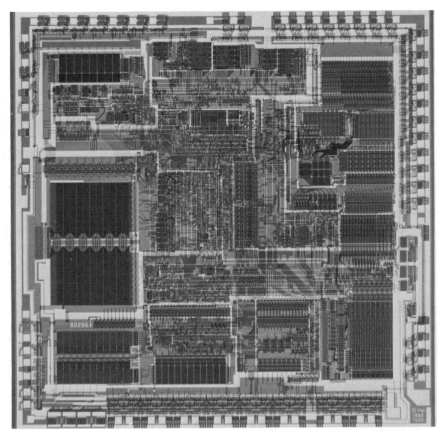

FIGURE 1.3. An Intel 80386 integrated circuit. Over 250,000 transistors are integrated into this silicon chip. (Courtesy Intel Corporation.)

microcomputer system has arrived in time for the challenges of the scientific laboratory.

New central processing units (CPUs) are being developed for data communications, calculations, data acquisition, signal processing, machine and instrument control, among other specialized applications. The CPUs that are used in personal/laboratory computers, however, are usually less specialized.

1.2. THE MODES OF COMPUTER OPERATION

The computer can be used to process the data and control the parameters of a scientific experiment in several ways. If we can avoid excessive rigidity at this stage, three modes of an experiment–computer relationship can be identified. As will become obvious, these modes are linked closely with the historical development described in the previous section.

1.2.1. Off-Line Operation

Schematically represented by Figure 1.4, off-line operation of a computer is a mode in which the computer is utilized almost exclusively as a computational element. Data are collected from the experiment by a manual process, without the aid of the computer. Panel meters, strip-chart recorders, and oscilloscope photos are common readout devices from experiments. Later, the data must be transferred manually to a computer for computations.

For example, consider the study of a transient phenomenon such as a rapidly changing chemical concentration during a reaction. The measurement usually requires that a chemical quantity (e.g., concentration) be changed to an electrical quantity (possibly by using a phototube to measure transmitted light) that can be recorded as a function of time. A permanent graphical record can be made with a strip-chart recorder or an oscilloscope and Polaroid camera. Time and the chemical information are encoded on the recordings as distance. That information is then converted with the aid of a ruler into digital values to be recorded in the lab book.

Typically, a scientific investigator finds that the data analysis model is too complex and/or the data set is too large to be handled by pencil and paper, slide rule, or hand calculator. A machine exists of enormous computational power and capacity to handle these very data sets. In order to use it, the data must be prepared for the computer by reencoding them on punched cards, on tape, or at a keyboard terminal to be delivered to a central site, the "Computer Center."

The term "submit" is used conventionally when programs and data are presented to a large computational facility, and the data are submitted to the computer in several senses of the word. The time interval between submission of this job and receipt of the results (turnaround time) is almost independent of either the computer's execution speed or the user's needs. Rather, it is determined by backlogs, priorities, maintenance schedules, and, some would allege, the phase of the semester, fiscal calendar, or moon.

The results of this computation are returned to the investigator, and, after interpretation, they are applied to the design of the next experiment. This interpretation may simply require reading only a single number (and possibly some statistical information that describes the justifiable confidence in that number) to determine what the new parameters should be. Typically, a plot

FIGURE 1.4. Off-line computer operation. Note that the experimenter is part of the data path.

of the fit of the data to a model is required so that the quality of the fit can be evaluated visually.

In some instances, the efficiency and accuracy of the off-line method have been improved without altering the overall concept. For example, output data are sometimes automatically digitized and recorded on a medium which can be carried or otherwise transmitted to the central computing facility; that procedure avoids the tedium of card punching and the attendant transcription errors.

In any case, the availability of off-line computing was an enormous improvement, in spite of slow and uncertain turnaround, particularly for investigators whose data were to be analyzed with reference to complicated models.

1.2.2. On-Line Operation

The second mode, represented schematically in Figure 1.5, is the principal subject of this book. The position of the researcher has moved now, out of the direct line of the data flow, but in a position to control both experiment and computer. A link exists between experiments and computer that is the electronic interface, the boundary between the digital domain of the computer and the many types of domains in the experiment.

In such a mode, data move directly from experiment to computer, where the data are processed and decisions are made. These decisions are based on the program written by the researcher, on the information entered by the researcher at the time of the experiment, and on the processed data; resulting commands or data can be returned to the experiment for the purposes of control. The "turnaround time" for the control is now in the submillisecond

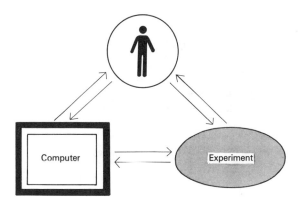

FIGURE 1.5. On-Line computer operation. The experimenter is no longer part of the data path, but can interact with both the computer and the experiment; information moves directly between the experiment and the computer.

regime, so that control can readily take place during the experiment as well as between experiments.

The principal enabling features of the small computer for on-line operation are the reduced size and cost, two factors that allow the computer to be drawn close to the site of the experiment and to be responsible to fewer users. Twenty years ago, when that cost was $50,000 to $150,000, the justification had to be quite compelling. However, both cost and size decreased and continue to drop rapidly to the present $200 to $20,000, so that this mode of operation can be extended to many more types of experiments.

1.2.3. In-Line Operation

The evolution of the computer-controlled system has gone one step further, again as a result of the shrinking size and cost of computers. Manufacturers of instrumentation saw the potential of placing the computer, now a microcomputer, directly *within* the instrument (Fig. 1.6). No longer does one view two entities—the instrument and computer are fully integrated. "Microprocessor-controlled" instrumentation now abounds. Therefore, a researcher can have many of the benefits of on-line computing—chiefly enhanced data processing and real-time control—without having to program the computer on the computer's terms (i.e., in a computer language such as assembler, BASIC, FORTRAN, etc.) and without having to deal with an external electronic interface.

This mode is not without its disadvantages. Not being *required* to program the computer often means not being *able* to program the computer, and frequently the very nature of the research requires that the instrumental measurement is to be carried out in a unique way not envisioned by the designer of the instrument. Furthermore, the exact nature of the integration, curve fit, or other calculation is not always revealed by the manufacturer, particularly when corrections for interferences or nonlinearities are involved. Some important information available to the instrument's microprocessor may simply be unavailable to the interface through the instrument company's oversight.

A good example of this problem is seen in a spectrophotometer marketed by a leading instrument company. A built-in interface makes instrument control simple. One needs only to send the correct code to the instrument,

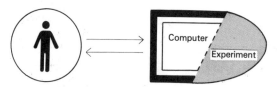

FIGURE 1.6. In-line operation. Information moves directly between the experiment and the computer, but the two are now indistinguishable.

and it responds with almost any action or datum that one could desire. However, an important part of this instrument, an automatic sample sequencer, is completely independent of the interface. Although the same microprocessor controls the sequencer as handles the interface, it is not possible for the computer to tell the instrument when to change samples or for the computer to discern the number of the sample that currently is being studied. The company finds it uneconomical to make the appropriate changes in the software, so that some users may find the advantages of a fine instrument and interface negated by a serious omission.

It should be noted that some progressive vendors, in addition to providing general-purpose "canned" routines, allow some programming in common languages, principally BASIC or Pascal, so that the user may make significant custom enhancements. However, this is still the exception, not the rule.

Although the in-line mode is not within the scope of this book, there are, nevertheless, occasions when it can be a useful approach to be used even for an "in-house-built" computer-controlled instrument, particularly when there are many users of varying computer skills. Furthermore, a commercial computer-controlled instrument (often advertised as "microprocessor-controlled") is invaluable as a timesaver when only routine operation is involved. In Chapter Four, the interface between the in-line computer-controlled instrument and the on-line computer will be considered.

1.2.4. Comparison of the Modes

Comparisons of the three modes can be made with respect to their operation in a scientific laboratory. Several factors are subjectively evaluated in Table 1.1 concerning characteristics that are important to laboratory applications.

Execution speed is not easily evaluated, particularly since each type of computer has its own special talents or aptitudes. However, it is safe to say that most large mainframes can perform large-scale computation faster than smaller computers suited for on-line work. There are three primary reasons:

TABLE 1.1. Comparison of Modes of Computer Operation

	Off-Line	On-Line	In-Line
Execution Speed	High	Moderate	Variable
Response time	Highly variable	Moderate-to-high	High
Language availability	High	High	Low
Application software availability	Moderate	Moderate-to-high	Low, specialized
Peripherals availability	Moderate	High	Low
Reliability	Low	Moderate	High

(1) the larger computer can usually handle more significant digits of a number in a single operation while the smaller computer requires more operations; (2) the mainframe often has more memory, so that it may not have to use mechanically operated storage devices (tapes, disks, etc.) as often; and (3) the mainframe can make use of more expensive specialized computer technologies: fast disks and data-transfer channels, fast arithmetic and array processors, and so on.

In-line computers, while possibly sharing the same processors as the microcomputers, can be optimized for certain computations that will be encountered often. For example, if trigonometric conversions are needed continually, a table of partially computed values can be stored permanently in a memory table for reference, thereby shortening the calculation that must be done as a series approximation.

Response time is an issue often insufficiently examined by mainframe users. It is one thing to have a mainframe that can compute in a second a calculation that takes up to an hour on a small computer, but if a day is required on the mainframes from job submission to collection of results, while the small computer stands idle, nothing has been gained. Most users of large machines are familiar with the situation that exists on university campuses when computer science students are near deadlines or when the management of the college or industry must process records: the response time for accepting single characters sent to the mainframe becomes long and outside the control of any scientific user.

Going one step further, when the computer can reliably give a turnaround time of milliseconds, it has the capability for control during the experiment. The results of single computations can be used to affect the remainder of the experiment. The mainframe must satisfy the needs of a large number of users and can seldom be allowed to adjust its time scale to the needs of single users.

The differences between large and small computers in the next several categories are largely matters of economics. Computers in general are limited by their physically available **memory**, but there are techniques to bypass these limitations: memory-management schemes, cache memory, virtual memory, etc. (See Chapter Two.) These methods are reaching microcomputers. Only in systems that are shared by many users has this been easily justified economically in the past, but plummeting memory costs mean that this distinction is beginning to blur. Sufficient memory to store several million characters is no longer uncommon in microprocessor-based computers.

Many **programming language** options are available both for large and for small computers. Because of the larger number of users, nearly every possible language is implemented on the large mainframes. However, all major languages useful for on-line computing also are available on small computers, although the expense of installing and maintaining this wide range of software in a small system for a single user is usually prohibitive. Language options will be compared in Chapter Three.

There are also **application programs** being developed that find their way into research labs: word processors, statistical packages, equation solvers, spreadsheets, database managers, and the like. The expanding market for microcomputers has generated a wider selection of software tools for microcomputers than was ever available for larger computers. Additionally, there are few types of software packages available on large machines that do not eventually find their way onto the small computer. Even many of those programs that are considered to be very large-scale computations have found their way onto the microcomputers of today!

A wide range of **peripherals** is made available to mainframe users. These include printers, plotters, and drives for magnetic disks and tapes. As the number of on-line installations has mushroomed in the last decade, newer and cheaper peripherals have been developed. Although the strides in this area have not equalled those of semiconductors, the cost of good printers and plotters has been cut at least in half, even more in constant dollars, and the capacity of storage devices continues to grow.

Reliability is critical to all computer users. How often does a computer "go down"? And once down, for how long does it stay down? Although statistics in the area are difficult to obtain, a reasonable guess is that fewer than 10% of computer malfunctions in previously reliable computers are due to failures of semiconductor components, except in power supplies. Consequently, simpler computers can be much more reliable. This discussion will be continued in Chapter Twelve.

1.3. IMPORTANCE OF ON-LINE OPERATION

It is clear that small computers do provide considerable raw computing power. However, this is not the primary reason for moving one into the laboratory. Rather, it is the property that all of the power and rapid responsiveness can be brought to bear on a single problem, enabling the computer to respond in time scales sufficiently short that the computer can be used for control. In this section, the reasons for doing so should become clearer.

1.3.1. Advantages of On-Line Operation

A consideration of these advantages is now possible. This will lead to the beginnings of a list of environments in which this mode can be profitably employed:

1. **Flexibility:** The operation of a computer is defined by a program. *Changes* in this program are made in the same way that the program is prepared: at a keyboard. Consequently, if the computer is involved both in data acquisition from and control of an experiment, then

changes in the sequence of operations, the processing of the data, or the time scale are made through changes in a program rather than in the hardware of the experiments.

2. **Speed:** Both the on-line control function and the computation function make use of the computer's speed. The question of "How fast is fast?" is not answerable in a general way since any small computer may have particular strengths and weaknesses different from the next computer. However, the computer will usually decrease the time required for control or computation compared to manual operation.

3. **Accuracy and precision:** Since all of its functions are carried out digitally under the control of a program, the result of a computation is always as accurate and precise as the program. Problems such as drift or instability in a result when the inputs are constant can only be due to either a malfunction or programming error.

   Furthermore, the approximations required when data processing is implemented in analog electronics can be forsaken since the most exact forms of data processing can be implemented. For example, in many applications, such as in spectrophotometry where Beer's Law defines the computation of absorbance, the logarithm of the ratio of two parameters must be measured. Without a computer or calculator, a logarithm is difficult to obtain, so several approximations have often been used. A very simple approximation was derived from the series expansion: When the log ratio is small, it is proportional to the difference between the parameters. The difference function is easily implemented electronically and was substituted in instrumentation at one time. The actual log ratio is more difficult, although still readily carried out, electronically; however, for the computer, it is just as easy as the approximation. Further corrections involving backgrounds, nonlinearity, derivatives, integration, or curve fitting are still simple to implement with the computer but difficult without it.

4. **Endurance:** There are occasions when the ability of the computer to provide results rapidly is of less importance than its ability to provide them continuously. Because it does not require coffee breaks or overtime pay and does not become bored or careless, experiments that are very long or need to be repeated often may be supervised by a computer advantageously

1.3.2. Range of On-Line Applications

On the basis of the previous discussions, it is now possible to identify certain types of activities that may be improved with the addition of a computer or cannot be done at all without a computer:

1. Experiments that occur *too rapidly* for the acquisition of the data by strip-chart recorder, but provide more information than can be re-

solved when recorded from an oscilloscope. A phenomenon that occurs as quickly as 1–50 ms can be followed by direct data acquisition with many small computer configurations.

2. Experiments in which rapid and/or continual *control* on the basis of data input, processing, and subsequent feedback is required in order to adjust one parameter on the basis of the measurement of several others.

3. Experiments in which more data are acquired in a short period of time than could be processed manually in a reasonable period if digitization, transcription, and calculation were to be done manually.

4. Very *slow experiments* in which data must be acquired, parameters must be controlled, and errors must be corrected or recorded over a long period.

5. Experiments in which the signal-to-noise ratio (SNR) is enhanced by use of *signal-averaging* a repetitive signal, such as making the same measurement repetitively and averaging the result.

6. Experiments that require substantial electronic hardware for counting, controlling, or determing experimental variables. A computer may replace much of that hardware at lower cost.

Now, if the computer is already in the lab for the reasons listed above, it can be used to provide several functions that, although essential for efficient operation, could be considered conveniences. The first is data handling, both with fixed algorithms and using interactive techniques. With the use of graphics depicting curve fits and error curves, the user can deal with the data most advantageously, combining innate pattern recognition abilities with the computer's speed and accuracy. Other functions include word processing, record keeping, modeling calculations, and report generation.

1.4. GENERALIZED ON-LINE COMPUTER STRUCTURE

So far, we have dealt with computers and their capabilities in a generalized way, and have begun to zero in on the small computer that is used in the laboratory. It is now time to carefully define this computer, determining the components necessary to obtain the benefits presented so far.

The system required to obtain the flexibility and power required for computer-assisted experimentation is shown in Figure 1.7. *Minicomputer* could be the term to properly describe this system; it lies somewhere between the large shared computer and a microcomputer that is dedicated to some particular function. In common usage, however, any computer whose CPU consists of a microprocessor is called a *microcomputer*, and most, if not all, small computers now built for on-line operation as we have defined it, are

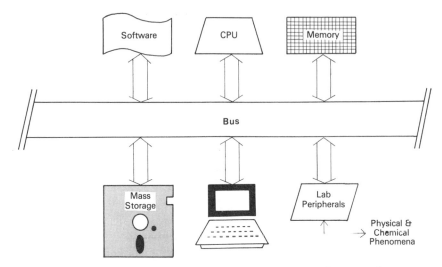

FIGURE 1.7. Components of a computer suited for on-line operation.

built around microprocessors. The term minicomputer is consequently of less value for this discussion. The term *microcontroller* will be applied to the system that is built around a microprocessor but is dedicated to a single function.

The definition of microcomputer that we choose is then an operational one, and the characteristics of this microcomputer match those of single-user minicomputers of the recent past. A microcomputer is a computer that

1. Is programmable by a user in one or more languages
2. Has facilities to communicate with a user for the purposes of program entry and information I/O
3. Has means of permanently storing data and programs
4. Is expandable to communicate with many different experiments, possibly, but not usually, at the same time
5. Was designed for the purpose of real-time interaction with its user or users

This definition is nearly synonymous with the on-line computer, although it allows some systems to fall "in the cracks" between off-line and on-line or between on-line and in-line.

For a microcomputer to be considered useful for experimentation, it must include the components listed. Each will receive more discussion in the following chapter.

1. The **CPU** is the part of the computer that interprets and executes instructions.

2. The **Memory** is a system of electronic devices that store instructions and data in a form directly accessible to the CPU.

3. The **Mass Storage** unit stores programs and data that can be moved into memory in sequential fashion; the media are magnetic (tape or disk) or punched paper (tape or cards).

4. Common forms of **user I/O devices** are the cathode ray tube (CRT) terminal, also termed video display unit (VDU), or the teleprinter.

5. In order for a laboratory computer to be flexible, it must have an open-ended architecture, and usually this means that it must have an available **bus**. That is, there must be a number of equivalent or nearly equivalent sockets for adding circuit boards that provide additional functions, especially different instrument interfaces.

6. In the laboratory, devices are necessary to encode and decode experimental information for the computer; these units will be termed the **laboratory peripherals**.

7. Finally, **software** is required at three levels. The first is the operating system (OS) for the computer, the programs that manage input, output, and the files on disk or tape. The second is the development software, often high-level languages such as BASIC, FORTRAN, or Pascal, which convert English-type statements into the code to be executed by the computer. The third is the application software, which is a program that is executed to solve a specific problem.

1.5. MULTIPLE USERS AND MULTIPLE COMPUTERS

For nearly any computer, programs can be written so that the resources of a single computer can be brought to bear on more than one problem simultaneously; the problems may have equal priorities and receive equal attention from the computer, or they may have different priorities, with the most important problem taking the bulk of the attention. Similarly, the resources of more than one computer can be utilized for a single problem; coordination between the computers allows each computer to handle a different portion of the problem. It is useful to consider the relative merits of these approaches, the advantages and disadvantages.

1.5.1. Time-Sharing

An example of time-shared computer operation is depicted in Figure 1.8, showing several experimental stations using the same computer simultaneously. When each station consists only of a user carrying out tasks such as preparing a program or performing computations, the operation is relatively simple. The computer is programmed to respond to each user's inputs and

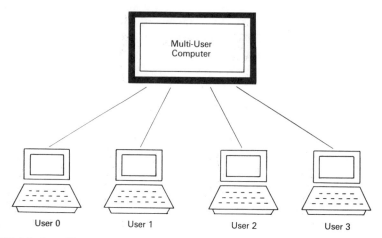

FIGURE 1.8. A multiuser computer with terminals. Four users share the use of a single computer system.

does so on a time scale that often *appears* to be instantaneous. In actual fact, the computer's processor switches its attention rapidly between users, the sequence being determined by the user's needs and/or a predetermined priority scheme.

It is quite typical, when interactive computers are employed, for the computer to spend most of its time waiting. This wait may be for the operator to enter commands or data or for the external devices (terminals, printers, etc.) to become ready to accept or provide another character. Even terminals operated at very high data rates still typically require the computer to spend 90–99% of its time in an unproductive loop when interacting with the user. In the situation where the computer is limited by its rate of I/O, it is considered **I/O bound**; it must idle between each byte received from or sent to an external device, and the speed of execution of the entire task is limited, not by the speed of the CPU, but by the I/O data-transfer rate. In the opposite case, where the computer is, for example, processing computations on an array of data, the computer is limited by the rate that it can operate on a task; it is then **compute bound**.

When compared with the alternative of using one computer per process or user, the reasons for a time-shared system are primarily economic; the actual cost of the hardware is that of a single computer (possibly with more memory than a single user would require) and a *shared* set of peripherals (printer, plotter, storage device, etc.), although a separate terminal is still required for each user. This sharing process may work well, since if a computer is I/O bound much of the time, other users can use that time without noticeable effect on the I/O-bound user.

Two methods of time-sharing, shown in Fig. 1.9, are common in small computers, foreground–background and round-robin. Two programs or processes are executed simultaneously in **foreground–background** operation,

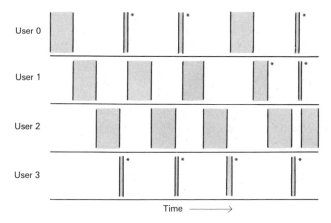

User 0

User 1

User 2

User 3

Time ———→

*Time slice relinquished while waiting for I/O

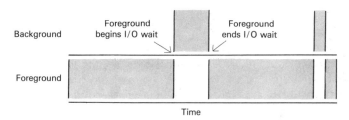

Background

Foreground
begins I/O wait

Foreground
ends I/O wait

Foreground

Time

FIGURE 1.9. Two modes of time-sharing. Above, each user's process is allowed an equal slice of the available time; however, if an active user is unable to use the time while waiting for I/O, it relinquishes its slice. Below, the process in the background is only given time when the foreground process is tied up with I/O.

with the foreground program always having priority; the background program operates only when the foreground program is waiting for some event (such as a keystroke) to take place. **Round-robin** operation usually implies that the computer's time is shared between the processes on an *equal* basis; the computer switches its efforts from one user to the next at regular intervals.

For example, a user might operate an editing program to enter and edit the code for a program that will be used for the real-time control of an instrument. While the user is editing, the computer must respond rapidly to every keystroke; otherwise, characters could be lost. Following the program preparation, the program is executed and the computer must respond to all of the instrument's needs. Later, data must be processed.

Particularly while the editor program operates, but also at many other times, the computer is often I/O bound. Consequently, some other program could operate at what appears to be the same time; the processor would actually switch operation between the two. If the second program was also

I/O bound, the user of the editor program would see little or no delay. The difference between the round-robin and foreground–background methods lies in the priorities attached to the programs that execute.

In the foreground–background case, the background program would *only* execute while the foreground program is waiting for the operator to enter characters or is waiting for the printer or terminal to be ready to accept the next character. The foreground program loses none of the computer's resources to the background program except memory space, since it always has priority. At the same time, all of the computer's time can be usefully employed.

In the round-robin system, all users may be equal. Unless priority schemes are implemented, the controlling program system, often termed an executive, simply switches the processor to different tasks after regularly timed intervals. However, if one user is I/O bound, that time is again wasted; consequently, when the executive finds an I/O-bound user, it immediately switches to the next user in the sequence. In this way, I/O-bound users have almost no effect on the processing speed of others; on the other hand, if all users are compute bound, the time for the completion of a single task on a dedicated computer must be multiplied by the number of users.

Other terminology also is used to describe types of time-sharing. Both of the methods described are called **multitasking** since more than one program may reside in the computer at any time. However, the round-robin systems is usually used in a **multiuser** system in which each task may be associated with a separate terminal. The foreground–background method is suitable for a single user.

When compared with a single-user system, time-sharing systems experience two possible drawbacks. The first is clear from the foregoing discussion; the CPU, the primary computer resource, must be shared. At most, *one* user can expect to receive submillisecond response continuously; in many cases, this response is available to none, and this could be disastrous for laboratory programs that perform data acquisition and control. Furthermore, when more than one process is compute bound, the speed of operation must decrease. The second drawback is that the multiuser microcomputer hardware is slightly more complicated and the software is much more complicated; furthermore, if a single system becomes inoperable, *all* users are brought to a halt.

The advantages, however, are that if the experiments are relatively slow on a continuous basis or only operate at high data rates in bursts, the hardware resources can be pooled advantageously. Either the costs of the peripherals (printers, plotters, and storage devices) are reduced by sharing, or peripherals of higher quality and/or capacity can be justified.

1.5.2. Networks

The network lies at another pole in computer operation. A computer network is a system in which two or more computers are able to pass information from any one to any other through an electronic link. These links normally are set up for three reasons: (1) so that two computers might operate on different aspects of the same problem simultaneously, (2) for communications, and (3) for sharing resources.

We will look at three general classes of network topologies useful for on-line computer systems: (1) the master–slave network, (2) the local area network (LAN), and (3) the long-haul network.

Master–Slave Network. With the dropping cost of computer hardware, the option of using more than one processor in a *master–slave* configuration (Fig. 1.10) is becoming increasingly prevalent. The processors involved may be identical or very different. In the case of a computer used for real-time data acquisition and control in an experiment, separate dedicated computers might be used to handle each of several parameters; for example, one might monitor and conotrol a gas pressure or fluid flow while another handles graphics. In most cases, one processor must be considered the master; it unloads certain tasks onto the slaves.

The slave processors may be similar to the master processor, but there are also cases in which the network includes a special-purpose processor, for example, one which is very good at performing certain computations, but poor at I/O or decision making. Many other special-purpose integrated circuits, CRT controllers for example, are, in fact, microprocessors specially designed and optimized for particular functions. Very often, the type of master–slave relationships described above requires the processors to be "tightly coupled"; that is, they must constantly be responsive to each other and often do this by sharing memory.

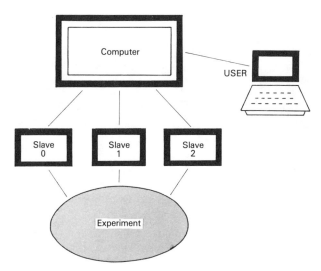

FIGURE 1.10. Master–slave computer operation. More power can be brought to bear on an experiment by use of slave processors, each of which supervise a portion of the task.

Local Area Networks (LANs). Local area networks serve two primary purposes: to provide communications between workstations or computers and to facilitate the sharing of resources. The communications application may be, for example, the transmission of a memo to a group of other people sharing the same network or it might be the movement of a program or data file to another computer. Shared resources might be information (i.e., literature files or mailing lists) or expensive peripherals such as a large-capacity disk, a high-quality printer, or a plotter. The peripherals are made available to any user while appearing to the computer program to be part of that computer; for example, disk drive "*F*" in the OS of one person's computer may actually reside on another computer, but it acts as though it were local.

Many arguments are made for various types of LAN topologies. One type of LAN, the **Bus** configuration illustrated in Figure 1.11, ties a group of equal workstations together; each is capable of taking control of the connecting link to send a message to any other, including those controlling printers or other devices. The Bus network does not depend on the reliability of any single workstation, but does require sophisticated methods of allowing many senders and receivers to share the same lines. The bus LAN is typical of office network systems.

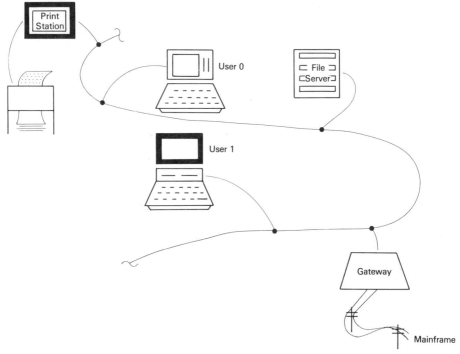

FIGURE 1.11. A LAN with a bus topology. All computers connected to the bus have rapid access to other devices connected to the bus.

A *Star* network (Fig. 1.12), in which all communications must pass through a single hub computer, is even better suited for sharing resources. It is, however, less suited for communications. Laboratory information systems will more typically use star topology than the bus topology; communication between workstations is less important than communication with a central hub. The hub computer can be made responsible for all information management, mass storage, printing, plotting, and communications with other networks or computers. In this system, no communications can take place when the hub fails, even though the workstations can still function.

In the future, when new science buildings are constructed, the conduits for computer-network cables will become just as common as the electricity, gas, and telephone distribution systems are at present. Movement of information will take on a similar priority.

Long-Haul Networks. Computers must still move information to and from distant sites, such as other universities, laboratories, and plant sites. For this discussion, we will describe the conferencing system and the store-and-forward system.

Most conferencing systems utilize single hub computers with which all of the participants communicate directly. Some are commercial "information utilities" such as CompuServe and the Source; these give access to a great deal of information, but also support electronic mail between participants, and conferencing. By conferencing, we mean that individual partic-

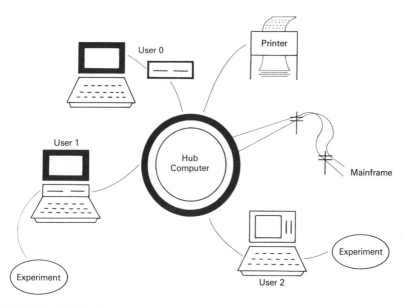

FIGURE 1.12. A LAN with a star topology. The access of each computer to other resources is routed through a centralized computer.

ipants may leave messages for any interested party to read and to which anyone can respond.

Store-and-forward systems typically link computer systems. For example, many computers used on academic sites are linked by a system known as BITNET. By correctly addressing the target system, any user can send a message to any other at any BITNET site. The BITNET software, running on each BITNET computer, finds a route to the destination, sending the message to successive sites, each of which forwards the message to the next site. No central control of this network exists. While electronic mail is easily accomplished in such a system, conferencing is more difficult to accomplish.

We can expect that in the not too distant future, most small computers will be linked into high-speed LANs, and the LANs will link to long-haul networks. It will be possible to move data across the room or across the continent with only a few commands.

1.6 THE SMALL COMPUTER AND THE REAL-WORLD INTERFACE

This chapter has attempted to place the small laboratory computer itself in perspective, compared with other types of computer systems. Typically, it

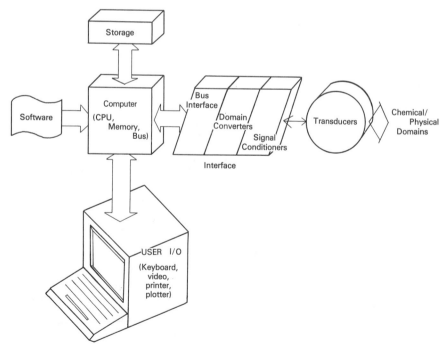

FIGURE 1.13. A laboratory computer system interfaced to a physical or chemical phenomenon.

is sufficiently small and inexpensive that it can be used for only one task at a time, although that task may take many other forms beside interaction with the experiment. The task that is most important to us now is data acquisition and control, laboratory automation, or (using a generalized terminology) Computer-Assisted Experimentation.

The interface, which controls the boundary between the computer and the phenomena involved in the experiment, is the heart of our study. It is shown in somewhat more detail in Figure 1.13 than was described earlier. The various components of this diagram and their interactions will be the subjects of the following chapters.

CHAPTER TWO

Computer Fundamentals

In the previous chapter, the general-purpose laboratory computer (Fig. 2.1) was defined and its place in the laboratory was identified. In this chapter, that general-purpose machine will be dissected; we will look at the "computer-hardware" component of the diagram of Figures 1.13 and 2.1. However, the approach will continue to be strictly from a user's point of view. We will begin by establishing how information is represented in a computer, so that we can prepare for a closer look at the units that were presented as integral components of the general-purpose machine in Section 1.4.

2.1. REPRESENTATION OF NUMBERS AND CODES

A unit of information can be electrically manipulated either in analog form or digital form. In analog form, the value of a datum is stored as a *continuous* value, for example, as a voltage on a capacitor; the precision with which that value can be read then depends on the electrical noise in the circuit or on the resolution of the readout device. On the other hand, digital logic systems describe information by using conditions or states that may be only **ON** or **OFF**; the representations '1' and '0' or **TRUE** and **FALSE** also are used to describe these states (Fig. 2.2). There is no representation for "partly TRUE" on "$\frac{1}{2}$ ON". When a datum is stored in digital form, it is encoded in *discrete* units having only integral values. Each unit of measurement can only have a fixed number of values: 10 in a decimal system and 2 in a binary system.

If the binary unit, having a value of '0' or '1', is to be represented and

27

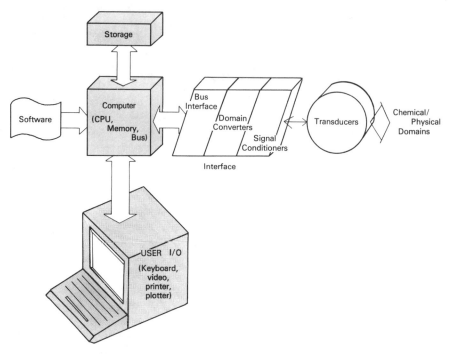

FIGURE 2.1. A laboratory computer system. The small computer's hardware is highlighted for study in this chapter.

stored as a voltage, many different definitions of how these states are electrically encoded could be invented. For example, voltages greater than $+3$ V can usefully represent a value of '1' while voltages less than 1 V can represent '0'. Since the region in between the '1' and the '0' is not included in the definition, there is no border line, and a slight uncertainty in measuring

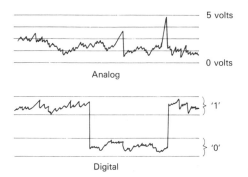

FIGURE 2.2. Analog and digital representations. Noise in the analog signal changes the value, which is continuous between 0.0 and 5.0 V. In a digital representation, the value can only be 0 or 1, and a small amount of noise has no effect.

these voltages is of no consequence. The certainty with which binary numbers can be represented is a primary reason why the binary numbering system is used internally by a computer. The details of the electrical representation will be saved for Chapter six.

The computer hardware is capable of manipulating these binary units of information only according to programmed sequences of instructions. Structures of information such as fractional values or alphanumeric characters are not inherent. Therefore, in order for a user to communicate with a computer, the computer's program must accept information encoded in a recognizable way. A standardized fashion of representing numbers, commands, data, and characters is necessary, and we will now consider these methods of representation.

2.1.1. Bits, Bytes, and Words

A single digit in a binary number is analogous to a decimal "dig-it," and is called a bit (b-it). When the computer manipulates numbers larger than 1, for arithmetic or storage purposes, for example, the numbers are organized into groups of 8 bits making up a unit termed a **byte**. As shown in Figure 2.3, the eight individual bits in a byte are numbered from the **Least Significant Bit (LSB)**, bit 0, to the **Most Significant Bit (MSB)**, bit 7; then each bit has the value of 2^n where n is the bit place. A byte can then contain digital values from 0 (in binary, $0000\ 0000_2$) to a base ten 255. (The subscript denotes the number base.) In binary,

$$1111\ 1111_2 = \sum_{n=0}^{7} 2^n = 1 + 2 + 4 + 8 + 16 + 32 + 64 + 128$$

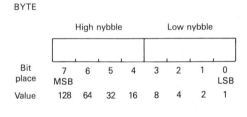

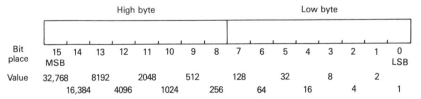

FIGURE 2.3. A byte and a word. A byte is composed of 8 bits. A word is the unit manipulated by a computer, in this case, 2 bytes.

A computer may be designed to manipulate one or more bytes in a single operation, and the number of bits that are manipulated at once is termed the **word length**. Some common microprocessors (including the Intel 8080/8085, the Zilog Z80, and the Mostek 6502) and computers based on these microprocessors usually manipulate a single byte at a time and are known as 8-bit microprocessors and computers. However, the current generation of microprocessors (including the Intel 80286, the Zilog Z8000, the Motorola 68000, and the National 16000 and 32016 families) and most existent minicomputers (e.g., the PDP 11 and Nova) can perform all operations on 2 bytes at once; these machines have 16-bit word length. Larger computers have 32-bit (4 bytes), 64-bit, and even 80-bit word lengths.

2.1.2. Integers

If data to be manipulated and stored in the computer are limited to 8 bits, there are only 256 possible values (2^8); this is clearly insufficient for handling data in the real world. Consequently, methods must be derived not only to handle larger integers but also to encode the sign and handle fractions.

For manipulation of **integers**, a word of 2 bytes is convenient for most problems. When only positive values can be expected, **unsigned integers** can encode the range of 0 to 65,535 (1111 1111 1111 1111$_2$). Signed integers are more typical (Fig. 2.4); 1 bit, the MSB, is used to encode the sign leaving

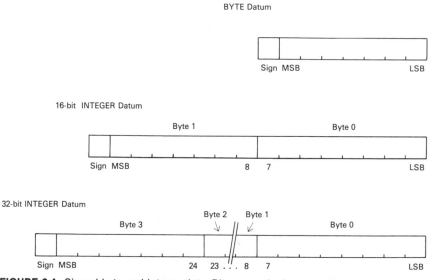

FIGURE 2.4. Signed byte and integer data. Bit seven of a byte encodes the sign; a byte datum encodes the range $-64 - 63$. Bit 15 of a 16-bit (2-byte) datum encodes the sign, leaving 15 bits for the magnitude. For a 32-bit (4-byte) integer, bit 31 encodes sign, and the remainder encode magnitude.

15 bits to encode the magnitude. The 15-bit magnitude has a positive range of $2^{15} - 1$ or 32,767 (111 1111 1111 1111_2).

Encoding negative numbers is somewhat less obvious. First, the sign bit is changed, but then the magnitude must be represented in a format compatible with the computer's operation. Since 0000 0000 0000 0000_2 represents the value of 0, the value of -1 must be represented by the bit sequence obtained if the count of 0000 0000 0000 0000_2 was decremented by 1. That sequence is 1111 1111 1111 1111_2. To obtain that representation by the subtraction of 1, there must have been a borrow of 1 in the bit 16 position, but that borrow bit is ignored.

This representation of a negative number is termed the **2's complement**. The 2's complement of any integer (negation) is produced by first performing a **1's complement** (all 1s become 0s and 0s become 1s) and then adding one. Therefore -1 is obtained from the 1's complement of $+1$ (1111 1111 1111 1110_2) plus 1. Table 2.1 shows some examples.

TABLE 2.1. Sixteen-Bit Integer Representation in 2s Complement Notation

Decimal	Binary	Octal	Hexi-decimal	BCD	Gray
1	00001_2	1_8	1_{16}	0000 0001	00000
2	00010_2	2_8	2_{16}	0000 0010	00001
3	00011_2	3_8	3_{16}	0000 0011	00011
4	00100_2	4_8	4_{16}	0000 0100	00010
5	00101_2	5_8	5_{16}	0000 0101	00110
6	00110_2	6_8	6_{16}	0000 0110	00111
7	00111_2	7_8	7_{16}	0000 0111	00101
8	01000_2	10_8	8_{16}	0000 1000	00100
9	01001_2	11_8	9_{16}	0000 1001	01100
10	01010_2	12_8	A_{16}	0001 0000	01101
11	01011_2	13_8	B_{16}	0001 0001	01111
12	01100_2	14_8	C_{16}	0001 0010	01110
13	01101_2	15_8	D_{16}	0001 0011	01010
14	01110_2	16_8	E_{16}	0001 0100	01011
15	01111_2	17_8	F_{16}	0001 0101	01001
16	10000_2	20_8	10_{16}	0001 0110	01000
17	10001_2	21_8	11_{16}	0001 0111	11000
18	10010_2	22_8	12_{16}	0001 1000	11001
19	10011_2	23_8	13_{16}	0001 1001	11011
20	10100_2	24_8	14_{16}	0010 0000	11010
21	10101_2	25_8	15_{16}	0010 0001	11110
22	10110_2	26_8	16_{16}	0010 0010	11111
23	10111_2	27_8	17_{16}	0010 0011	11101
24	11000_2	30_8	18_{16}	0010 0100	11100
25	11001_2	31_8	19_{16}	0010 0101	10100

These examples hold for a 16-bit format. If an 8-bit format is used for numerical representation, the sign bit will again be the high bit and the range will be from -2^7 to $2^7 - 1$ (-128 to $+127$). If integers of greater magnitude are used than can be accommodated in 16 bits, it may be convenient to employ a 4-byte **double-precision integer** format which handles the range -2^{31} ($-1,073,676,290$) to $2^{31} - 1$ ($1,073,676,289$).

We should note well that the chosen format of numerical representation results from the requirements of the data, not the word length of the computer. The precision with which a number may be stored or manipulated is strictly controlled in software and is not dependent on the word length of the computer.

2.1.3. Floating-Point Numbers

A difficulty arises when numbers that are not integers are to be represented in a binary system; there is no natural place for the fractional component. A similar difficulty appears when numbers are to be stored whose magnitude exceeds 2^{31} but whose precision is less than 31 bits; these numbers can be efficiently stored in an area of 32 or fewer bits if trailing zeros do not need to be retained explicitly.

In decimal notation, the problem is solved by the common use of scientific notation. The value 0.0098765 or $98,765/10^7$ is represented by *normalizing* the number so that the most significant digit is positioned adjacent to the decimal point; the result is 0.98765×10^{-2} or 9.8765×10^{-3} depending on the chosen format.

The same procedure can be applied in base 2. Consider the representation of the decimal 0.15625; this number is convenient because it also can be expressed as a binary fraction. As a decimal fraction, it is $\frac{5}{32}$. In binary, this could be written as $101_2/10\ 0000_2$, that is,

$$101_2 \times 2^{-101_2} = 1.01_2 \times 2^{-11_2}$$

This value is obtained from the $\frac{5}{32}$ fraction by noting that

$$5_{10} = 101_2 = 1.01_2 \times 2^2$$

and

$$32 = 10\ 0000_2 = 1.0 \times 2^5 = 1.0_2 \times 2^{101_2}$$

The representation with the "binary point" after the leading bit is the binary form of scientific notation, and, in that form, the fraction takes a form suitable for manipulation in a computer.

Unfortunately, the format for storing floating-point numbers in a series of bytes is less standard than the format for integers, and several different formats are in common use. For any floating-point representation, the following items must be designated:

1. The total number of bytes to be allocated.
2. The number and placement of the bits reserved for storing the exponent.
3. The number and placement of the bits reserved for the fractional portion called the significand or mantissa.
4. The method(s) for storing sign and magnitudes of both exponent and fraction.
5. The position of the binary point, that is, whether the digit before the binary point will be a 1 or a 0.

Some computers have circuitry or floating-point arithmetic instructions that operate directly on floating-point numbers. In those cases the format is determined by that circuitry or instruction set. The software must follow that protocol. Otherwise, the floating-point operations are implemented by a series of steps using integer instructions, and, in this case, the programmer that writes the arithmetic software determines the format.

In order to relieve the possible confusion, the Institute of Electronic and Electrical Engineers (IEEE) developed a standard that is in use both by circuitry and software developers. It is this standard format that will be described here. The IEEE format is *not*, however, used in the larger computers produced by DEC or IBM, or in the software from some prominent software houses. In most cases, all of the computation and storage of floating-point numbers are handled by the computer language in use, and so long as there is consistency, the user need not worry about these internal formats. However, some knowledge of floating-point representation is useful for an understanding of certain errors and limitations.

The floating-point number, in binary, follows the form $\pm 1.M \times 2^{\pm E}$. In small computers, the floating-point number is stored to a first level of precision (single precision) in 4 bytes, 32 bits. The binary point is, by convention, fixed in a standard position that immediately *follows* the MSB of the mantissa. Since for any nonzero binary number the MSB *must* be equal to 1, it can be represented by implication, without actually allocating a bit of memory.

In the IEEE proposal shown in Figure 2.5, 23 bits (bits 0–22) store the mantissa, (M), and provide 24 bits of precision; the MSB, always being a 1, does not need to be represented explicitly. Bits 23–30 store the exponent, E, with a bias of 127; that is, the number that represents the exponent is $E + 127$. That is a simple 1's complement of the 8-bit exponent. Bit 31 stores the mantissa's sign, MS. Then the number, F, can be represented by the formula

$$F = (-1)^{MS}\, 2^{(E - 127)}\, (1.M)$$

There is one special case. Because of the implied bit, there is no way to represent 0. Therefore, a number in which the 8 exponent bits are all 0 is reserved for the representation of "0". The decimal values supported are

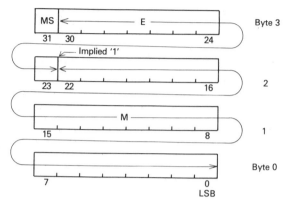

FIGURE 2.5. The IEEE floating-point 4-byte format. The mantissa is encoded in 23 explicit bits plus one implied bit with the sign in bit 31. The signed exponent is encoded in bits 23–30.

in the range $\pm 1.17 \times 10^{-38}$ to $\pm 3.40 \times 10^{38}$; the numbers are precise to approximately seven significant digits.

To return to the representation of 0.15625_{10}, the base 2 floating-point representation is $1.01_2 \times 2^{-11_2}$. The representation of the exponent E will be

$$(-3 + 127)_{10} = 1111\ 1100_2$$

The mantissa is 1.01_2, and since the leading 1 is implied, 01 is stored. The entire representation works out to

0111	1110	Byte 3
0010	0000	Byte 2
0000	0000	Byte 1
0000	0000	Byte 0

Alternative methods of representing floating-point numbers differ in the arbitrary assignments listed above. To expand the range that a format could handle, another bit could be added to the representation of the exponent, but that bit would come at the expense of the precision of the mantissa.

Data that require more than seven decimal digits of precision or have absolute values greater than can be represented in single precision can be stored with 64 bits in **double-precision** format. This format is shown in Figure 2.6, by the formula

$$N = (-1)^{MS}\ 2^{E-1023}\ (1.M)$$

This double precision format will represent decimal numbers in the range $\pm 2.22 \times 10^{-308}$ to $\pm 1.80 \times 10^{308}$. Numbers are valid to approximately 16 significant decimal digits.

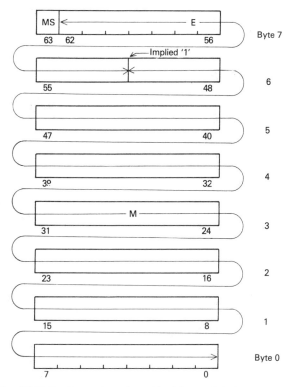

FIGURE 2.6. The IEEE floating-point 8-byte (double-precision) format. The mantissa is encoded in 52 explicits plus one implied bit with the sign in bit 63. The signed exponent is encoded in bits 52–62.

Even when a number can be adequately represented in a given format, errors due to round-off within a computation can easily occur. Consider the integer calculation of the length of one side of a right triangle whose hypotenuse is 10 (1010_2, 4 bits) and one side of which is 6 (110_2, 3 bits). The answer is, of course, 8 (1000_2, 4 bits). However, to calculate that result, we must calculate the square root of $10^2 - 6^2$ (i.e., $110\,0100_2 - 10\,0100_2$), which require 7 and 6 bits, respectively. If the intermediate result is rounded to 4 bits, the maximum length of the representations of the input values and answer, the result is $(96 - 36)^{1/2}$ which is 7 plus a remainder of 11; converted to floating point, it is 7.75.

To avoid any problems with rounding errors in intermediate results of a computation, extra bits might be added temporarily. The Intel 8087 math coprocessor performs all internal calculations in an 80-bit format with a 63-bit mantissa and a 15-bit biased exponent, regardless of the actual precision of the input numbers.

2.1.4. Special Representations and Codes

There are several other representations of numbers that either are used to represent binary numbers or are alternatives to binary encoding methods. An explanation of those methods here will make later discussions easier.

Binary, Octal, and Hexadecimal Codes. The visual representation of a 16-bit number as a string of 16 ones and zeros is inconvenient. The interpretation of that pattern is, for humans, time-consuming and error prone. Unfortunately, the decimal numbering system that predominates among humans owes its popularity to the number of digits commonly found on two human hands and has little relevance to computers; it is difficult to take a decimal representation of a large integer and determine which of the bits in the binary representation are 1s and which are 0s. Consequently, when represented visually, a data word is commonly segmented into subgroups that can readily be understood.

Two methods for this grouping are useful: **octal** and **hexadecimal (hex).** In the former, the byte or word is segmented into 3-bit groups having eight possible values; in the latter, 4-bit groups having 16 possible values are used. In order to represent 16 possible values in the single character, our numbering system must be extended past 9; the letters *A–F* are used. Table 2.2 illustrates the various systems by providing the equivalents.

The octal system has the advantage that all eight possible codes can be represented by symbols from the decimal system; however, dividing a byte (8 bits) into octal digits creates two octal digits with 3 bits and one with only two; the concatenation of two octal bytes into a 16-bit word can be quite confusing. The hex system avoids this problem and is used much more frequently.

In order to avoid confusion, numbers can be provided with a subscripted suffix to indicate the numbering system; the suffixes for binary, octal, decimal, and hexidecimal systems are 2, 8, 10, and 16, respectively. (Suffixes *B, O* or *Q, D,* and *H* also are used to indicate binary, octal, decimal, and hexidecimal, respectively.)

Binary-Coded Decimal. There are other coding systems that are also alter-

TABLE 2.2. Numbering Codes

Decimal	Binary	Hexidecimal	Octal
32,767	$0111\ 1111\ 1111\ 1111_2$	$7FFF_{16}$	$77,777_8$
1	$0000\ 0000\ 0000\ 0001_2$	0001_{16}	$00,001_8$
0	$0000\ 0000\ 0000\ 0000_2$	0000_{16}	$00,000_8$
-1	$1111\ 1111\ 1111\ 1111_2$	$FFFF_{16}$	$177,777_8$
-32768	$1000\ 0000\ 0000\ 0000_2$	8000_{16}	$100,000_8$

nate representational systems. These are of interest in certain special applications. The first is **Binary-Coded Decimal (BCD)** code. The BCD code is, as the name implies, a method of encoding decimal numbers in a binary format while maintaining a correlation between groups of bits and decimal digits. Four binary bits are used to encode each decimal digit, 0–9, in the binary codes 0000 to 1001; the codes 1010 to 1111 are not allowed. That is,

$$0000\ 1001_{BCD} + 1 = 0001\ 0000_{BCD}$$

One drawback to the BCD system is the waste of storage space; note in Table 2.2 that to encode the decimal value 25 in binary required 5 bits whereas the BCD scheme required 6. If one attempted to encode the number 25,000 in BCD, 18 bits would be required ($10\ 0101\ 0000\ 0000\ 0000_{BCD}$) compared with 15 in binary ($110\ 0001\ 1010\ 1000_2$).

There are advantages to the BCD encoding scheme, however. It is useful in systems in which a substantial fraction of the effort requires decimal I/O, such as in handling displays of numbers. Software math packages can be written to perform BCD rather than binary arithmetic, and the processors themselves can be constructed to carry out all operations in BCD without sacrificing significant speed or accuracy. BCD arithmetic is often used in packages for business applications, where more significant figures are required without roundoff (accountants wish to be accurate to the penny), but magnitudes less than 0.01 or greater than about 10^{10} are seldom encountered (except when dealing with the national debt).

Gray Code. This coding system does not conform to the rules of numbering systems in which each digit has a set value. This is a special scheme that allows only 1 bit in an entire number to change upon an increment or decrement. It is therefore valuable where a value is counted mechanically. The gray code eliminates ambiguity that might occur during a transition such as that from 3 (011_2) to 4 (100_2); a transient 7 (111_2), 0, or any number in between could appear if the change in each bit was not precisely simultaneous (see Fig. 2.7). The value of this sequence will be illustrated in Section 7.5.2.

2.1.5. Alphanumeric Codes

One very important set of data that must be manipulated in the computer is text (letters, numerals, punctuation, etc.). Computers must be able to accept typed input and must provide printed or displayed output. The format in which this is done must be compatible with the wide variety of text I/O devices, particularly printers and CRTs. The predominant code is the **American Standard Code for Information Interchange (ASCII)**. A similar code, the **Extended Binary-Coded-Decimal Interchange Code (EBCDIC)**, is used by IBM for its larger computers, but it has been nearly abandoned in small computers in the quest for compatibility.

The ASCII code, listed in Appendix One, is seven bits in length which

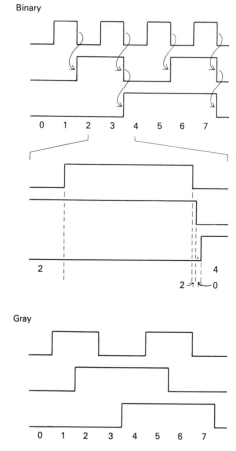

FIGURE 2.7. The gray code and ripple error. In the binary counting sequence, the count may ripple through the count bits causing transient error when the count changes from three to four; the glitches can be seen where the counting waveform is expanded. Since only one bit can change at a time in the Gray code, these glitches do not occur.

allows the representation of 2^7 or 128 characters. Besides upper and lower case letters, the standard includes the important punctuation symbols but few of the special characters of importance to scientists such as Greek letters, the integral sign, or the square root symbol.

The 32 codes below 20_{16} are control codes, which are nonprinted characters. Control codes are used to control terminals or send signals to a computer. Two examples are the carriage return and line feed characters (codes $0D_{16}$ and $0A_{16}$). Another example, the code 13_{16} (named DC3 and often written ^S which is pronounced "control-S"), can be generated by a keyboard by depressing the "S" key while holding down the CTRL key. Control-S is often used by a terminal operator to signal a computer to stop sending alphanumeric information to the terminal. The code 11_{16} (DC1 or

ˆQ) is frequently used to signal the computer to resume. Control-C (ˆC) is often used in computer languages to signal a program to abort.

2.2. THE CENTRAL PROCESSING UNIT

At this point, we can move ahead to descriptions of the various components of the small laboratory computer, beginning with the **CPU**.

The CPU is the heart of any computer, containing the logic for fetching instructions from memory, decoding them, and executing them. Over the past decade, continual improvement in the compression of transistors into integrated circuits has led to increasingly powerful CPUs for small computers (Fig. 2.8). The success of the electronics industry in increasing the numbers of transistors in a single package is truly spectacular.

However, of greater importance to the laboratory user of small computers than the number of transistors on a chip is a particular CPU's ability to execute a task. Fundamentally, this ability is dependent on the **architecture** of the CPU. The architecture is the general operating scheme of the CPU, how it handles memory, I/O, and the way in which it executes instructions including jumps and calls to subroutines. The architecture developed by the designer of a CPU is naturally a compromise between the capabilities that

FIGURE 2.8. An 80386 integrated-circuit CPU. The CPU functions and some other general-purpose functions, such as DMA, are integrated into a single "chip." (Courtesy Intel Corporation.)

would be useful and those that can be accommodated within the constraints of producing the integrated circuit.

If we compare several CPU designs, we will usually find that one CPU will excel at one task and another at a different task. That complicates comparisons, but we must still try. A program used to make a comparison of the execution speed of different CPUs in handling a particular task is termed a **benchmark**, and benchmarks are often performed for tasks such as a bubble sort (placing a series in sequence, alphabetically or by value), a mathematical operation such as a square root, a character string comparison such as finding the occurrence of a word in a document, and a table look-up operation.

The speed with which a CPU can perform a task depends on several factors, some of which are readily identifiable and quantifiable, but others of which are not. This makes the task of actually preparing and performing a benchmark for a particular application nearly impossible to do with unambiguous results. Examination of the factors affecting the results is useful nevertheless.

The first factor contributing to the execution speed of a computer is the **cycle time** of the CPU. Any CPU requires an oscillator or clock, and the execution of instructions is synchronized with that clock. The clock frequencies of common microprocessors vary over an order of magnitude, from less than 1 MHz for early 8-bit CPUs to as high as 20 MHz for some of the recent entries, so this factor alone has a great effect on the speed with which a task can be performed.

Obviously, the execution time may be reduced by the simple expedient of increasing the clock rate. However, there are limits to the speed with which both the CPU and its associated circuitry may execute. If those limits are exceeded, errors will occur in the processing of instructions. Furthermore, if the CPU can be operated at a higher clock rate than that at which the other components of the system can operate, errors will occur in data transfer.

If the CPU can outpace the memory or other computer components, it is possible for those components to ask for a **wait state**. In such a case, the request signal from the CPU *to* the memory or I/O circuitry will trigger a "not ready" signal *from* those components which will cause the CPU to idle while the other circuitry "catches up." For example, when the memory is unable to provide a datum requested by the CPU in the time allowed by the CPU, it may signal the CPU to wait for 1 or more clock cycles. This slows down the overall computer execution, but the delay is much less than that produced by slowing down the clock of the CPU.

Related to the cycle time is the **efficiency** with which the time is used. Most instructions require more than one tick of the clock (cycle); the number of cycles varies with the complexity of the instruction. Later versions of the same microprocessor often require fewer cycles per instruction. Early microprocessors also follow a completely linear sequence when executing instructions. First, the instruction is "fetched" from memory. Then it is interpreted and executed before the next instruction is fetched. Advanced microprocessors use techniques such as **pipelining**, whereby successive instructions are fetched from memory and stored internally in the CPU *while* the earlier instruction is being executed. The throughput is greatly improved, since, the fetch and execute processes occur concurrently.

The second factor, **word length**, has significant effect on efficiency. All other factors being equal (they never are, of course), a CPU having a word length of 16 bits will execute functions involving 16-bit data *more* than twice as fast as an 8-bit processor, since some of the overhead in handling two separate bytes rather than one word may be eliminated. For example, incrementing a 16-bit word by one requires the former CPU only one instruction. The latter CPU may first increment 1 byte, then check for overflow; if a carry is noted, then the high-order byte is incremented.

It should be noted that few CPUs use only a single word length for all operations. For example, the 8-bit Zilog Z80 and Intel 8080 CPUs have a few instructions for moving and adding 16-bit words, and the 16-bit Motorola 68000 CPU has many 32-bit manipulation instructions. On the other hand, the 8088 used in the IBM PC computer is an example of a microprocessor usually placed in the 16-bit category that lacks some of the characteristics associated with 16-bit CPUs; it transfers data to and from other circuitry only a byte at a time although, internally, it can carry out any operation on a full 16-bit word.

The third factor is the **efficiency** of the instruction set. A more powerful instruction set will allow a programmer to do in one instruction a task that may take many instructions in a less powerful CPU. In most cases, this is an elusive factor to quantitate since, as was mentioned earlier, it is difficult to make benchmarks that cover all types of applications. In one example, however, it is clear: the Zilog Z80 instruction set is an extension of the instruction set of the Intel 8080; that is, all of the 8080 instructions were retained and many useful new ones were added. Similarly, the NEC V20 and V30 are extensions of the Intel 8088 and 8086. A microprocessor that executes *more* instructions is probably more powerful that one which has fewer instructions, if other factors are more or less equal.

In some cases, CPUs have been designed particularly for specific languages; consequently, such CPUs may be efficient in that language, but poor in others. Carefully written benchmarks are usually required to make valid comparisons. General-purpose small computers are unlikely to use specialized CPUs, however.

It has been argued that if the trade-off is necessary, execution of simple instruction sets quickly is more powerful than executing complex instructions more slowly. Certainly more flexibility would be gained.

The fourth factor determining the time required to execute a task may be the most important factor of all, and that is the **programmer's skill**. Clearly, if a clever and skillful programmer can write a program that requires only half as many instructions to complete a task as the program written by another, that effort will contribute significantly to increasing the execution speed.

For the average computer user who programs in a higher level language (BASIC, FORTRAN, Pascal, etc.), the most important time savings are to be made by using a language which, both inherently and through effective implementation, can generate efficient instruction sequences; this means that care must be taken in purchasing that software. There is a wide range of execution times for the same program when interpreted or compiled by the development software from different software houses. It is not unusual to see a single benchmark program written in the same language and executed on the same computer have execution times which vary by an order of magnitude for different companies' compilers. When the application software is written in assembly language, the responsibility lies entirely with the applications programmer. The importance of the skill of the writer of both the development and applications software cannot be overestimated.

Certain deficiencies in the instruction sets of CPUs, particularly the inability to handle mathematical functions, can be circumvented in several ways. For use with 8-bit microprocessors, **math processor peripherals** are available; these are integrated circuits to which the CPU can send data and commands. The math processor then makes available to the CPU the results of the desired operation in a small fraction of the time the CPU would require on its own. For example, the AMD9511 can perform floating-point multiply, divide, and transcendental functions rapidly and thereby both increase the execution speed of the microprocessor and decrease its memory requirements (by eliminating the need for space for subroutines to perform those functions).

A similar but even more efficient approach is taken by several 16-bit microprocessors for which **coprocessors** are available that operate closely coupled with the CPU. The coprocessor is not an I/O device to which the CPU sends data and commands and receives results; instead, it is a device that executes an extended instruction set for that CPU architecture. Coprocessors for the 8086 microprocessor are the 8087 for performing math functions and the 8089 for supervising I/O. Coprocessors are being developed for other specialized tasks such as searching for a string of characters in an area of memory and handling graphics functions.

The coprocessor approach provides a nearly ideal path for upgrading CPU performance, but before jumping too quickly at the coprocessor possibility, one should evaluate the availability of software that support math peripherals or math coprocessors. Very little development software ever became available which could support the 9511, although the outlook for the 8087 coprocessor is much brighter.

Finally, certain computational tasks lend themselves to help from **array processors** (Fig. 2.9). If the same operation is to be carried out on an entire array of data (i.e., multiplying by a constant, multiplying two arrays, or

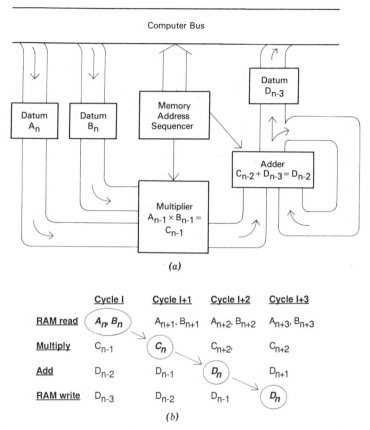

FIGURE 2.9. A pipelined array processor. (*a*) While data are fetched, the previous data are being multiplied; the data before that are being summed, and results are being returned to memory. (*b*) The matrix shows data moving through the pipeline in four cycles.

inverting an array), the operation can be done in a special pipelined processor. The computer's CPU is either placed on hold or performs other operations. Then *while* one number is being multiplied (by a hardware multiplier which can do the multiplication in a single cycle of the computer clock), the number multiplied previously is returned to memory, *and* the following number is being fetched directly from memory.

Advantage is gained only when the computation deals with *arrays* of numbers, but in such cases, a tremendous increase in speed may be obtained. The most common example is the computation of the Fast Fourier Transform (FFT). Fourier transform instruments (for nuclear magnetic resonance and infrared absorption measurements) are increasingly using microprocessor-based computers with array processors.

For routine laboratory applications, one should be aware of the CPU performance factors which have been presented, but a detailed examination should often be subjugated to a consideration of factors influencing ease-of-use and software availability.

2.3. THE MEMORY

The **main memory** is required to store both the instructions for the programs to be executed by the CPU and the data on which those programs will operate. In most computers, regardless of their word length, the memory is organized in bytes, and each byte has a unique **address**. Whenever information is to be transferred to and from memory, the CPU (or any other device having the capability to make that transfer—see DMA) presents the binary address and a READ or WRITE signal, and the memory must respond by either sending or accepting the datum. The process of sending a datum *to* a memory location is termed a **memory write** and the process of fetching a datum *from* a location is termed a **memory read**.

Memory may be characterized in several ways. The read–write capability is first, that is, its ability to not only store instructions or data but also to be modified. Secondly, it is categorized by its ability to retain information when the power is removed. That which loses its information is volatile, whereas that which retains its information is nonvolatile. Finally it is characterized by its **access time**. Memory units do not accept and read out data instantaneously; a finite amount of time must elapse between the time the memory location is addressed and the time that the data can reliably be read from that location. If memory is not sufficiently fast, it may be accommodated by adding wait states as was discussed in Section 2.2.

The areas in which technological improvements are being sought by the industry are primarily in density (number of bits per chip) and speed. Secondarily, efforts are made to develop fast and inexpensive nonvolatile memory.

A very important nontechnical characteristic of memory also should receive some attention, and that is its cost. Whereas at one time the cost of memory was such that it was worth the expenditure of significant amounts of time to reduce the number of bytes in a program, the cost of memory has in a decade dropped considerably, by nearly two orders of magnitude. The tradeoff between saving programming time and reducing program memory now usually falls heavily on the side of saving time.

Memory devices that can be addressed directly may be divided into three categories, according to both the technology by which they are made and operated and the way that they are used. A discussion of these categories will follow.

2.3.1. Random Access Memory (RAM)

The term RAM is generally used to include semiconductor random access *read–write* memory. In general, it is inexpensive but volatile. Semiconductor RAM now is the dominant form of memory in small computers. RAMs may be either **static** (S-RAM) or **dynamic** (D-RAM); the latter requires that each memory bit or cell be recharged or **refreshed** every few milliseconds. The refresh requires additional circuitry which in simpler designs may cause occasional delays in the access to a memory cell. This disadvantage is more than offset in most cases by significantly lower cost, higher density (more bits per chip), and lower power consumption. As a result, most RAM used in computers having 32 kbytes or more memory is dynamic.

In addition, D-RAM has often been considered less reliable, but this reputation stems from poor designs in the early days of microcomputers and is not a major factor now so long as the memory board is obtained from a reputable manufacturer. Early dynamic RAMs could be affected by alpha particles from the atmosphere or plastic case, but current devices include protection from such error sources.

Some memory, particularly that manufactured using the **Complementary Metal-Oxide Semiconductor** (**CMOS**) process, uses so little power that operation from dry cell batteries over long periods of time is possible. Since mass storage (disk and tape) is available for laboratory computers, battery operation is not often necessary for storing data but is useful in portable "lap-size" computers or when operating in environments where the power is not dependable.

2.3.2. Read-Only Memory (ROM)

ROM is also random access memory, but as the name suggests, ROM can only be read; for memory, this would be a dubious distinction except for the fact that ROM is **nonvolatile**. As such, it is particularly valuable when the power is first turned on. At that time, there is only random data in the RAM, but a simple program in the ROM can load the operating software from the disk or tape. This initial program is called the **bootstrap loader** or just the boot. A few computers maintain user-accessible software, such as the BASIC language, in ROM, so it is always available with or without the mass storage device. Also highly useful, although not at all universal, is a monitor, a program in the ROM that allows the user to modify, test, and display memory from the keyboard without having to load a program from the mass storage unit. This allows the computer to be tested even when the disks or part of the RAM are faulty.

ROMs differ in how the stored code is entered and in whether or not there are mechanisms for changing the code. The mechanism for entering information into the ROM is given the same term as the process for writing a program, and that term is **programming**. The first category of ROM is programmed at the factory when the ROM is manufactured; when a large quantity of ROMs is required, the data is sent to the ROM manufacturer who prepares a mask to be used in the production process. These **masked-ROMs** can never be modified but, in large quantity, can be manufactured at the highest density and lowest cost of all types of memory.

In order to satisfy the need for ROMs that could be programmed at the user's site, **Programmable Read-Only Memory (PROM)** is available. One type uses fusible links which can be "blown" by a special circuit to create permanently programmed bits. Others can be erased and reprogrammed.

The **Erasable Programmable Read-Only Memory (EPROM)** (Fig. 2.10) can be easily programmed by a user but has the advantage that it also can be erased and subsequently reprogrammed. In order to program an EPROM, a 25-V pulse is applied to a memory cell; at that high voltage, a conduction barrier is broken down and a charge is forced onto the unconnected gate of a transistor, thereby causing that memory cell to read a **0** or LO. The only way that the charge may be removed is by illuminating it with ultraviolet light, which again renders that barrier conductive and the charge bleeds off. A quartz window is placed over the circuit for this purpose. Circuit boards are available for many computers that simplify this ROM-programming process.

Having to expose the EPROM to ultraviolet light is certainly a disadvantage to those who would like nonvolatile storage of programs or data that could be changed frequently. Consequently, the **Electrically Erasable Read-Only Memory (EEPROM)** is available. The programming operation for an

FIGURE 2.10. An erasable programmable ROM chip. The silicon chip is mounted under a quartz window, and the memory can be erased by illumination with ultraviolet light. (Courtesy Intel Corporation.)

EEPROM is basically the same as that of the EPROM; however, since an electrical pulse can be used to erase the former, it is possible for the computer to handle the entire operation. The erase operation of an EEPROM clears all cells of the memory unit simultaneously. Some variants, electrically alterable PROMs (EAPROMS), permit erasing and reprogramming single bytes. Although that sounds like ordinary read–write memory, the erase–write cycle is far too slow for the memory to be used as RAM. The semiconductor industry is working to eliminate from both versions the two chief liabilities, higher cost and lower speed.

2.3.3. Core and Circulating Memories

Two other types of memory are useful for specialized applications; most examples are magnetic.

The term **core** refers to magnetic RAM. The storage units are discrete ferrite rings, one for each bit, arranged in an $X-Y$ plane (Fig. 2.11). A current in the drive wires passed through the rings magnetizes the rings in one of two directions according to the direction of the current; the *direction* of the magnetic field, detected by the sense circuitry, encodes the binary information. Since the direction of the magnetic field can be changed only by the current pulse, the information is retained when the power is removed; it is the nonvolatility of core that is its chief advantage.

Semiconductor memory is now available at a fraction of the cost of core. Since core has a longer access time, it is now used only in a few critical situations which demand nonvolatility; even that justification is seldom applicable since low-power memory can be supported for long periods of time with batteries. However, core was used almost universally before LSI. The term is often mistakenly used to refer to memory in general.

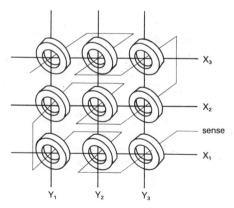

FIGURE 2.11. Core memory. The bit is stored in the direction of magnetization of the ferrite ring.

Circulating Memories. Two types of circulating memories find limited application. Access is not random but sequential, since the device must shift the bits in a loop in order to access any given bit. Consequently, the time required to read or write a given bit depends on its current position in the loop; as a result, circulating memories are not suitable for use as directly accessed memory, but have their application as solid-state mass storage devices. A projected but uncertain advantage is low cost.

Two types of circulating memories are in use: **bubble** and **charge-coupled device**. The former are based on circulating microscopic magnetic domains (bubbles) and as such are nonvolatile, whereas the latter are based on circulating electrical charge packets and are volatile. It had been thought that both of these technologies would provide extremely high-density, low-cost, temporary storage media. The bubble memory also claims a nonvolatility advantage. The cost projections have so far failed to hold up, and only the bubble memories have seen significant numbers of applications.

Bubble memory continues to have application where rugged, nonvolatile, solid-state mass storage devices are needed, for example, when data must be collected and stored at a remote location.

2.4. INPUT—OUTPUT PORTS

A computer must not only move data to and from memory in the execution of a program, but it must also move data in and out of the computer, and this requires I/O ports. An output port is basically a data latch that can acquire a datum when signaled, plus the logic that is required to detect a unique latching signal from the CPU. Synchronized by that signal, the output port can "catch" the datum at the precise instant that it becomes available.

An input port must provide a datum to the CPU at the correct instant. The timing comes from logic that decodes signals from the CPU. The input port also may have the assignment of detecting that some external device is signaling the presence of new data. The direction of the data transfer is always, as in the case of memory, with reference to the CPU: data is *input to* the CPU and *output from* the CPU.

Although there are many different ways that the I/O ports can be organized to accurately transfer data (these will be discussed in later chapters), there are two primary methods of communicating with the port. Each method has its advantages, disadvantages, proponents, and opponents.

When I/O ports are **memory-mapped**, the relationship of the port to the CPU is identical to the relationship of memory to the CPU. If, for example, the port handles data from the computer terminal, the byte resulting from a keystroke is obtained from the input port using any of the same repertoire of instructions as are used to obtain a byte from memory. Conversely, the instructions used to send a character to the terminal are identical to the instructions used to store the byte in memory. The only difference is that one group of addresses are allocated to I/O rather than memory.

The term **data channel** is left over from the earlier days of computers when each port had its own logic, circuitry, and even its own chassis; the I/O was completely separate from the memory. This concept is used in several modern CPUs (such as the 8080, Z80, and 8086–8088). The primary difference between the data channel and memory-mapped port is that separate address space is used for I/O from that used for memory. When the CPU issues an address, it also indicates with a separate signal whether it is making a transfer with memory or an I/O port. In the 8080 and Z80, there are only 2^8 addressable I/O ports compared with 2^{16} memory locations. Consequently, it is easier for an I/O port to detect its own address than for a memory byte to detect its address. Data-channel I/O requires that machine-level instructions must be provided for I/O separate from those provided for memory.

Memory-mapped I/O has the attraction of more efficient software, since the more powerful CPUs have a wide variety of instructions for storing data in and retrieving it from memory whereas CPUs using data-channel I/O typically have only a single input and output instruction, and seldom have more than a few. On the other hand, the primary argument *for* the data channel is that I/O port addresses do not come at the expense of memory; this competition for address space is a definite limitation in such processors as the LSI-11, 6800, and 6502. However, the current generation of microprocessors, the LSI 11/23, 68000, etc., provide the means to directly address so much memory that the competition is less severe.

2.5. MASS STORAGE

One special set of I/O devices with which the computer communicates is used for mass storage. This type of storage, in contrast to memory, can be permanent and can have a huge capacity compared to the computer's main memory. Mass storage devices, principally magnetic disks and tape, are essential to the task of efficiently developing application software, storing experimental data, and manipulating that data. The process of manipulating these devices should not be the "rate-limiting step" in the effective use of the laboratory computer. In this section, we will consider the key specifications that apply to mass storage and review the primary technologies presently in use with small computers.

One of the key specifications of a mass storage device is the time required to store or retrieve a set of information, and that depends on the mode of access. In contrast to the computer's main memory, which was termed *random access* because all locations could be accessed rapidly and with equal or nearly equal ease, mass storage devices are either **sequential access** or **semirandom access** devices (Fig. 2.12). In the former case, the example of

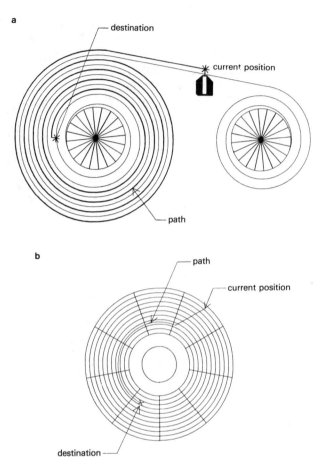

FIGURE 2.12. Sequential and semirandom access storage. In (*a*), a magnetic tape, it is necessary to move the entire tape over the read–write head, from the current position to the destination, in order to access the desired datum. In (*b*) the access path of a disk is semirandom since the head can move directly to the desired track, but must wait for the disk to rotate to the correct position.

which is a magnetic tape system, the tape must be advanced sequentially to find the desired data or record; the worst-case access time is the time required to advance a tape from the beginning to the end. On the other hand, another group of magnetic mass storage devices known as **disks** are organized somewhat in the fashion of an LP record, as a series of concentric, although noncontiguous, tracks. In order to access a particular spot on the rotating disk, the head can move directly to the desired track and then wait for at most one revolution; the worst-case access time is then the time required to move from the outermost to the innermost track *plus* the time for a single complete revolution.

Once the desired spot on a medium has been located, the second specification is applicable, the **data-transfer** rate. This is the number of bytes that can be transferred to or from the computer per second. The transfer rate is usually limited by some mechanical part of the system except for certain high-speed disk and tape systems.

The third important specification is the total **capacity** and **cost** per kilobyte of a storage unit. These prices change almost daily, so they are difficult to compare.

All mass storage devices consist of five components:

1. The medium (tape, disk, paper tape, cards, etc.)
2. The read–write system (tape drive, disk drive, paper tape reader/ punch, card reader/punch, etc.)
3. The controller, the electronics controlling the read–write system
4. The interface which matches the signals of the computer with those of the controller; often the interface and controller are placed on a single printed circuit board.
5. The software driver, programs used to format data and transfer it to and from the interface.

In the following discussions, the *system* will be considered to be the combination of medium, read–write system, controller, and interface. The software driver is usually part of the host computer's OS software.

2.5.1. Magnetic Disk

Previously, the magnetic disk was likened to an LP record; it is different in that the data are recorded *magnetically* and are located in concentric **tracks** rather than in a spiral. Disks also rotate at a much higher rate, from 300 to 3600 rpm. Disks are semirandom access devices since, in contrast to tapes, the drive can move directly to the desired data track without reading intervening tracks. Consequently, disks are the devices of choice for use in systems in which both reading and writing of files occurs within the same program. This includes most small computer applications.

The access time for a disk is the time for the head to move to the correct track plus the time for the disk to rotate into the correct position. These times are the **head seek** time and the **rotational delay** or **latency**. The worst-case seek time is that required to move from the first to the last track, and the average seek time can be shown to be one-third the worst case. The worst-case latency is the rotational period, and the average latency is one-half the worst case.

Disks are classified as **flexible** or **hard**; the former are commonly nick-named **floppies** and are removable. They represent the low-unit-cost, low-performance end of the disk spectrum. The hard disks may be either removable or nonremovable. They have much higher capacities and higher performances.

The three primary categories of disks, flexible, hard nonremovable, and hard removable are considered in greater detail in following sections.

Data Organization. When a disk is manufactured, a track is not marked into subdivisions except to determine a reference location, the beginning of each cycle. The disk must be **formatted** into 8–50 **sectors** which are subdivisions of each track as shown in Figure 2.13; a file or part of a file may then be addressed by its track and sector. The process of formatting the disk involves writing a few bytes of identifier and error-detection information at the beginning of each sector.

The required sector size is determined by the OS software and may vary, usually between 128 and 2048 bytes. Smaller sectors can make better use of disk space when the data files are small, since an incompletely full sector will usually leave only a small unused area. If the sectors are small, more space is required for formatting, and the usable capacity of disk is reduced.

Another important issue when formatting *removable* disks is that of standardization; not only might the size of the sector vary, but the information in the header and the sequence of using sectors may vary. This has led to difficulty in standardizing high-density flexible disks in particular.

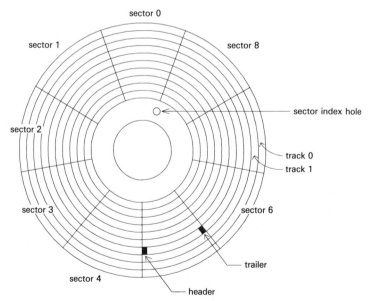

FIGURE 2.13. The organization of a flexible diskette. the diskette is organized into tracks and sectors. The beginning of sector 0 is determined by detection of the sector index hole.

The OS for the computer is responsible for reading files from and writing files to the disk. In order to organize the disk usage, a **directory** is maintained. For example, when a program is to be loaded into memory from the disk, a search of the directory for the file name leads to information regarding the location of the file. In many cases, the file is not stored in contiguous areas, and the directory must keep track of all segments. When a file is stored, the directory is used to locate available space. If a file is to be erased, it is only necessary to remove the name from the directory; the data is not actually lost from the disk until new data is written over that space.

Flexible Disk Systems. In terms of the number of units in service on small computers, the flexible or floppy disk systems dominate. Compared with the paper tape and magnetic tape systems of earlier minicomputer generations, they offer a major leap in speed and capacity. Compared with the cartridge disk systems of larger minicomputers, they offer a dramatic decrease in both initial and operating cost.

Floppy disks are available in a variety of sizes and capabilities. Originally, the 8-inch disk was developed by IBM; on one surface, approximately 250 kbytes can be encoded in what is now termed *single-density* (IBM 3740) format; each of 77 tracks is divided into 26 128-byte sectors. Later, the industry moved to reach two new goals simultaneously. The first was to develop a lower performance, smaller unit: the $5\frac{1}{4}$-inch drive. A series of even smaller drives, typically $3\frac{1}{4}$ inch, are also manufactured and now achieving common use. The second goal was to increase the capacity of the drives. Changing the encoding scheme doubled the capacity, and placing a read–write head on either side of the diskette so that both surfaces could be used also doubled the capacity. The density limit has not been reached; "quad density" floppies are now in widespread use, and floppies with greater density are to be found in some computers.

Consequently, there are several important variables in floppy disks:

- *Density*: single, double, quad, or higher
- *Number of sides*: one or two
- *Diskette size*: 8, $5\frac{1}{4}$ or sub-5 inch.

Physically, the floppy disk medium, the *diskette* (occasionally spelled discette) consists of a round sheet of mylar plastic coated with iron oxide. A carboard jacket with appropriate holes protects the surface from dust and fingerprints (see Fig. 2.14); the cardboard is lined with a soft material that continuously wipes the surfaces. When placed in the drive and the door is closed, the diskette is clamped to the spindle and rotates at either 360 rpm for 8-inch diskettes or 300 rpm for 5-inch diskettes.

The starting location for positioning the sectors is determined photoelectrically using a light-emitting diode light source and a phototransistor detector on opposite sides of the diskette to detect the rotating index hole.

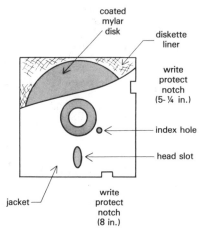

FIGURE 2.14. Flexible diskette construction. Diskettes of this type come in 8- and 5¼-inch styles; the 3½-inch diskette is mounted in a hard plastic sheath.

The head slot provides access by the read–write head. In 8-inch drives, when read–write operations are desired, the computer directs the drive to activate a solenoid which allows the head to ride directly on the surface of the diskette; when the operation is complete, the head is lifted, so that the operator hears the clicking of the head during disk access operations. The heads remain continuously in contact with the media in smaller drives, but the rotation motor usually stops when the drive is not in use.

Single-density 8-inch diskettes, formatted according to IBM 3740 specifications, are customarily sold preformatted. However, only de facto standards exist for double-density or 5¼-inch disks; it is rare that a double-density diskette written by a system of one manufacture will be readable by a system of another manufacture even if the same OS is used. Consequently, software that runs on diverse computers but under the same OS, CP/M for example, can be distributed on single-density 8-inch disks. The de facto standards for 5¼-inch disks are usually named for the computer in which they are used: Apple, IBM, NorthStar, KayPro, etc.

Floppy diskettes are vulnerable to accidents and abuse, which can damage or destroy them. An error can lead to the loss of part or all of a file, and an error in the directory can make it impossible to find any of the files on the diskette.

Errors also may occur if the head or diskette is dirty or if electrical noise distorts the signal; if an error in reading a sector occurs, the OS normally requests that the sector be read again, typically up to ten times. If the error persists, it is probably due to a damaged diskette and is a "hard" error. Some hard errors can be corrected if they are due to dirty read–write heads or misalignment of the heads. Dirty heads should be cleaned with a cotton swab and isopropyl alcohol or a head-cleaning kit, but the alignment of the drive requires a special diskette, an oscilloscope, and some patience. Fortunately, realignment is not often required.

Winchester Hard-Disk Systems. In order to significantly decrease the access time and increase the storage capacity of a disk surface, improvements must be made in the uniformity of the surface and the disk must be rotated at a significantly higher rate. The use of heavy aluminum platters rather than plastic for the medium provides a better surface. The head also had to be redesigned; allowing the read–write head to ride directly on the surface as it does on the floppy diskette is precluded at high speeds. The development of the air-bearing head enabled high rotational rates while keeping the head in close proximity to the disk surface; the head rides aerodynamically on a cushion of air while rotating at 2000 to 3600 rpm. This type of disk is commonly called a **Winchester**; (it is said that the name comes from IBM prototypes which had two 30-Mbyte units and were called 30–30s as were certain Winchester rifles).

The distance of the head from the surface, or flying height, is small even when compared with particles of tobacco smoke; a high degree of cleanliness is mandatory if damage to the head and surface are to be avoided (see Fig. 2.15). The simplest means of maintaining that cleanliness is to either seal the disk compartment from the atmosphere or to allow the compartment to breathe through a filter. Under these conditions, the flying height can be reduced to under 0.5 μm, so that over 42 Mbytes can be stored on three 8-inch surfaces. Up to five double-sided platters can be mounted on a single spindle. The transfer rate may exceed 1 Mbyte/s, and, typically, the average access time is 5 ms.

Winchester technology disks are becoming the dominant form of mass storage in this decade. They are usually nonremovable and are manufactured in 3-, 5-, 8-, 10-, and 14-inch diameters with the drives of the smaller sizes designed to be the same size as the floppy disk drives that they are to replace. The *unit* cost of Winchester disk drives is substantially greater than that of

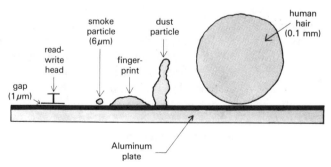

FIGURE 2.15. Head height of a Winchester disk compared with dirt sources. The importance of a sealed environment is illustrated. Flexible disk heads ride on the surface at lower speeds and are affected to a lesser degree.

floppy disk drives, but the cost per kilobyte, particularly with larger units, is about a tenth of that of the floppy.

The main disadvantage with this type of disk is the inability to store data off-line, and a separate mass storage device is required for backup. The devices of choice are the streaming cartridge tape and the floppy disk. The cartridge tape is in many ways ideal, since when the information on the disk is dumped in a single operation, the tape can be moved rapidly and has high capacity. There are two advantages to using the floppy disk for backup, however, (1), on many laboratory computer systems, the floppy drive was already in place before the Winchester was added, and (2), the floppy diskette is commonly used for the interchange of software and data between machines.

Hard errors do occasionally occur on hard disks in spite of the various precautions. Since the media cannot be repaired or replaced outside a factory clean room, "spare" tracks are provided. When a hard error is detected, a track can be substituted either by a utility program or by the controller itself, so that the bad track effectively disappears.

Removable Hard Disks. Two types of hard disks are available in which the media can be removed and stored; these actually preceded the Winchester by almost two decades. The first is the **cartridge disk**, a single platter housed in a plastic shell which can be inserted in the drive from the side. The second is the **disk pack**, a unit containing up to 12 platters on a single spindle which is mounted on the top of the drive. Both types are presently used in larger and in older computers and are seldom cost-effective on smaller computers when compared with the combination of a Winchester drive and either a floppy disk or streaming tape drive.

2.5.2. Magnetic Tape

The concept of storage of information on magnetic tape is quite familiar since the audio tape system is quite ubiquitous. However, the recording and reproduction of digital data involves problems different than those encountered in the recording and reproduction of sound. As is well known from elementary physics, when a magnet is moved past an electrical conductor, a current is only induced in the conductor when a *change* in the magnetic field is experienced; this matches the requirements of reproducing sound since audio information is encoded in the *frequency* domain. However, digital data is electronically represented as voltage *levels*, and voltage levels cannot be directly recorded; a conversion is required.

There are several methods of representing digital data in a form suitable for magnetic tape recording, but these are outside the scope of this study. It is sufficient to point out that special circuitry in the interface is required for digital data recording and that where *audio* tape recorders are employed in computer systems, the binary levels must be converted to frequencies in the audio range, a severe limitation.

The types of magnetic tape systems may conveniently be categorized according to the method of handling and the physical package. The three major types, shown in Figure 2.16, are cassette, cartridge, and reel-to-reel, each having several important subclassifications.

Cassette Tape Systems. The (Phillips) tape cassette package commonly used in home tape recorders represents a convenient method for low-cost tape packaging. Recording is generally made on a single track so the data are stored in a *byte serial* format. That is, not only do the bytes follow each other sequentially, but the bits in each byte also are encoded serially.

Audio recorders have been employed as very inexpensive storage devices by many entry-level personal computers. They are inexpensive, but very slow alternatives to disks.

In order to take advantage of the very low cost of portable audio tape recorders, coding methods had to be developed which would overcome the problems of the limited frequency range of the read–write heads and the poor speed regulation of the drive motors. Typically, due to the difficulties of precise calibration, tapes produced on identical systems may not be interchangeable, even though they work on the systems on which they were recorded.

Digital–Data Cartridge Tape Systems. In order to retain the packaging convenience of the Phillips-type cassette but to obtain better performance and reliability of the tape system, the **digital–data cartridge** was developed by the 3M Company. By mounting the reels on a rigid aluminum plate, protecting the tape from dust, driving the tape with the capstan rather than the hubs, and using superior quality components throughout, high speeds and capacities can be obtained.

Cartridge systems are best suited to applications in which the data are

FIGURE 2.16. Magnetic tape. The 9-track tape is shown on the left. A Phillips-type cassette is located in the lower center. The cartridge on the right may be the only form useful for small computers.

continuously recorded, since the greatest stress is placed on the system during starts and stops. The cartridges were originally developed for use in data-acquisition systems, for continuous monitoring, or for recording events such as bank transactions. However, the fastest growing application is now for backup or archival storage. A fast access device such as a Winchester disk may be used for general on-line storage of programs, data, libraries, and so on; however, in order to safeguard the information in the event of malfunction, the contents of an entire 20- to 30-Mbyte disk can periodically be copied to the cartridge for safekeeping. Since in backup applications, the drive does not start and stop, it is often called **streaming tape**.

Reel-to-Reel Tape Systems. Typical tape reels used for high-density data recording hold ½-inch magnetic tape with a typical length of 2400 feet on a 10½-inch diameter reel. In order to both increase the data density and decrease the access time, nine read–write heads are used to record the data in parallel, laterally across the tape; a character or byte can then be stored in parallel using nine tracks; 1 bit is available for error detection. The data density varies up to 6250 bits/in., and the speed of the drives varies from about 12.5 to 200 in./s. The higher speeds are quite expensive and clearly out of the range of the small computer user unless compatibility with other equipment is mandatory.

2.5.3. Optical and Other Mass Storage Systems

Optical disk storage is an emerging technology. At this writing, optical disks are available for video reproduction and as "compact disks" for digital sound reproduction, and some of these are adaptable for computer-compatible digital data. The technology promises to have a large impact on the scientific user, but the way that optical disks will be used depends on further developments. We can expect to see 10-Gbyte (10^{10} bytes) removable disks, capable of storing on the order of 500,000 pages of text and figures with very low error rates.

We can make three classifications of the technologies:

1. At present, most technologies are *read-only*; the writing must be done at the factory, but duplication costs are only a few dollars per copy in high volume. One can envisage all types of databases being made available on-line; an entire encyclopedia is now available on a single platter.
2. A few are developing *write-once* processes. These will be very valuable to the pharmaceutical industry, among others, that must keep copious amounts of raw data for regulatory agencies. Some large research labs have to keep many megabytes per day, filling warehouses with tapes that must be periodically maintained. These disks will greatly reduce the space and maintenance requirements.

3. *Read–write* (erasable) capability is still in its infancy. If fully developed, these optical disks could replace today's magnetic technologies.

Typically, an optical disk is derived from a glass substrate coated with a multilayer photoresist. A modulated laser beam either burns pits or raises bubbles in the photoresist less than 1 μm wide, which encode the information. The read process uses a low-power laser to detect those pits. The bubbles produced in some types of photoresists can be removed by annealing them with a low-power defocused laser.

Circulating memories, considered in Section 2.3, have certain applications in which moving parts cannot be tolerated. Bubble memory systems are often designed to appear to the computer's OS like a disk drive. They have the advantages of moderate speed and reliability, making them excellent for ruggedized systems. However, these units still must be backed up by floppy diskette or tape. They will probably remain expensive when compared with disks as long as the production volume stays small.

Memory drives are blocks or banks of memory, from 64 kbytes to several megabytes, which the OS uses like a disk. Consequently, transfer is very fast, although the contents are volatile. Memory drives are useful for storing intermediate results during program compilation or experimentation. As memory has become less expensive, memory drives have become more popular, particularly in 16-bit computers that can address much more memory than is typically used by the programs.

The discussion of mass storage should not end without mention of systems that either have limited application or are relics of the past. The most significant relic is the punched **paper tape** system. A descendent of the ASR 33 Teletype teleprinter with tape reader and punch, the later paper tape systems were developed with the capability to punch and read paper tape at speeds up to several hundred characters/s. Punched tape has the psychological advantage that the data are visible, encoded as a series of holes, and invulnerable to electrical or magnetic hazard. (Stories exist, probably apocryphal, of entire cabinets of data on magnetic tape or disk being wiped out by the field of a passing floor polisher). However, the bulk (1 Mbyte requires 1½ miles of tape) and the inconvenience have nearly eliminated it.

2.6. THE BUS

Five hardward components have been identified as parts of the computer: the CPU, memory, I/O ports, operator interface peripherals, and mass storage peripherals. We have seen that the CPU communicates with each of the other components by first specifying that component's memory or I/O port address, and then accepting or sending data. There are typically 8 or 16 bits of data, and 16, 20, or 24 bits define the address. In order to define whether the CPU wants a memory or I/O address and whether it wants to send or receive the data, it also must generate *control* signals.

In microprocessor-based computers and other modern computers, each of the main components is a separate circuit, often on a separate printed circuit module. In order for a memory or I/O circuit to recognize that it is being uniquely addressed, a circuit with the responsibility of decoding the address and control signals is required. When the computer's separate functions are placed on separate printed circuit modules, the decoding responsibility is placed within that module. Consequently, computer systems can be designed in which a series of equivalent sockets are provided, each having all of the address, control, and data signals available. When new modules are added, each has access to all the address, data, and control lines required for its function; it also depends on that socket for its power. The group of conductors carrying these signals is termed the **system bus**, and a computer that contains equivalent sockets having all of the bus signals is said to be bus-oriented (Fig. 2.17).

In the history of computer engineering, the system bus is a relatively new development. In earlier computers, which had thousands of discrete components making up the CPU and a limited number of peripherals with which it communicated, each data channel or memory unit was simply wired directly into the CPU; a common set of address, data, or control signals could not be identified. The development of the bus-oriented computer, principally

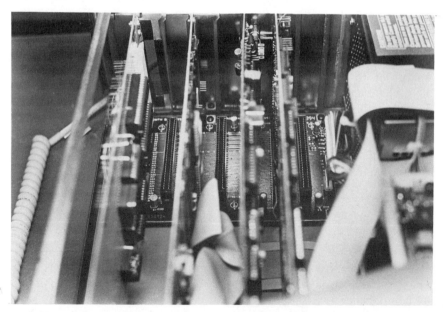

FIGURE 2.17. A computer bus. Seven equivalent slots on an IBM PC bus can be observed. Four of them contain cards, but three more are available for expansion of the computer.

by DEC in their PDP-8 and PDP-11 series computers, led to increased flexibility. At first, the user needed to purchase only as much memory or I/O as required at that stage, but later expansion required simply plugging in additional modules. A second source of flexibility follows when manufacturers other than the original vendor produce new circuit modules for bus-oriented computers that can be accommodated as long as spare sockets exist; the benefits lie in both price and choice.

One should be aware that the term "bus" is used at several levels: an *internal bus* is found within the CPU and is not accessible to the user, and each module may have its own set of signals, buffered from the system, that might be termed a *local bus*. Furthermore, many popular small computers have no external system bus; all of the functions are placed on a single printed circuit board, and any expansion requires use of the *expansion bus*, a single connector containing the bus signals. The latter approach has great advantage in that it lowers cost dramatically by (1) eliminating one of the most expensive components—connectors, (2) increasing reliability by eliminating a source of problems, again the connectors, and (3) allowing the designer to use power supplies with only the minimum required capacity. The Macintosh, Commodore-64, and TRS-80 series of computers are prime examples as are most home computers priced under $1000. However, the lack of a system bus also eliminates one of the prime advantages of laboratory computers, the flexibility in expansion.

Standard Buses. System buses generally are first proposed by a manufacturer, which places a product in the marketplace. If that product is accepted, other manufacturers may provide products that are compatible with it, and a de facto standard results. In certain cases, the independent manufacturers also develop entire computers that use the same bus, and in such a case, it is often necessary for a committee to define the logic and timing of each signal on the bus, since those definitions could vary among manufacturers.

A case of the former exists with two buses developed by DEC, the "Unibus" and the "Q-bus," the latter developed for the LSI-11 members of the PDP-11 family, and the former for all but the largest members of that family. The bus signals were well defined, and dozens of other manufacturers have moved to duplicate each of DEC's offerings as well as new functions. A similar case exists at present with the IBM PC and the Apple II.

The necessity for a standard arose during the development of the IEEE 696 (S-100) bus. The bus was initially defined by the manufacturers of the Altair 8-bit computer; almost immediately after its introduction, other companies emerged as alternate sources of both the component modules and the entire computers. Soon, new signals were needed to handle unforseen functions and later to accommodate the arrival of 16-bit microprocessors. Each manufacturer did as it thought best until the oft-used phrase, "S-100 bus-compatible" became nearly meaningless. Finally, in 1981, a committee of the IEEE developed a standard, the IEEE 696. This standard defines the activity of each of the 100 lines of the connectors including function and timing; levels of compatibility also can be specified.

Several other buses developed for microprocessor-based computers that have a wide following are the "STD" bus from ProLog Corporation, the "Multibus" (I and II) by Intel, and the Versabus and VME buses from Motorola.

2.7. USER-INTERFACE PERIPHERALS

The computer is an efficient device for moving and manipulating information that can be encoded in a digital format. One of the primary sources and recipients of that information is the human user. Although most people are accustomed to thinking of information being *typed in* or *read out* on screen or paper, we can now go much farther than that.

Reading is one of the least efficient methods of obtaining information. By developing capabilities to draw "pictures" and other information in graphic form, we can dramatically increase the efficiency of information transfer to the operator. In certain cases, significant advantage can be gained by providing the operator with a degree of mobility; voice output is then useful.

Similarly, the abilities to *enter* information by techniques other than typing are attractive. Information can be directly entered graphically, by voice, or direct from a printed page. The types of peripherals that provide these capabilities are listed in Table 2.3. A brief discussion of each follows.

2.7.1. Terminals, Printers, and Optical Character Readers

This group of devices deal with character information. On a terminal, we enter, character-by-character, information that has been converted to character strings. Optical Character Readers move information from the printed page. Printers return that information to the printed page.

The simplest approach to I/O is the use of a terminal. The terminal consists of two components that, from the vantage of the computer, are independent, but are usually designed to be a single unit, physically: a keyboard provides input from the user, and a display unit provides the output. If the output

TABLE 2.3. User Peripherals

Form of Information	Input Peripherals	Output Peripherals
Alphanumeric	Keyboard	Printers and VDUs
Graphic	Digitizers	Plotters, Graphics VDUs
Speech	Voice Recognition	Voice Output units

display unit is a printer, the device is a **teleprinter** and if the device uses a **CRT** or a similar nonprinting output, the device is a **VDU**.

The independence of the keyboard from the output is an important concept. In a small computer environment, when a key is struck and the corresponding character appears on the output unit, the code for the character is sent from the keyboard to an interface where it is read by the computer; then the computer will echo the character by sending it to the output interface, which relays it to the display unit.

Keyboards. The keyboard consits of a matrix of momentary switches and the digital logic to generate the ASCII code corresponding to the depressed key(s). The designs of the keyboards in use in various computers involve a myriad of techniques for the detection of a key depression including electrical contact, change in capacitance, and Hall effect (a magnetic phenomenon). The trade-offs involve the feel, the cost, and the reliability.

Keyboards also vary in layout. The standard typewriter layout (called the QWERTY arrangement, named for the top row of alphabetic keys) is nearly always used for the placement of letters and numbers. However, the locations of the other characters and even the RETURN, TAB, and CONTROL keys are not standard, so that it is often frustrating to move between terminals.

Office managers have often lamented that the typewriter keyboard was intentionally designed to keep the operator from typing too fast. The keys were placed so that common combinations would be inconvenient, reducing the likelihood of a jam of the keys of a manual machine. Tests have shown that even random key placements would be more efficient, and several designs have been developed and tested that optimize finger movements. One example is an efficient one-hand sphere utilizing finger combinations. Another, for which keyboards are actually on the market, is the Dvorak arrangement. However, the enormous investment in machinery and training behind the present layout has killed any chance for improvement.

Teleprinters. The combination of keyboard with printer is sometimes a lower cost alternative to using a VDU and independent printer. The printing unit must have some speed constraints since it must be able to start and stop for each character printed; nevertheless, printers of remarkable speed have met this challenge, so that speeds up to 180 characters/s are possible. Even so, teleprinters are less desirable in small computer systems than VDU–printer combinations.

Video Display Units (VDUs). VDUs are more efficient devices for displaying text since most can display data at rates up to 1000 to 2000 characters per second. The simplest VDU simply displays the characters as they arrive from the computer and moves them up a line (scrolls) when a linefeed character arrives.

Nearly all new VDUs employ microprocessors to manipulate the characters in an array of their own memory, which is electronically mapped onto the screen. A number is deposited in a given memory location, and the VDU will display its ASCII representation at the equivalent position on the screen.

Since the ability of the terminal to perform other functions on the incoming information rests with the internal microprocessor, it becomes a relatively simple matter to add functions that are collectively considered "intelligent" functions. Application programs such as spreadsheets and word processors require abilities to vary the intensity, move the cursor, insert or delete lines, and manipulate bodies of text smaller than the full screen. In the laboratory, these functions will become essential as well, as programmers learn the value of making data entry routines operate through a screen that looks like a form, or provide menus that can be made to disappear and reappear over a graphical image of the data.

Although most VDUs employ CRTs, there are applications in which these are not suitable because they are either too fragile or too bulky. Plasma panels are an expensive alternative. Efforts are underway to produce VDUs with a flat panel display that is also inexpensive; liquid crystal, electrochromic, and light-emitting diode (LED) panels are possibilities. New lap-size computers are already using these large liquid-crystal displays although visibility and contrast presently compare poorly to CRTs.

Printers. Most small computer systems require a printer or at least access to a printer. These come in widely varied price and speed ranges. In many environments, a user has access to both types: an inexpensive printer attached to the computer, and a top-quality printer available through a LAN.

Characters are printed by two principal methods. The first produces *fully formed* characters. A hammer strikes a wheel or thimble on which each character has been engraved. Fully formed character printers are often termed "letter-quality," since the characters have the quality of typewritten characters; the detail depends on the engraving quality, and fonts can be changed by manually changing the wheel or thimble.

Dot-matrix printers form characters by sweeping a print head with pins in 7–29 positions across the page. A character is formed by discharging individual pins. The dots are produced either by impact of the pins on a ribbon, by ejecting microdroplets of ink, by thermal effect on treated paper, or by electrostatic effect on treated paper. In laser printers, the laser beam creates the image by a xerographic process.

The quality of dot-matrix print depends on the size of the matrix; simpler printers form characters in a 5×7 matrix that severely limits the detail; often, the "tails" or descenders of the lower case leters, g, j, p, q, and y do not descend below the line and can be difficult to read. These descenders are printed by units that can print a larger matrix.

Newer dot-matrix printers, using up to 24–29 dots in the vertical dimension of the matrix, form characters with a resolution rivaling those of the fully formed character printers. In addition, they are faster, and of great importance to scientific users—character fonts can be defined in software. Laser printers also print by placing dots at high resolution, typically 300 dots per inch. The laser printer hardware has the capability of mixing an infinite number of type-styles with Greek letters, symbols, even graphics.

Dot-matrix printers can be given additional capabilities by their manufacturers without adding to the hardware; in particular, many have the capability to operate in a "dot-addressable" mode. Instead of sending an ASCII character, which the printer converts to the corresponding series of pin discharges for the appropriate character matrix, a character sequence can be sent that allows the user to control each individual pin. The user can then define the character matrices if special characters or character sets are desired. Reasonably high-resolution graphics can be produced; if the printer prints 80 characters across an 8-in. line, and each character is represented in a matrix with a width of 12 dots/character, the horizontal resolution in a dot-addressable mode would be about $\frac{1}{120}$–$\frac{1}{300}$ inch, and the typical vertical resolution would be at least $\frac{1}{72}$ and up to $\frac{1}{300}$ inch. Letter-quality dot-matrix printers may produce graphics with a resolution in both dimensions of about an inch.

In order to satisfy the graphic and text needs of scientific document preparation, printers that generate a matrix of dots will eventually dominate the scientific market. The hardware is in place to create a page using a microcomputer with mixed text and graphics and with various fonts and symbols. The result would be virtually indistinguishable from typeset print on a printer of sufficient resolution (300 dots per inch or more).

Optical Character Readers. Although this technology is not immediately germane to most laboratory environments, there are cases in which direct character input would be valuable. Optical Character Readers (OCRs) are capable of taking printed text and converting it to an ASCII file.

OCRs are valuable to publishers such as the Chemical Abstracts Service, which must publish some previously printed abstracts. Conversion of older manuscripts to a form that can be manipulated by machine may be useful in individual laboratories, but the technology may be more important for record-keeping: for example, reading labels on x-ray films.

OCRs operate by first digitizing the image of a single character, and then matching it with tables. An OCR may need to deal with varying fonts, spacing, character size, and print quality. Consequently, a low error rate is difficult to achieve.

The simplest OCRs, often used in department stores and libraries, operate only on a font optimized for the purpose. OCR print wheels are available for many printers and deliver the famliar "technical-looking" font. The most complex OCRs can scan a page and "learn" new fonts using sophisticated pattern recognition techniques.

2.7.2. Graphics Devices

The importance of graphics devices to the efficient use of the computer system is difficult to underestimate. The adage "a picture is worth a thousand words" is again applicable; the time saved in obtaining the information in easily assimilated graphic form instead of attempting to gage trends in data from printed tables will very often save what appears at first to be significant cost.

Besides being able to readily understand the data, the operator is also able to manipulate data rapidly when doing so interactively with the computer. In many cases, the innate ability of the user to recognize patterns in complex data is much more powerful and easier to develop than a software algorithm. However, that ability is only put to use if the user can readily display the data at will over any desired range or possibly as differences, derivatives, or integrals. Each entity—user and computer—is able to do what it does best, efficiently.

A complete graphics configuration for a computer could contain four components: a VDU with graphics display capability, a graphics hard-copy device, a digitizer for entering data that is already in graphical form, and software that takes advantage of the hardware capabilities. A detailed discussion will be saved for Chapter Nine.

2.7.3. Voice Peripherals

This section, at the time of publication, may seem to be either only wishful thinking or unnecessary ornamentation. However, the technologies are available for both producing speech under computer direction and accepting spoken commands, and both capabilities can aid the scientist. A speech unit would have definite application in a computer system which aids research, which must be carried out in total darkness, such as is required in certain optical experiments. Just as frequently, an operator might need to accept information while manipulating an experimental parameter away from the console. In other cases, the experimental sequence might be previously entered in a disk file so that the computer can announce steps at the correct times. In the same types of situations, the capability of entering information by voice could be useful, although this capability is much more difficult to provide. In the following sections, a brief description of voice I/O units will be provided.

Speech Output Units. There are several techniques for the computer generation of natural speech. An obvious technique is **waveform synthesis**. Carefully spoken words and phrases are digitized at a rate twice the normal speech bandwidth of about 4 kHz; the data for each sample are stored sequentially to be reformed into an analog signal via a **digital-to-analog converter (DAC)**. Unfortunately, even quantizing into only 8 bits, 8 kbytes of memory are required per second of speech. Although the resultant speech can be very pleasant and natural, the memory requirements are prohibitive.

Careful study of the sounds produced in the vocal tract has led to a technique for their reproduction called **linear predictive coding (LPC)**. The particular sound is synthesized by feeding a series of periodic impulses and white noise into a programmable 10-stage filter; for highest quality speech output, control of the impulse frequency, energy, and the filter requires up to 50 bits, but that sound can be sustained for a frame length of at least 25 ms. The memory requirement is reduced to 1–2 kbits/s depending on the required quality.

Modeling speech still further, English, as spoken in the United States can be divided into 42 building-block sounds called "phonemes" and further divided into 128 "allophones." A library of allophones and a set of rules can be placed in ROM for direct "text-to-speech" conversion; the result is not entirely pleasant, but is understandable. Alternatively, spoken speech can be edited for "analysis synthesis" with an interactive computer, and the LPC data for each word placed in ROM; the result is a limited vocabulary but a pleasant sound. Inexpensive subsystems can be obtained for either technique, with the trade-off of large vocabulary in text-to-speech systems for speech quality, with only 100 or so words in ROM for analysis synthesis systems.

Voice Recognition. Recognition of speech is somewhat more complicated than generation. Voice recognition techniques must store reference waveforms and cannot easily account for differences in voice between users, unless substantial computing power is applied. At this writing voice recognition cannot be classified as an inexpensive addition, although the capability is available on some small computers.

"Smart" keyboards are available that accept spoken words and convert them to the equivalent or predetermined character strings. A training period is required in which each word is spoken into a microphone to be stored by phoneme along with a character string that may be the spoken word or any command string.

Problems still exist. Distinguishing "HI" and "BYE" is difficult. After a user has "trained" the unit, it might not recognize the same person if he or she has a cold. Background sounds might be recognized as words or interfere with recognition. Complicated systems of thresholds are necessary to determine how close an utterance must be to the reference in order to be recognized.

Laboratory applications are easy to envisage, but a greater degree of error checking is required just as is the case when spoken communication between people is used instead of written communication.

CHAPTER THREE

Software:
Systems and Languages

Software is included as a central component of the computer system, as shown in Figure 3.1, since effective use of the computer requires quality software. Three types of software introduced in Chapter One were the OS (with utilities), the development software, and the application software. To reexplain these in a few lines: the application software consists of the programs that solve the user's *specific* problems; the development software is the group of languages and other development aids used in the production of application software; the utilities and OS manage the computer hardware, simplifying and making the computer process more efficient.

The software goal of the laboratory user is to have effective application programs. It may be that these programs are available commercially or it may be that they have to be written locally. If these programs are to be written by or specially for the user, the programmer must have efficient tools for preparing and debugging the programs.

Two questions concerning software are central to using the small computer in the laboratory. First, is sufficiently capable application software already available ("off-the-shelf")? When the answer is yes, the fastest and easiest solution may be at hand. When the answer is no, a second question follows: What is the most efficient path for the generation of that software? This question leads to development software, a topic that will be discussed in the remainder of this chapter.

3.1. SOME BACKGROUND AND DEFINITIONS

To begin to understand software needs, we must understand the type of instructions that the computer actually executes. If a single line in a computer

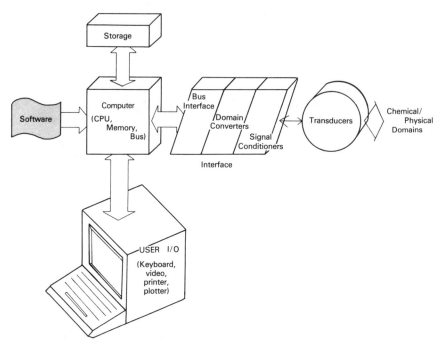

FIGURE 3.1. A laboratory computer system. The small computer's software is highlighted for study in this chapter.

program specifies that "if *n* is greater than 5 then print *n* on the screen," many primitive machine instructions must be executed to make that comparison and to perform the appropriate actions. The CPU follows a program composed from its own repertoire of instructions, which are discrete **operation codes (op codes)**. The OP codes are relatively primitive so that even a simple task may require several hundred steps.

After learning the set of op codes for a particular CPU, it is *possible*, with sufficient care and patience, to manually prepare an application program by assembling the codes for a sequence of machine instructions to complete a task. Great attention to detail is required: the absolute memory address of every operation code and data storage location must be fixed by the coder. The programming time required by this method is prohibitive for programs longer than a few score instructions.

The process is improved with the use of a software development tool called an **assembler**. Each machine instruction is represented by a **mnemonic**, and the collection of mnemonics makes up an **assembly language**. The programmer prepares a list of mnemonics that represent the program instruction sequence; mnemonic labels are used to represent locations in memory. This list is stored on a storage medium such as a disk. The assembler program reads the stored code and produces a listing of the code that is executed by the CPU by interpreting the mnemonics and determining the absolute memory addresses for each of the labels.

The assembler approach, though much better than manual assembly, retains several disadvantages. Foremost, program development is slow, since so many instructions are required for a given operation. Second, since each CPU has its own instruction set, a new skill must be developed each time a programmer faces a new computer. Furthermore, the program written for one type of computer cannot be executed on another type, that is, it is not *transportable*. Third, the mnemonics are not usually English language words and are therefore tedious to learn. Finally, there are few programming constraints in assembly language programming so that many more kinds of mistakes are possible than with higher level languages.

The computer user needs a language with a syntax that makes the programming job easier. Assembly language generally requires one written instruction for each executed instruction; it is *machine-oriented*. However, the *user* comes to the computer with a specific problem and needs development software that can be applied easily to that problem. The ideal environment would allow the user to state the problem in terms with which he or she is familiar and to ask the computer to translate that to the computer's language. The best available answer is a language that is *problem-oriented*; that is, the programmer is allowed to write statements in the language of the problem instead of the language of the machine.

A **high-level language** results from an attempt to provide a problem-oriented programming aid. The program **source code** can be written in English-like phrases that are relatively easy to learn; the language program package will convert the source to the executable **object code** at the machine level. Typically at least ten to a few thousand machine instructions are required for each command in high-level code.

Many experts say that a programmer will produce a relatively constant number of lines of code per day whether that code is written in high-level language or assembler. A high-level language allows the objectives of the programmer to be met more rapidly than would be possible with assembler alone. Although languages prepared specifically for problems in the areas of, for example, spectroscopy or statistical mechanics may not be readily available, languages have been developed specifically for instrument control, computation, or information sorting. The right language choice can greatly facilitate application program development. Furthermore, when a program is written in a given language, it can be converted to machine code for any of a variety of computers with minimal difficulty.

There is a price to be paid for this improvement in affairs, but usually it is acceptably small. If one compares programs written in a high-level language with those in assembler for the same task (both written by skillful programmers), the high-level method is less efficient in terms of both memory usage and execution speed. A fixed development tool simply cannot translate from source code to machine code and optimize the result as well as a skilled programmer.

This penalty may provide assembler level programming a place in the laboratory when high-speed applications or memory-limited applications are required. However, memory is normally inexpensive, and a programmer's time is normally expensive. Therefore, because of the penalty in preparation time, it is usually prudent for persons whose primary objective is experimentation rather than computational efficiency to accept the slightly slower execution time. Assembly level programming has been placed largely outside the scope of this book.

Because there are many types of problems and varying opinions of how best a problem can be encoded in a program, high-level languages abound. They differ from each other in several ways, two of which are important to the user. The first difference is in commands and functions: What operations can be done with the commands that are provided? (For example, can the tangent of an angle be computed with a single command, or would the programmer using that language have to write the routine to compute a tangent using only multiply–divide functions?).

The second way in which languages differ is the stage at which the source code is translated to object code. There are **interpreters** that make this translation line-by-line as the program is executed, **compilers** in which the source code is subjected to a single translation (compile) step, and **threaded-code** languages that compile each line as it is entered at the computer's console, and store it, in machine-language form, for future use. Each language has its own strong advocates, and it is often difficult for a user to determine which is best to apply to the problem at hand.

In this chapter we will begin the study of software with the component that is responsible for general "housekeeping," the OS. Following that, several language options will be explored and compared. Next, since readers with previous programming experience on larger computers may not be familiar with methods for controlling interfaces directly from high-level languages, the special instructions for that purpose will be explored. Finally, some useful types of application programs will be introduced.

3.2. OPERATING SYSTEMS

Even though a program is prepared for a given CPU according to that CPU's instruction set, there is yet another level of nonstandardization that the program must face. Each type of computer has its own combination of peripherals (terminal, printer, disk, etc.), and the software required to communicate with the interface varies with the hardware. Therefore, it is useful to employ a separate module of software that may be customized for a given hardware configuration through which the application program can communicate in a generic fashion with each of the specific devices. Consequently to the writer of software, many different computers will appear to be the same. That software is the OS or, since the mass storage unit is usually a disk, the *disk operating system* (DOS).

The first assignment of the OS is to handle the I/O with respect to the user and the mass storage; that is to control the terminals, printers, disks, etc. Any development or application software that is written to be compatible with the OS of a particular computer need not concern itself with the details of the I/O procedures, but can make service calls to specified routines that pass data to and from the OS which in turn handles the physical I/O.

The OS has at least two other components. One, the set of **utilities**, is a group of commands and/or programs that can handle special housekeeping functions, most of which are related to the disk. For example, the OS al- locates several tracks of the disk for the directory which contains the names of all the files and information about their absolute locations on the disk. Consequently, a DOS includes commands such as one that causes a list of files as read from the directory to be printed on the console, possibly with information about the sizes of the files, dates, and their locations on the disk. Other commands are used to copy files from one disk to another, to rename files, to edit files, or to erase files. (The erase command normally does not actually erase the file from the disk surface, but removes the information about the file from the directory, so that the area on the disk can later be overwritten.) A utility not directly related to files might be one that sets or prints the time of day or date.

The other component, the **monitor/executive** section of the OS, handles the user's interaction with the console and the scheduling of the computer in multiuser systems. The monitor sends a prompt to the console that the user may answer by requesting a utility command or by requesting that a particular program be executed. The executive handles the internal timing clock, often based on pulses derived from the line frequency (50 or 60 Hz) in order to compute the time of day, and it switches between users in a multiuser system.

There are several commercial OSs that are sufficiently common to merit mention in a general discussion. A "Microcomputer Control Program," CP/M-80 (Digital Research) was first designed for computers using the Intel 8080 microprocessor; it is also upward-compatible with the Intel 8085 and Zilog Z80 microprocessors. Since each different computer that employs those microprocessors may have a different combination of peripherals, the OS is distributed with guidelines for customization for a particular configuration; the manufacturer of the computer system is usually responsible for the complete system customization.

When 16-bit microprocessor-based systems became available, the same set of commands and the same disk organization was implemented to produce an OS for the Intel 8086 microprocessor, CP/M-86, and a similar version for the Motorola 68000. As the demand for software supporting multiuser systems grew, Digital Research developed Concurrent CP/M, a foreground–background version (see Section 1.5.1.) of CP/M-80, and also MP/M-80 and MP/M-86, the multiuser analogs.

Since CP/M-80 is implemented on many different computers using 8080 and Z-80 microprocessors, a programmer can write a CP/M-80-compatible program that uses the CP/M-80 I/O and disk functions and expect it to run on many different computers with different types of disks, terminals, and other peripherals.

The Intel 8086/8088 has become very important in personal computing, particularly with the advent of the IBM PC, the PC-AT, the DEC Rainbow, the Zenith Z-100, and other similar and compatible computers. With IBM support, MS-DOS (and variants PC-DOS and Z-DOS) became the most important OS for 16-bit personal computers. Again variants developed which provide multitasking and multiuser environments.

Another well-known OS developed when programmers at Bell Laboratories gave careful consideration to the utilities that should be available in an OS and the methods for moving information between tasks performed by a computer. The result was UNIX, an OS that has probably been implemented on more CPUs than any other. Many "serious" computer users prefer UNIX, enjoying its powerful abilities to organize and direct information. UNIX is popular as a multiuser system for program development and many aspects of computer science research.

UNIX is powerful and as such consumes a lot of the computer's resources. Practical implementation on single-user computers suitable for the laboratory are beginning to become available as the memory capacity and power of microcomputers grow. However, the particular needs of real-time on-line experimentation were not considered in the development of UNIX, and it has not been important for lab computers, but its most valuable features have been included into other OSs.

One special OS feature sometimes found on larger minicomputers, and increasingly in 16- and 32-bit microprocessor-based systems, is **virtual memory**. In the hardware and OS that supports virtual memory, the application program operates as though the computer has a much larger memory than actually exists; sections of the mass storage are treated as though they are actually RAM. When memory is referenced which actually exists on the disk, that segment is transferred from the disk to a transient area of the actual memory; although this slows the execution significantly, it makes possible the execution of extensive programs.

Simplified alternatives to virtual memory, which commonly are implemented on small computers, are the use of **overlays** and **banked memory**. With overlays, when a program is too big to fit entirely in memory, the program is segmented by the programmer so that sections that are not used simultaneously can overlay each other, being swapped from the disk during program execution. In this case, the size of the program is limited only by disk size, although the size of data memory may still be limited. A relatively simple hardware approach for larger programs requires using some hardware feature to select one of several physical banks of memory, so that different segments of the program or data can be accessed; the programming must be specific for the particular machine. In a true virtual memory system, the programmer seldom needs to consider memory size or how that memory space is achieved.

3.3. INTERPRETIVE LANGUAGES

An interpreter is nearly always the first type of high-level language to be implemented on a newly developed small computer, since it is the most convenient to use, and often, the easiest to implement. When an interpreter is in operation, the user can both enter and modify program statements (edit) and execute the program directly. The source code statements are entered by typing in the commands. Execution of the program is normally invoked by simply typing "RUN." As each instruction is reached in the program sequence, it is both converted into machine code and executed.

The ease of moving between the editing mode and the execution mode gives the user of an interpreter an immediacy that makes the development and debugging of small programs easier than is the case in languages that require several steps in the preparation. The program is always "live," able to respond to the RUN command. If an error in execution is encountered, the interpreter reports it and returns immediately to the command mode for program modification.

In order to achieve this live operation mode, the interpreter must remain resident in memory at all times; it contains all of the code that is necessary to execute any legal statement. This requires a rather large area of memory even before the user's application program is entered.

The penalty for this ease of use is execution speed. Each command must be interpreted into the executable program code each time it is executed, no matter how many times it is encountered. The extent of the penalty depends on the program. A program that is primarily computational will spend a much larger proportion of its execution time in the mathematical subroutines and a smaller proportion interpreting. Consequently, the penalty may only be a small increase in execution time compared with a compiled program. However, a program that is used for sorting information and is heavy on branching statements may be slowed greatly by the interpretation process. Examples will be provided later.

In most interpreter implementations, words that are used as instructions are compacted for storage in memory or disk; the shorter representation, called "tokens" can be interpreted somewhat more rapidly. For example, on the disk and in memory, a single byte can be used to represent commands such as PRINT. Some versions of interpreted languages claim to be "semi-compiled," but often that operation goes no further than the tokenizing operation.

3.4. COMPILERS

A compiled language separates the various steps in the development of the executable program code. First, the entire program or at least a self-contained section of it is written into a source-code file with the aid of editor software. The editor has little or no regard for the nature of the text that is being typed, but simply edits the characters that form the syntax of the program and stores the source code. In order to convert the text to machine code, the **compiler** program is executed; it accepts as input the file of source code and prepares as output a file of **object code** and, optionally, a printable file with a listing of the source annotated with line numbers, error or warning messages, and lists of variables and subroutines.

The object file incorporates the machine-level instructions, but without resolving the absolute memory address locations for the instructions and without including the code for the procedures and subroutines that are called by the main program. That requires the final step, which is the link process: a program called a **linker** searches the main program's object file to see what procedures and subroutines are necessary. These procedures may be mathematical, such as floating-point addition or logarithms, or they may be operations such as the execution of a loop, reading from a file, or printing. The subroutines are other program modules previously written by the same or other programmers that were previously compiled. The linker then searches the **libraries** of subroutines (as requested by the operator) in order to find the subroutine code that must be added to the main program module. Then the absolute addresses are determined. The result is an executable program that is self-contained except for those functions that it calls from the OS.

One advantage of this process is quite obvious: the compilation of the program is done once and for all so that the execution can proceed at a much higher speed. A rough measure of this speed advantage can be obtained from the following exercise: two programs were written in BASIC (the language

```
        Program A              Program B
    10  PRINT "Start"       10  DEFINT A-Z:PRINT "Start"
    20  FOR I=1 TO 5000     20  DIM A$(50): NWORD = 50:                    REM # of words=50
    30       X=SIN(I)        30  FOR X=1 TO NWORD: READ A$(X): NEXT X:  REM Read word list.
    40  NEXT I               40      SWITCH=0: FOR X=1 TO NWORD-1:         REM Go through list,
    50  PRINT "Done"         50          IF A$(X) < A$(X+1) THEN GOTO 70: REM & test if any 2
                             60          SWAP A$(X),A$(X+1): SWITCH=1:     REM are out of order.
                             70      NEXT X
                             80  IF SWITCH=1 THEN 40:                      REM Are we done?
                             90  FOR X=1 TO NWORD:PRINT A$(X),:NEXT:       REM If so, print list.
                            100  DATA seed, when, said, have, been, pace, sign, cost
                            110  DATA seem, then, know, next, many, year, made, hope
                            120  DATA seen, seep, seek, sine, mail, hone, home, that
                            130  DATA test,testing,tested,boy,girl,wise,secret,arch,book
                            140  DATA children,child,childless,a,story,star,single,verbatim
                            150  DATA elephant,omnivore,omnivorous,everyman,happiness
                            160  DATA chromatography,spectroscopy,spectrophotometry
                            170  DATA spectrometer, at
```

FIGURE 3.2. Sample programs for computation and sorting, written in BASIC. The programs enable comparison of the effects of compilation on computation-intensive and nonintensive programs. The results are provided in Table 3.1.

TABLE 3.1. Timed Results for Execution of Programs in Figure 3.1

Program	Execution Time Interpreted	Execution Time Compiled
A	30.2 sec.	28.5 sec.
B	5.9 sec	0.8 sec.

will be described later) and executed both interpretively and after being compiled. Although BASIC is seldom optimized for speed, this language provides a unique opportunity for this experiment since both BASIC interpreters and compilers are available that operate on the same source code. The two programs are listed in Figure 3.2; one is computation intensive, and the other, a sorting algorithm, is not. The results are listed in Table 3.1. They demonstrate that the speed advantage of a compiler is most significant when the program spends a smaller proportion of its time in computation.

The other side of the execution speed factor is the preparation time factor. For a user to prepare the compiled program, an editor program must be invoked in order to enter or modify the source code; a compiler must convert to machine code, and a linker must combine the code with subroutines and procedures in libraries. This edit-compile-link cycle typically requires 10–20 min each time the programmer wishes to make a change, greatly increasing the "debug" time compared with that needed when programming in an interpreter.

3.5. THREADED-CODE LANGUAGES

Another group of languages has been developed that contains some of the advantages of both interpreters and compilers. Using a self-contained editor, one enters the lines of the program just as in an interpreter. The program is built up by defining a "dictionary"; the procedures are defined and entered into this dictionary, and following procedures, build on the preceding procedures until the result is simply a single procedure. Hence the name, threaded code. Since each line defines a complete entity, the source can be compiled a line at a time as it is entered; consequently, these languages also are called **incremental compilers**.

Probably no other group of languages has developed a more dedicated following than the threaded-code languages since they are extremely powerful tools for real-time control applications; the programmer has the immediacy of the interpreter with the execution speed of a compiler. However, the fact that threaded-code languages have not conquered all implies that there might be drawbacks. They present a considerably greater barrier to mastery than do other languages since the code is difficult to read, and they are not well suited to data processing. More will be stated later in the discussion of the primary threaded-code representative, FORTH.

3.6. HIGH-LEVEL LANGUAGES—A SURVEY

Having considered the nature of interpreters, compilers, and threaded languages, we go on to consider the languages that are available. High-level languages can be characterized as being problem-oriented. Consequently, since there are many kinds of problems amenable to computer assistance, we should not be surprised that a wide variety of languages exist. Over 150 are in use today with many others already dead. This discussion will consider only a few of those that are useful in the context of laboratory computing. Some languages that are important to other areas of computing, COBOL (*CO*mmon *Bu*siness-*O*riented *L*anguage), for example, can be disregarded safely.

A high-level language should, as was already suggested, allow the programmer to express the problem easily. A language without multidimensional arrays is a poor choice if matrix algebra is used to describe the problem. Where character strings such as text are important, a language that only allows variables to contain 32-bit floating-point numbers would be unsuitable.

A language also should address one of the nagging problems in the use of computers: the fact that programs written by one programmer tend to be difficult for another to follow. Indeed, it is frequently difficult for that same programmer to decipher previous work unless a great deal of commentary was added to the source code. The result is a huge investment in time to write programs that are difficult to debug and impossible to revise later.

In order to deal with this latter problem, **structured programming** has emerged which constitutes both a philosophy and a technique. This approach was in reaction to the typical program that "grew" as the programmer wrote it without preplanning, and then was patched until it worked. If lines are drawn on the printed listing of such a program to show the central path taken by the program, the result resembles spaghetti.

Structured programming is a technique that *requires* an orderly approach; tasks in the program become independent entities with definite beginnings and ends. Then, the program executes in time in much the same way that it appears on a page: execution progresses in sequence; self-contained tasks are sidetrips (subroutines) that, on completion, return execution to the points from which they left.

The best-known hallmark of structured programming is the abolishment of the infamous *GOTO* statement because of its tendency to disorder a program. That statement allows the program to jump from any location to any other location, and that brings disorder.

With structure, program listings can take the form of a diagram on the printed page. A loop is indented, and a loop within a loop is further indented. The result is a program that is much easier to design, read, and debug.

Note that in this discussion of structured programming, there has been no mention of efficiency in execution. The philosophy suggests that the reduced number of errors and reduced development and maintenance time greatly outweigh any improvements in execution speed that may accrue from taking short cuts.

Programs can be written in any language in a structured way. However, some languages are called structured languages because they facilitate logical and systematic structure by including certain capabilities, such as IF . . . THEN . . . ELSE commands, and by not including JUMP or GOTO commands. These languages can then enforce structured principles absolutely.

For the following discussion of languages, a sample algorithm has been given in order to provide a feel for the programming style. This program will compute the slope and intercept for the best fit of a straight line ($y = bx + a$) to a data set by the method of least-squares. The computation assumes that all of the data are equally reliable, so the *unweighted* form is used. The structure of the programs is illustrated in Figure 3.3. The user of the program is asked to enter the X and Y values. When no more data are available, the user types in a negative number for X or Y; therefore, negative values cannot be entered as data.

In these examples, data input, data output, conditional branching, iterative loops, and the use of subroutines are illustrated. The coding of the examples has been done in the simplest way using what is hoped to be the most standard version of the language. "Clever" methods that could make the program more elegant or efficient have been avoided, and some simplifying changes to the algorithm have been avoided in order to illustrate more of the language.

3.6.1. BASIC

Among the high-level languages used on microcomputers, Beginners' All-purpose Symbolic Instructional Code (*BASIC*) dominates. The language was first developed at Dartmouth College as a teaching tool so that "Dartmouth BASIC" is sometimes deemed the standard version. It is available for nearly all computers as an interpreter. In some cases, a compiler is available that will compile code already written and tested using the interpreter. Some "semicompiled" versions of BASIC are also in popular use, but these have application primarily in business environments.

Except for the person who learns to program in a high school or college computer science class, interpretive BASIC is commonly the first language learned by small computer users. This is because of the following several persuasive advantages:

1. Interpretive BASIC is almost completely self-contained. The user with only the one package has immediately the functions of an editor to write, print, and modify programs and the ability to execute that program immediately. This characteristic also frees the beginner from

Comments:

The task of the demonstration programs is to compute the coefficients a and b in the equation,

$$y = bx + a$$

First, we enter all of the data at the keyboard, terminating by entering a negative number. Then, we use the method of Least Squares to fit a line to the data where the x_i and y_i are the n independent and dependent data, respectively. Then, where weighting procedures are not used, the coefficients are

$$a = \Delta^{-1} \left(\Sigma x_i^2 \Sigma y_i - \Sigma x_i \Sigma x_i y_i \right)$$

$$b = \Delta^{-1} \left(n \Sigma x_i y_i - \Sigma x_i \Sigma y_i \right)$$

$$\Delta = n \Sigma x_i^2 - \left(\Sigma x_i \right)^2$$

Variables:

XArray and *YArray* are the arrays of up to 100 x and 100 y data.
XInput, YInput are individual entries of x_i and y_i.

SumOfX, SumOfY, SumOfXX, SumOfXY, and *NumbData* represent
$\Sigma x_i, \Sigma y_i, \Sigma x_i^2, \Sigma x_i y_i,$ and n.

Delta, CoefA, and *CoefB* represent Δ, a, and b.

Program structure:

As long as the last *YInput* and *XInput* are positive
    Call a subroutine to get *XInput, YInput*.
    If neither *XInput* or *YInput* is less than zero, do the following:
        increment *NumbData*,
        put *XInput* into *XArray*, and
        put *YInput* into *YArray*.
Repeat this sequence.

Set the summation variables (*SumOfX, SumOfY, SumOfXX, SumOfXY, NumbData*) to zero.

Compute the required sums of the NumbData members of the arrays:
        add *XInput* to *SumOfX*,
        add *YInput* to *SumOfY*,
        add *XInput* squared to *SumOfXX*,
        add *XInput* times *YInput* to *SumOfXY*.
Continue until done.

Compute the results from the equations above as follows:
    Delta is *NumbData* times *SumOfXX* minus *SumOfX* squared;
    take *SumOfXX* times *SumOfY* and subtract *SumOfX* times *SumOfXY*;
        that result divided by *Delta* is *CoefA*;
    take *NumbData* times *SumOfXY* and subtract *SumOfX* times *SumOfY*;
        that result divided by *Delta* is *CoefB*;
Print the results on the screen.

FIGURE 3.3. Structure of a programming problem to be solved in various languages.

having to interact directly with the OS, which is often a less "friendly" experience.

2. Individual BASIC instructions can be executed as keyboard commands. A datum can be input from a port and printed or be output to a port with simply a single keyboard command. The operator also can interrupt the execution of a program, change a variable value, and resuume execution.

3. It is easy to learn; the immediacy of the interpreter provides the feedback long known to be of pedagogical value. Also, tutorials abound.

4. The English-like key words make it easy to understand.
5. There is a large body of application programs written in BASIC for everything from *Star Trek* games to Fast Fourier Transform numerical analysis.

The disadvantages are just as significant:

1. There is little standardization, in spite of the continued efforts of the American National Standards Institute (ANSI) committee X3J2. The only standard thus far approved is for "Minimal BASIC" that, although released for microcomputer implementations in 1978, is now obsolete because of the unforeseen development in semiconductor capabilities. Cheap memory allows nearly every computer to execute an extended version of BASIC. Powerful implementations of BASIC have been tailored for the particular configurations of popular computers, adding clever and powerful ideas. However, these improvements also result in reduced transportability.
2. With a few exceptions, it is not possible to make use of independent subroutines written in BASIC or other languages without the tedium of making all the variable names and line numbers conform to those of the main program. Furthermore, in most common BASIC implementations, all subroutines muust be addressed by a line number. (Use of an English-like label is much easier to use—*CALL SORT* to call a subroutine instead of *GOSUB 32500*). These drawbacks can be severe in large programs, particularly if more than one worker is collaborating on a project.
3. The execution of programs written in BASIC is slow. This is inherently true for interpretive versions, but it also tends to be true of the compiled versions when compared with compilers for other languages.
4. Some implementations do not have the extensions required for direct I/O to and from I/O ports or for direct modification of absolute memory locations. These facilities are required for interfacing the computer to experiments. However, nearly all versions for microcomputers are so equipped.

Overall, BASIC is an extremely useful language for real-time laboratory computing. In the interpretive versions, single commands can be entered that directly operate on interfaces, and that makes interaction with the circuitry easier. The immediacy of the interpreter also makes the program easier to debug. In some cases, when the debugging is complete, the program can be compiled, and the benefits of a compiler are obtained.

A preliminary set of standards for BASIC have been released by the ANSI committee, eliminating several previous defects. For example, independent subprograms are included; a variable name in the subprogram will not need to be the same as the corresponding one in the main program, and external libraries of subroutines can be used. Real-time commands are to be included and used with a clock, and plotting commands will be available. Matrix operations are being included. Finally, many structured programming features are being added.

The example BASIC program is provided in Figure 3.4.

3.6.2. FORTRAN

The name FORTRAN is an acronym for *FOR*mula *TRAN*slator, which gives a clue to the type of problem for which it was designed. FORTRAN was originally developed for scientific computing by batch processing. (By batch, it is meant that no user interaction takes place during the execution of a program.) It was the first language to be widely implemented, and it represented a major leap from assembly-level coding, so much so that the authors in 1954 claimed that it ". . . should virtually eliminate coding and debugging. . . ."

FORTRAN has been standardized by ANSI, the common version being FORTRAN IV. A recent revision, FORTRAN 77, is an attempt at a structured version. All versions for small computers require extensions (additional commands) in order to interact directly with I/O hardware or to take advantage of the interactive capabilities of small computers.

Although many computer science-oriented programmers harbor strong feelings against FORTRAN, it does have the following advantages:

1. The strongest argument for the continued use of FORTRAN is the existence, in the scientific community, of a massive library of programs and subroutines written in FORTRAN. Although other high-level languages are vying to replace FORTRAN, it will be some time before that software is translated. Because FORTRAN allows (some say, encourages) disorganized programming, the process of translating a FORTRAN program to another language is not always simple.

2. A second argument for retaining FORTRAN is the wealth of skill present in computer users who learned to program in FORTRAN before coming to the small computer. (On the other hand, although learning a new language represents a significant activation energy barrier, the other side of the barrier may bring a net improvement in programming productivity.)

3. Because the goal of a FORTRAN program is usually to accomplish large-scale computations, FORTRAN is often optimized more toward execution efficiency than most other languages. However, this feature is highly dependent on the particular implementation.

4. Standard FORTRAN has complex number arithmetic capability.

For a computer language to have lasted so long in an ever-changing field

```
*****************************************************************************************************

1000 REM  Computation of the best fit of a straight line to
1010 REM  an array of data by the method of unweighted Least Squares.
1020 REM
1030 REM          K. L. Ratzlaff, Kansas University, Lawrence, Kansas
1040 REM          Date begun: 7/4/85;  Last updated 7/13/85
1050 REM
1060 REM        Data in the form of X-Y pairs are entered via the keyboard
1070 REM  until a negative X or Y value is entered.  Then the coefficients
1080 REM  a and b are computed for the equation
1090 REM          y = bx + a.
1100 REM
1110 REM  Reference:  Bevington, P.R. Data Reduction and Error
1120 REM      Analysis for the Physical Sciences, 1969, McGraw-Hill.
1130  REM
1200 REM  Dimension the array that is to be used.
1210          DIM XARRAY ( 100 ), YARRAY ( 100 )
1300 REM  Set to zero the initial number of data, and the input variables.
1310          NUMBDATA = 0
1320          XInput = 0: YInput = 0
1330  REM
1400 REM     Now loop to enter data.  (Some readers will note that the
1410  REM          more modern WHILE-WEND construction could be used.)
1420  REM  First call a subroutine to get a data pair.
1430          GOSUB 3000
1440  REM  Test if data entry is complete.
1450          IF (YINPUT < 0) OR (XINPUT < 0) THEN GOTO 1600
1460  REM  If the data are valid, put them in the array and try again.
1470          NUMBDATA = NUMBDATA + 1
1480          XARRAY ( NUMBDATA ) = XINPUT
1490          YARRAY ( NUMBDATA ) = YINPUT
1500          GOTO 1400
1510  REM
1600 REM     Data entry is complete; now compute the sums.
1610 REM     First initialize the summing variables to zero.
1620  REM  Note that SUMOFXX is the sum of the squares of X, and SUMOFXY
1630  REM  is the sum of the products of XY for each data pair.
1640          SUMOFX = 0: SUMOFY = 0: SUMOFXX = 0: SUMOFXY = 0
1650 REM
1800 REM  Compute the sums for all NUMBDATA members of the arrays.
1810      FOR I = 1 TO NUMBDATA
1820          SUMOFX = SUMOFX + XARRAY ( I )
1830          SUMOFY = SUMOFY + YARRAY ( I )
1840          SUMOFXX = SUMOFXX + (XARRAY (I))^2:          REM Add the square of the X input.
1850          SUMOFXY = SUMOFXY + XARRAY ( I ) * YARRAY ( I ) :      REM Add the product of X and Y.
1860      NEXT I
1870  REM
2000 REM     Data entry is complete; now compute the results.
2010      DELTA = (NUMBDATA * SUMOFXX) - (SUMOFX)^2
2020      COEFA = (1/DELTA) * ( (SUMOFXX * SUMOFY) - (SUMOFX * SUMOFXY) )
2030      COEFB = (1/DELTA) * ( (NUMBDATA * SUMOFXY) - (SUMOFX * SUMOFY) )
2040 REM   Print the results.
2050      PRINT: PRINT "The best fit is": PRINT "y ="; COEFB; "* x +"; COEFA
2060      END
2070 REM
3000 REM   Data entry subroutine.
3010      PRINT "What is the independent variable (X)";
3020          INPUT XINPUT
3030      REM  Note that if XINPUT was negative, YINPUT is not needed.
3040          IF XINPUT < 0 THEN RETURN
3050      PRINT "What is the dependent variable (Y)";
3060          INPUT YINPUT
3070      RETURN
*****************************************************************************************************
```

FIGURE 3.4. Solution of the problem of Figure 3.3 in BASIC.

suggests that it was very well planned for its time. However, standard FOR-TRAN has a number of important weaknesses:

1. The syntax does not encourage structured programming or well-documented programming.
2. Most implementations handle character strings poorly.
3. I/O to the peripherals uses complicated and restrictive *FORMAT* instructions.
4. Most implementations have poor error-handling facilities.
5. The capability for bit-for-bit Boolean mathematics is not provided.

Many of the available extensions are intended to overcome these weaknesses. However, many FORTRAN implementations are also mere subsets of standard FORTRAN, usually deleting complex number capabilities.

We should not leave the impression that FORTRAN is a dead language. (Witness the move to develop structured FORTRAN). Despite its limitations, there is no doubt that FORTRAN will remain in the scientific laboratory for some time to come.

The FORTRAN programming example is given in Figure 3.5.

```
C          Computation of the best fit of a straight line to
C    an array of data by the method of unweighted Least Squares.
C
C    K. L. Ratzlaff, Kansas University, Lawrence, Kansas
C    Date begun:  7/4/85;  Last updated 7/13/85
C
C          Data in the form of  X-Y  pairs are entered via the keyboard until a negative
C    X  or  Y  value is entered.  Then the coefficients a and b are computed for the equation
C          y = bx + a.
C
C    Reference:  Bevington, P.R. Data Reduction and Error
C          Analysis for the Physical Sciences, 1969, McGraw-Hill.
C
C  First thing is to dimension the array.
          DIMENSION XARRAY ( 100 ), YARRAY ( 100 )
C
C  Next begin execution,
C    by setting to zero the initial number of data, and the input variables.
          NUMDAT    = 0
          XINPUT    = 0
          YINPUT    = 0
C
C  Now enter the members of the array until either value is zero.
C          Get the two input values.
100       CALL GETINP ( XINPUT, YINPUT )
C  Test the input variables.
          IF (YINPUT .LT. 0) GOTO 200
          IF (XINPUT .LT. 0) GOTO 200
C  Put the values into the arrays.
          NUMDAT = NUMDAT + 1
          XARRAY ( NUMDAT ) = XARRAY
          YARRAY ( NUMDAT ) = YINPUT
C  Get data again.
          GOTO 100
C
C  Data entry is complete; next process the data.
C  First set summing variables to zero.
```

FIGURE 3.5. Solution of the problem of Figure 3.3 in FORTRAN.

```
200   SUMX  = 0
      SUMY  = 0
      SUMXX    = 0
      SUMXY    = 0
C
C Compute the desired sums, Counting from 1 to NUMDAT.
      DO 300 I = 1, NUMDAT
            SUMX  = SUMX + XARRAY ( I )
            SUMY  = SUMY + YARRAY ( I )
            SUMXX = SUMXX + ( XARRAY ( I ) )**2
            SUMXY = SUMXY + XARRAY ( I ) * YARRAY ( I )
300   CONTINUE
C
C We are ready to compute the final results.
      DELTA = (NUMDAT * SUMXX) - (SUMX)**2
      COEFA = (1 / DELTA ) * ( (SUMXX * SUMY) - (SUMX * SUMXY) )
      COEFB = (1 / DELTA ) * ( (NUMDAT * SUMXY) - (SUMX * SUMY) )
C
C Print the final results.
      WRITE (6,1000) COEFB, COEFA
C Done.
1000 FORMAT ( ' The best fit is',/,' y = ', F7.3, ' * x + ', F7.3 )
C
      END


      SUBROUTINE GETINP ( XENTRY, YENTRY )
C This subroutine returns XENTRY and YENTRY; however, if XENTRY is negative
C   then YENTRY will not be acquired.
C Print out a prompt for the datum.
      WRITE (6,1000)
C Accept the result from the keyboard.
      READ  (6,2000) XENTRY
C Test if it is less than zero.
      IF ( XENTRY .LT. 0 ) GOTO 100
C Print out a prompt for the Y datum and get it.
      WRITE (6,1010)
      READ  (6,2000) YENTRY
C Return to the main program.
100   RETURN
C
1000  FORMAT (' What is the independent variable (X)? ')
1010  FORMAT (' What is the dependent variable (Y)? ')
C The format for the datum that is to be typed in:
2000  FORMAT ( F10.4 )
      END
```

FIGURE 3.5. (*continued*)

3.6.3. Pascal

Unlike most other language names, the name Pascal is not an acronym; the language was named for the French mathematician Blaise Pascal. It was introduced by the Swiss computer scientist Nicklaus Wirth in 1968. Pascal is somewhat English-like and very much a structured language.

The nature of Pascal's advantages and faults can be understood when one realizes that it was devised and first implemented as a teaching tool to train programmers to use structured programming concepts. Consequently, source code for application programs is usually readable, even self-documenting (i.e., the statements themselves can fully describe their own actions). Therefore, Pascal programs tend to be easy to debug and possible to maintain. In most universities it has replaced FORTRAN as the beginning language for computer science students.

Pascal is usually considered in the compiler category because that is the most common implementation. However, Pascal also has been developed as an interpreter with all the attendant advantages, and has been implemented in an intermediate form called a *"p-version"* at the University of California, San Diego. There, a level of standardization and transportability was achieved by writing a series of compilers for a large variety of CPUs by incorporating this intermediate concept. The compiler generates intermediary p-code which executes on a hypothetical "p-machine" (p represents "pseudo"). Then a program that *emulates* the p-machine interprets the p-code as it is run. The concept is somewhat between that of an interpreter and a compiler. The result is both transportability of the source code and transportability of the p-code between all computers using this type of Pascal system. Because of the source of this implementation, the p-version is often referred to as "UCSD Pascal". In order to further guarantee this uniformity, the "p-machine" emulator software is often contained *in* an OS rather than executing under some other OS.

Another feature of Pascal is that it is strongly typed. That is, there are many more types of data variables than just the few described in Chapter Two; for example, 1-, 2-, 4- or 8-byte integers might be employed. Each variable used in the program must be defined to hold values appropriate to one and only one of those types. Orderly structures of data can be defined and manipulated in ways that reduce the probability of programming error.

In the lists of advantages and disadvantages, many will appear in both categories according to the bias of the person making the list. Among the advantages

1. Pascal is highly structured. This should result in fewer programming errors.
2. Pascal is strongly typed, a feature that also should lead to fewer errors.
3. More science students are now being taught programming using Pascal as a programming language than are learning any other language.

Among the disadvantages

1. The high degree of structure and error checking *tend* to lead to slower execution speed and larger programs.
2. I/O support was not developed well by Wirth, and tends to be very implementation dependent.
3. Some programmers find that the strict data typing gets in the way of more than just error-producing short cuts.

An extension of Pascal, **Ada**, was defined by the U.S. Department of Defense as the standard language for military applications. Ada was named for Augusta Ada Lovelace Byron (as mentioned in Section 1.1.1). The definition of Ada is an attempt to incorporate the useful features of FORTRAN, PL/I, and even COBOL, as well as Pascal; consequently, it is very large and difficult to implement on small computers. However, subsets of Ada for small computers do appear. Its popularity derives simply from the voracious military appetite for computer software, and the tight military control on the definition which guarantees transportability. The size and complexity of the language may limit application in other areas; many feel that the same factors threaten the reliability of Ada code in critical military applications.

Another language that grew out of Pascal, **Modula-2**, is likely to become a practical replacement since it does standardize those functions which are nonstandard in Pascal and is oriented more toward practical applications programming.

The example coded in Pascal is provided in Figure 3.6.

3.6.4. C

C is a compiler whose development was closely related to the development of the OS UNIX at Bell Laboratories. Using some earlier definitions, C can be considered a structured language but not an English-like language. That is, programming in C requires structured techniques, but the syntax is extremely terse. This may have the advantage of requiring less "keyboarding" but also has the disadvantage that the terseness can make C somewhat more difficult to learn.

C is often described as a systems language. That is, C is used in place of assembly language for writing software related to the operation of the computer, OSs, word processors, communications packages, and even other compilers.

The following advantages of C support that description:

1. C has direct and easy access to I/O hardware and the bit-manipulation instructions. This eases the task of writing character-manipulation software and makes the handling of interfaces and peripherals easier. In fact, the OS UNIX and most C compilers are written in C.
2. A large body of utilities and subroutines exists. Functions for handling screen windows or database manipulations can be inserted into a program relatively easily.
3. Because of its relative youth, there has not been time for the development of disparate versions before standardization.
4. The resultant code, while retaining the portability of a higher level language, also has the performance necessary for writing systems applications.

Several disadvantages help to define the limitations:

1. The mathematical capabilities are often limited, although this depends

on the implementation; floating-point arithmetic is not standard (there may be good reasons why it should not be).

2. The high performance necessitates a minimum of built-in error checking.

3. The terse syntax makes it somewhat difficult to learn, although it has a loyal following among the subgroup of "serious programmers."

```
Program LSQ;
{
Computation of the best fit of a straight line to an array of data by the method of unweighted Least Squares.

        K. L. Ratzlaff, Kansas University, Lawrence, Kansas
        Date begun: 7/4/85;  Last updated 7/13/85

        Data in the form of X-Y pairs are entered via the keyboard until a negative Y
value is entered.  Then the coefficients a and b are computed for the equation
            y = bx + a

Reference: Bevington, P.R. Data Reduction and Error
        Analysis for the Physical Sciences, 1969, McGraw-Hill.
}
        { First thing is to define the variables. }
Var
        XArray, YArray           : array [1..100] of Real;
        NumbData, Index          : Byte;
        XInput, YInput           : Real;
        SumOfX, SumOfY, SumOfXX, SumOfXY  : Real;
        Delta, CoefA, CoefB      : Real;

        { Next we must define procedures that will be used in the main program.}
Procedure Get_input (var XEntry, YEntry: Real);
        { This procedure will return values for XEntry and YEntry; however, if
            XEntry is negative, YEntry will not be acquired.}
        begin
            Write ('What is the independent variable (X)? ');
            ReadIn (XEntry);                    {Enter XEntry from the keyboard.}
            If XEntry >= 0 then                  {If XEntry is OK, then get YEntry.}
            begin
                Write ('What is the dependent variable (Y)? ');
                ReadIn (Yentry)
            end
        end;                                    {of procedure Get_input}
{
Now to begin execution.
}
begin
        {First set to zero the initial number of data and the entry variables.}
            NumbData := 0;
            XInput  := 0;
            YInput  := 0;
        {Enter the members of the array until either value is zero.}
        While (YInput >= 0) and (XInput >= 0) do
        begin
            Get_input (XInput, YInput);          {This is the call to the subroutine.}
            If (YInput >= 0) and (XInput >= 0) then
            begin
                NumbData := Numbdata + 1
                XArray [NumbData] := XInput;
                YArray [NumbData] := YInput
            end
        end
```

FIGURE 3.6. Solution of the problem of Figure 3.3 in Pascal.

```
{
Data entry is now complete.  Next set the summing variables to zero.
}
        SumOfX    := 0;
        SumOfY    := 0;
        SumOfXX   := 0;
        SumOfXY   := 0;
{
Then compute the desired sums.}
        For Index := 1 to NumbData do
        begin
                SumOfX  := SumOfX + XArray [ Index ];
                SumOfY  := SumOfY + YArray [ Index ];
                SumOfXX := SumOfXX + Sqr (XArray [ Index ]);
                SumOfXY := SumOfXY + XArray [ Index ] * YARRAY [ Index ]
        end
{
We are now ready to compute the final results.
}
        Delta := (NumbData * SumOfXX) - Sqr (SumOfX);
        CoefA := (1 / Delta) * ( (SumOfXX * SumOfY) - (SumOfX * SumOfXY) );
        CoefB := (1 / Delta) * ( (NumbData * SumOfXY) - (SumOfX * SumOfY) );
{
And print the final results.
}
        Writeln ('The best fit is');
        Writeln (' y = ', CoefB:3:3, ' * x + ', CoefA:3:3)

end.
```

FIGURE 3.6. (*continued*)

In summary, C can be considered a "medium-level assembly language," but unlike assembly language, it is structured and the code is highly transportable.

The example program written in C is shown in Figure 3.7. In studying this example, one should bear in mind that C was not designed for mathematical data-processing applications.

3.6.5. FORTH

Threaded-code languages, of which FORTH is the primary example, are difficult to describe. Both the programming structure and the code produced by FORTH are substantially different from other languages.

Like other threaded-code languages, FORTH is based on the concept of a *dictionary*. Entries are either primitives—procedures written in assembly language—or secondaries, produced by combining primitives and/or other secondaries. The writing of either secondaries or a complete program consists of referencing (or threading) the dictionary entries. In order to implement that concept efficiently, a Last-In/First-Out (LIFO) buffer or *stack* is used; adding another element to the stack is called a *push* and removing an element is called a *pop*. One of the results of this approach is the need to use a formula entry notation different from other languages: Reverse Polish Notation (RPN), best known for its implementation on certain hand-held calculators. The code to compute the sum of 3 plus the product of 5 and 7 (in BASIC or FORTRAN, 3 + 5 * 7) is written

$$3\ 5\ 7\ *\ +$$

Numbers are pushed onto the stack as they are read; when an operator is read, two values are popped, the operation is executed, and the result is pushed back onto the stack.

```
/*
Computation of the best fit of a straight line to an array of data by the method
of unweighted Least Squares.

                Tom Peters, Instrumentation Design Laboratory
                The University of Kansas  8/85.

Data in the form of X-Y pairs are entered via the keyboard until a negative X or Y
value is entered.  Then the coefficients a and b are computed for the equation
      y = bx + a

Reference: Bevington, P.R. Data Reduction and Error Analysis
  for the Physical Scientists, 1969, McGraw-Hill.
*/
#include "stdio.h"                    /* definitions and variables required by the 'C' environment */

main()
{
/* first thing is to define the variables */
        float XArray[ 100 ];          /* array of real values */
        float YArray[ 100 ];          /* array of real values */
        float XInput;                 /* entry for x_i */
        float YInput;                 /* entry for y_i */
        float SumOfX;                 /* sum x_i */
        float SumOfY;                 /* sum y_i */
        float SumOfXX;                /* sum x_i squared */
        float SumOfXY;                /* sum x_i sum y_i */
        float Delta;                  /* Delta */
        float CoefA;                  /* a */
        float CoefB;                  /* b */
        char NumbData;                /* number of data points entered */
        char Index;                   /* index thru the array of values */

/* program execution starts here */
        /*
            First set to zero the number of data and the entry variables.
        */
        NumbData = 0;
        XInput = YInput = 0.0;
        /*
            Enter the members of the array until either value is negative.
        */
        do
            {
              /* call a subroutine to get XInput, YInput */
              get_input( &XInput, &YInput );

              /* check whether both XInput and YInput are positive. */
              if ( ( (XInput >= 0.0) && (YInput >= 0.0) ) )
                 {
                 /*
                   YES THEY ARE, so save each value and increment the number of points.
                 */
                 XArray[ NumbData ] = XInput;
                 YArray[ NumbData ] = YInput;
                 NumbData++;
                 };
            }
```

FIGURE 3.7. Solution of the problem of Figure 3.3 in C.

```
                 /* check both XInput and YInput and repeat the operations if necessary */
                 while( (XInput >= 0.0) && (YInput >= 0.0) );
                 /*
                        Set summation variables to zero.
                 */
                 SumOfX = SumOfY = SumOfXX = SumOfXY = 0.0;
                 /*
                        Then compute the desired sums.
                 */
                 for ( Index=0; Index < NumbData; Index++ )
                        {
                        SumOfX += XArray[ Index ];
                        SumOfY += YArray[ Index ];
                        SumOfXX += (XArray[ Index ] * XArray[ Index ]);
                        SumOfXY += (XArray[ Index ] * YArray[ Index ]);
                        };
                 /*
                        We are now ready to compute the final results.
                 */
                 Delta = (NumbData * SumOfXX) - (SumOfX * SumOfX);
                 CoefA = ( (SumOfX * SumOfY) - (SumOfX * SumOfXY) ) / Delta;
                 CoefB = ( (NumbData * SumOfXY) - (SumOfX * SumOfY) ) / Delta;
                 /*
                        print the results on the screen
                 */
                 printf( "The best fit is:\n" );
                 printf( " y = %f6.3 * x + %f6.3\n", CoefB, CoefA );
                 exit();
}
/*
        get_input returns values for XEntry and YEntry.
        Note on the C language:
        These variables must be pointers, since values are being returned through them
        to the calling program.  This corresponds to the Pascal requirement that the
        variables be declared as type var parameters.
*/
get_input( XEntry, YEntry )
float *XEntry;        /* address of the XInput value */
float *YEntry;        /* address of the YInput value */
{
.        printf( "What is the independent variable (X) ?  " );

        /* enter XEntry from the keyboard */
        scanf( "%f", XEntry );

        /* check if XEntry is OK. */
        if ( *XEntry >= 0.0 )
                {
                /* YES it is, so get YEntry */
                printf( "What is the dependent variable (Y)?  " );

                /* Enter YEntry from the keyboard */
                scanf( "%f", YEntry );
                };

        return;

}
```

FIGURE 3.7. (*continued*)

The advantages of this are that FORTH has the immediacy attributed to interpreters and the execution speed of compilers; the method of threading and using dictionaries allows it to operate with very small memory systems.

FORTH was invented by astronomers faced with the problems of controlling huge telescopes. In FORTH, such problems can be solved in a short period of time. However, its mastery also has been shown to be valuable

```
•••••••••••••••••••••••••••••••••••••••••••••••••••••••••••••••••••••••••••••••••••••••••••••••••••••
(
   File LSQ.FTH                    8/4/85
              Eugene H. Ratzlaff
              IBM Instruments
              Danbury, Connecticut

      Linear least squares fit.

      Execute LSQ and enter data as x,y pairs:  end data entry by entering
      negative number.  Floating-point data format requires entry with
      "E" notation.  All data and calculations in 64-bit precision.
                                                                          )

( make a "defining" word to allow use of floating-point vectors)
: FVECTOR
              CREATE          ( n --> )      8 * ALLOT                   ( allot 8 bytes per entry)
              DOES>           ( n addr1 --> addr2 )   SWAP 8 * + ;
                 ( multiply index by 8 to calculate offset into array and add to a;
                     indexing is 0. . .n)

100    FVECTOR  XARRAY           ( create a 100 element vector for X data;
                                      indexing from 0 so last element is 99 XARRAY)
100    FVECTOR  YARRAY           ( create a 100 element vector for Y data)
              VARIABLE NUMBDATA  ( create a variable for # of x,y pairs)

( convert string input to FP # on stack with flag indicating good/bad )
: INPUT#       ( --> r f )
          PAD 30 0 FILL                   ( fill PAD with zeros)
          PAD 1+ 25 EXPECT                ( get string into PAD, with room for byte count)
          SPAN @  PAD                     ( put byte count on stack; get PAD address)
            BEGIN   1+ DUP C@ 32 <>        ( search string for first nonblank)
            UNTIL   1-  SWAP OVER C!       ( then go back one, store count there)
          FNUMBER? ;                      ( convert string leaving FP # and flag)

( get FP string; leave # on stack with true flag if ok,
                                   else drop from stack and leave false flag)
: REAL_GET   ( --> r true OR  --> false )
            INPUT#                        ( get number and flag onto stack)
            ?DUP NOT IF                   ( duplicate flag if true, make false flage true)
                    FDROP 0               ( if number bad, drop it from stack and flag false)
                  THEN ;

( ask for X input until good, then echo input number
                                   for confirmation and leave on stack)
: X_GET        ( --> r )
          BEGIN
            CR ." What is the independent variable (X)?            ( do <CR> and ask )
            REAL_GET                                               ( input real # )
          UNTIL                           ( if input number bad, retry, else continue:)
          2 SPACES FDUP  E. ;             ( type 2 spaces, duplicate number on stack,
                                               and type top number out)

: Y_GET        ( --> r )
          BEGIN
            CR ."What is the dependent variable (Y)? "             ( do <CR> and ask )
            REAL_GET                                               ( input real # )
          UNTIL                           ( If input number bad, retry, else continue:)
          2 SPACES  FDUP  E. ;            ( type 2 spaces, duplicate number on stack,
                                               and type top number out)

( zero the data-pair counter; fill the vectors with input numbers)
: GET_INPUT  ( --> r )
          0 NUMBDATA !                                             ( store zero in data counter)
          100 0 DO                        ( loop 100 times)
              X_GET                       ( get X)
              FDUP I XARRAY F!            ( duplicate it and store it in array -
                                              if negative, don't worry, it won't be used!)
              FDUP F0<                    ( duplicate it again, is it < . ?)
              IF
                  LEAVE                   ( if number is negative leave loop)
              ELSE                        ( otherwise,)
                  Y_GET                   ( get Y)
                FDUP I YARRAY F!   ( duplicate it and store it in array)
                FDUP F0<                  ( DUPLICATE IT AGAIN, IS IT < 0.0 ?)
```

FIGURE 3.8. Solution of the problem of Figure 3.3 in FORTH.

92

```
                 IF
                    LEAVE                     ( if number is negative leave loop)
                 ELSE                         ( otherwise, if both numbers positive: )
                    1 NUMBDATA +!             ( Then increment data counter, and )
                    CR                        ( do a <CR>)
                 THEN                         ( End of if y<0)
             THEN                             ( end of if x<0)
         LOOP ;                               ( end of loop )

( define summation variables )
FVARIABLE      XSUM         ( sum of Xs)
FVARIABLE      YSUM         ( sum of Ys)
FVARIABLE      XXSUM        ( sum of X^2s)
FVARIABLE      XYSUM        ( sum of X*Ys)

( set the summation variables to zero)
: ZERO_SUMS ( --> )
         0.0E0                          ( put floating-point zero on stack)
         FDUP       XSUM F!             ( duplicate it and store at XSUM)
         FDUP       YSUM F!             ( duplicate it and store at YSUM)
         FDUP       XXSUM F!            ( duplicate it and store at XXSUM)
                    XYSUM F! ;          ( consume by storing at XYSUM)

( go thru the vectors calculating the sums)
: SUMS_MAKE      ( --> )
         NUMBDATA @ 0                   ( put data count, 0 on the stack, in order to: )
         DO                            ( loop NUMBDATA times)
             I YARRAY F@                ( get nth Y element using "I"- the current loop index)
             FDUP     YSUM F@           ( duplicate it and get current Ys sum)
             F+ YSUM F!                 ( add them together, and store new sum)
             I XARRAY F@                ( get nth X element)
             FDUP XSUM F@               ( duplicate it and get current Xs sum)
             F+ XSUM F!                 ( add them together and store new sum)
             FDUP FDUP F *              ( duplicate nth X element - previously left on stack-
                                          twice, and form X^2 by multiplying together)
             XXSUM F@                   ( get current X^2 sum)
             F+ XXSUM F!                ( add them together, and store new sum)
             F*                         ( multiply copies of Y and X previously left on stack)
             XYSUM F@                   ( get current X*Y sum)
             F+ XYSUM F!                ( add them together, and store new sum)
         LOOP ;                         ( loop until all vector elements have been summed)

( define variables for intermediate results )
FVARIABLE    DELTA
FVARIABLE    ACOEF
FVARIABLE    BCOEF


( calculate DELTA )
: DELTA_CALC ( --> )
         NUMBDATA @ S>F                 ( get number of data pairs, and convert to FP)
         XXSUM F@  F*                    ( get X^2s sum, and square it)
         XSUM F@ FDUP F*                 ( get Xs sum, and square it)
         F- DELTA F! ;                   ( now take difference, and store as DELTA)

( calculate ACOEF )
: ACOEF_CALC ( --> )
         XXSUM F@                        ( get X^2s sum)
         YSUM F@ F*                      ( get Ys sum, and take product with X^2s sum)
         XSUM F@                         ( get Xs sum)
         XYSUM F@ F*                     ( get XYs sum, and take product)
         F-                              ( take difference between products)
         DELTA F@ F/                     ( get DELTA, and divide into difference)
         BCOEF F! ;                      ( store result as BCOEF)

( write the resulting equation out)
: .EQN   ( --> )
         CR
         ." The best fit is" CR
         ." y = "
         BCOEF F@ E.
         ." * x + "
         ACOEF F@ E.
         CR ;
```

FIGURE 3.8. (*continued*)

```
( execute the entire process of linear Least Squares fit)
: LSQ  ( -> )
                GET_INPUT
                ZERO_SUMS
                SUMS_MAKE
                DELTA_CALC
                ACOEF_CALC
                BCOEF_CALC
                .EQN    ;
```

FIGURE 3.8. (*continued*)

in laboratories facing challenging problems in real-time instrument control. Chemists have developed a variant, CONVERS, and biomedical researchers wrote STOIC for scientific laboratory applications. However, the mathematical operations are usually quite limited in most implementations.

With such impressive advantages, why is FORTH so seldom used? The answer to this question lies in the difficulty of its use; FORTH, at the primary and lower secondary level, is the antithesis of a natural language, and the syntax is radically different from that of most other languages. However, after the lower level secondaries have been prepared, a FORTH program can read very naturally.

The example program, written in FORTH, is shown in Figure 3.8. Floating-point arithmetic is not standard, but is included in this example.

3.7. FEATURES FOR REAL-TIME OPERATION

The languages used on small computers are, with the exception of FORTH, extensions of languages devised for use on large computers. However, on mainframes, the integrity of the entire system is paramount: The system must ensure that no single user can interfere with the operation of the system for the other users. Consequently, certain types of operations cannot be tolerated. In particular, precision timing cannot be supported, and commands that could directly read to and write from I/O channels cannot be tolerated.

If one is to use a high-level language to develop the software portion of the interface to one's experiment, there must be extensions to the language that allow this direct access. Therefore, high-level languages for small computers usually have commands for reading and writing to absolute memory locations anywhere in memory, and commands for I/O channels if I/O channel architecture is used.

For channel I/O, commands of the type

$$I = INP (J)$$

and

$$OUT J,I$$

are typical where J is the port address and I is the value of the datum. Similar syntax usually exists for memory-mapped systems where the analogous commands are *PEEK* or *IPEEK* for reading a datum and *POKE* or *IPOKE* for writing a datum. In either set of commands, the user must be wary of an easily included error, that of failing to convert the address to decimal from the hexadecimal (or octal) printed in the hardware documentation.

A command similar to the *IN* and *OUT* commands is the command WAIT, found in some versions of BASIC. *WAIT J,M,N* tests the status of a device by simply stopping the execution of the program until the value input from port *J* becomes *TRUE* aftter being *AND*ed with *M* and *OR*ed with *N*.

Testing status of a device often requires that the condition of a single bit be tested. A Boolean algebra function may be required, but the execution of these logical commands, such as the Boolean *AND* and *OR*, may differ from one language implementation to the next. In a real-time environment, Boolean functions must be carried out bit-wise; for example, 14_{10} *AND* 5_{10} should yield 4 (1110_2 *AND* $0101_2 = 0100_2$.) In some BASIC interpreters and most FORTRAN compilers the result is simply **TRUE** since single-bit mathematical operations were not anticipated.

A second useful extension is seldom included, and that is a facility for software support of interrupts. Digital Equipment Corporation's implementation of FORTRAN for use with RT-11 includes the procedure, INTSET, which specifies any subroutine as the completion routine to be entered from a particular interrupt vector address; such a command plus commands to turn interrupts on and off would be a useful addition to many microcomputer languages. An extended Pascal for MS-DOS OSs, Turbo Pascal, also supports the function.

Most (but not all!) interpreters and compilers, whether or not they have the real-time instructions listed above, do have the facility to call a subroutine written in assembler or another high-level language. Commonly C, and sometimes Pascal and FORTH, allow a programmer to switch to assembler and back within a procedure; this is termed "in-line" coding. These techniques allow the acquisition of a burst of data at a very high rate from within a slowly executing program such as one written in BASIC.

The reader should understand that in most cases, excepting those in which very fast I/O or interrupts are required, all of the programming for real-time interfaces *can* be done exclusively in high-level language; a *very* rough estimate would be that data can usually be acquired and stored in memory by an interpreter at rates up to about 60–100 Hz and by a compiled program at rates approaching about 20 kHz. Given the two order-of-magnitude variation in small computer performance, these estimates should not be applied without a benchmark test using the specific computer and program routine to be used in the execution of the final application program.

3.8. PRODUCTIVITY TOOLS

The name Productivity Tools can be used to describe the class of application programs that are not necessarily prepared for scientific–experimental applications but which can be used to aid the scientist in the management of other tasks. These tools may well have had a greater impact on use of small computers by laboratory scientists than the on-line capability itself. Increasingly, the computer is found useful for the production of reports, manuscripts, and other documents, for management of experimental data and literature files, and for manipulation of arrays of data.

In this section, these types of application programs will be considered briefly. The goal of discussion is primarily to suggest the uses for and the impact of these tools for scientists.

3.8.1. Word Processing

Small computers are as well-suited for the manipulation of arrays of alphanumeric characters (documents) as they are for the manipulation of numerical data. A word processor is a program that can accept text input, can facilitate manipulation of that text through interaction with a CRT display, keyboard and/or mouse, and in most cases can direct the formatting of the text on a printer. Word-processing capability allows even a minimally competent typist to prepare a document and edit that document with greater ease than could be done by writing it in script and asking a secretary to type and correct it.

Before word-processing software arrived for small computers, entry of text required *editor* software utilities. The generally accepted distinction between word processors and editor utilities is that only the former are page-oriented. That means that only word processors allow the display of a page with direct access to any location on the page. Editors are line-oriented, editing a line at a time in a manner developed for printing terminals. That word processors can format edited text is another distinguishing characteristic.

Typically, a word processor accepts "free-form" text input. That is, the typist need not be concerned with carriage returns at the ends of lines. The only carriage returns that are entered are found at the ends of paragraphs. Consequently, paragraphs can be automatically reformed after words or phrases have been changed. Text can be moved, modified, or deleted at will without any paper copy being generated.

Most word processors also allow the insertion of characters that direct the printing process. These commands may send commands to the printer to change print density (for boldface), fonts, change character width and spacing, line height, and page size.

Of particular interest to scientists is the ability to print special symbols and equations. Most word processors for small computers are deficient in this area. The reasons are two-fold. First, few small computers possess CRT displays with the capability of displaying characters which are superscripted, subscripted, or not part of a standard character set. Obviously, Greek and other scientific characters are required for technical writing. Even if the printer can print these characters, seldom does the CRT display allow a user to "see what you get." Second, the printing of equations and special characters may be possible on a given printer, particularly on dot-matrix printers as shown in Figure 3.9, but most word processors do not support the necessary manipulation of line spacing and control of individual dots.

The reason is in part the limitation in ordinary screen resolution making detail difficult to display. The cost of computers such as the Xerox Star and the Apple Lisa and MacIntosh, which possess moderately high-resolution screens, is coming down steadily. These machines include word-processing functions that allow the operator to move text around on the screen and to change fonts and font sizes on screen. The accompanying printers are dot-matrix types, either impact or laser printers. The combination allows the writer to "see what you get" before it is committed to the printer. Although early models of these examples are poor for laboratory applications, the appearance of a computer that includes most all desirable features is a reasonable expectation.

One cannot presently predict the ways in which word processing will be used, but one might expect that most scientists will enter all drafts of memos, reports, and other documents using a word processor. Drafts of reasonable quality can be produced on a low-cost dot-matrix printer. Using a network, likely to become available in most laboratories, a document can be transmitted electronically (or on diskette) to coworkers and secretaries for modification, to the addressee(s) directly, or to a shared high-quality laser printer. With the laser printer facility, a simple command embedded in the text will determine whether the output will appear typewritten or typeset.

A word of warning might be in order. The process of revising and reprinting is nearly painless, an advantage not lost on teachers of technical writing. However, it can be all too easy to let a manuscript go through just one more revision before sending it, thereby losing the savings in time.

However, as important to effective science as the written word is, one can only expect use of this productivity tool to become incresingly widespread.

3.8.2. Database Management

Scientists must maintain databases. Those involved with routine testing will need to keep track of samples and testing results. Since the format of this

$$a = \frac{\int_{t=0}^{t} exp\,(bt^{2n})}{(2\pi sin\theta)^2}$$

FIGURE 3.9. Equation printed on a dot-matrix printer. A photographic enlargement was made of an equation printed with a low-cost printer.

data remains static for many samples or tests, the data are amenable to organization by a computer. Those involved in research may need to maintain that same type of data, but even more universally will need to maintain files of references to the scientific literature. In the latter category, card files containing several thousand entries are not uncommon for individual investigators. The task of sorting them to find an entry is formidable indeed, even with a well-developed indexing system, but again this data should be amenable to computer-assisted organization.

A **Database Management System (DBMS)** is a tool for organizing the storage of and access to information. Information takes many forms, and a general-purpose DBMS must be adaptable both to the type of data to be stored and the type of interaction that the user would like to have. The DBMS then must possess a high-level syntax for expressing a solution to a problem, and that sounds very much like an application-oriented *language*. DBMSs are indeed considered languages for the development of programs for acquiring and managing lists of numbers and character strings.

A DBMS developed expressly for literature file management must have a few additional characteristics that can set it apart from a general-purpose DBMS. It should possess the best characteristics of the large computer-search systems available by telephone at large technical libraries. (Most commercial DBMSs for small computers have been developed for the needs of the business community and are not entirely suitable for the needs of searching library data.) Suitable DBMSs must allow search by *combinations* of keywords that may be technical terms, authors' names, or journal names. For example, one may need to study horse's aortas knowing that P. Johnston has been another worker in that area since 1977. One might wish to start with a search with the boundaries that the resultant references should contain the words "horse" *AND* "aorta" *OR* the author "P. Johnston." To save time, the search should be limited to publications since 1977.

Suitable DBMSs also must have convenient methods of data entry. The user might be presented with a table including blanks to be filled in; an additional area might be included for notes. This feature requires some internal word-processing capability.

As more scientific literature becomes available electronically, we might envision a small computer connected by telephone to a host computer that has the tables of contents for scientific journals. Moving the cursor to a particular entry and sending a command to the host could produce an abstract, and further pursuit might give the text. DBMSs with the capability of "capturing" and formatting that information will greatly extend the productivity of scientific investigators.

3.8.3. Equation Processors

Programs have been developed for small computers that have the capability of performing some of the most tedious parts of modeling, the solving of algebraic equations, both numerically and analytically.

The software package TK!SOLVER is probably the most successful. A system of equations can be entered as "rules," and values for variables can be entered singly or in tables. An additional work area contains relationships between disparate systems of units. Upon command, TK!SOLVER seeks a numerical solution, directly if possible, but otherwise by iteration. Complex kinetic processes or steps in the equilibration of physical or chemical parameters can be modeled interactively.

Algebraic equation solving has been less successful, but improved tools for obtaining exact solutions to systems of equations and integrals can be expected.

3.8.4. Spreadsheet Software

The spreadsheet is a program developed to automate an important task for the business community. It is thought to have done more to stimulate sales of small computers than any other type of application program.

In its simplest form, the spreadsheet is a grid of "cells" displayed on the screen. Information can be entered into any cell of that grid, creating a table. However, the usefulness of the spreadsheet stems from the immediate computational capability afforded by the computer. Cells can be defined in terms of the mathematical combinations of other cells so that the result is observed instantly.

As a simple business-related example, assume that one needs to project the cost of a certain laboratory study. The total cost may require a fixed overhead for the laboratory, plus a cost per experiment. The latter would be calculated from a formula involving the time per experiment, an hourly wage, materials, and other overheads. Entering all of that data into an equation would, of course, lead to the single final number. However, entry into a spreadsheet allows the manipulation of the data. What would be the effect if the materials cost increased by 10% or the time required per experiment by a better worker was reduced by 15% while paying that worker 20% more? The result could be visualized immediately by moving the cursor to the cell containing the datum to be changed, making the change, and observing the other cells changing in response. When any entry is changed within the grid, the result changes immediately.

Applications in the nonmanagerial work in the laboratory are not so obvious, but several can be suggested. The computer with a spreadsheet on the lab bench provides a convenient manual data entry facility. If a simple density measurement experiment is undertaken, a column for weight and a column for volume might be prepared. Additional cells might be programmed to provide the ratio and the standard deviation of the measurement; "running totals" and measures of the measurement quality (standard deviation) would always be on screen.

The spreadsheet also can be used as a data manipulation tool. If the program to acquire data from an experiment stores that data on disk in a format that can be read by a spreadsheet program, the user can interactively experiment with the data. The corresponding values of the dependent and independent variables might be placed in columns *A* and *B*. If the relationship between these variables is linear when the reciprocal of *A* is plotted against the log of *B*, columns *C* and *D* should be defined as reciprocal *A* and log of *B*, respectively. Then by defining other cells as variables in least-squares fitting equations, the slope and intercept are computed immediately. If an "outliner" in the data is spotted, the result of removing it can be determined immediately. (This can easily lead to the interactive "doctoring" of data, and should therefore be used with caution.)

Those who also teach in a system requiring evaluation may find spreadsheets ideal for keeping scores with immediate readout of statistical parameters.

3.8.5. Integrated Software

Recently, several application software functions have been integrated in software packages. In the business market, spreadsheet, data management, and advanced graphing capabilities have been combined in a single program. A computer user can move easily between spreadsheet and graphing, for example, in order to immediately see the results of various data manipulations. By integrating this function further with data-acquisition programs, a very powerful system for data handling and interpretation results, and these systems are now becoming commonplace.

Similarly, laboratory software functions have been integrated into easy-to-use packages which bring together the data-acquisition drivers, graphing capability, and data management; often the package interfaces with a general-purpose integrated software package. Typically the data-acquisition drivers will support popular data-acquisition subsystems, allowing the user to set up the acquisition of a burst of data with only a few data entries. That data can be interactively manipulated using built-in transformations (such as logarithmic, inverse, or Fourier transformations) or the data can be displayed graphically. Data from conventional experiments can be acquired, manipulated in the spreadsheet, displayed, and stored without the need for programming.

As with all application programs, one must accept some inflexibility. Some of these packages support a limited number of hardware configurations or do not allow any real-time interaction with the experimental parameters. Frequently these limitations are a small price to pay for the power and ease-of-use that integrated laboratory software provides.

CHAPTER FOUR

High-Level Interfaces and Instrument Interfacing

Interfacing a computer to phenomena in the real-world requires an I/O interface module which is a *bridge* between the computer's bus and the transducers of the real-world phenomena with which the computer must communicate. The position of this bridge is illustrated in Figure 4.1. The interface must move information accurately and efficiently between the computer bus and the external phenomena.

In this scheme, two general components are identified which are necessary to acquire the information: the transducer and the interface. A **transducer** interacts with the chemical–physical domain, transducing information from the chemical–physical domain to the electrical domain and vice versa. (Some definitions would have transducers operate only *to* the electrical domain while "actuators" operate *from* the electrical domain; we will use only the single term.) Transducers are the subject of Chapters Seven and Eight. The electrical interface moves electrical domain information (encoded either as analog or digital) to the bus of the computer.

The path between the bus and the phenomenon is depicted in more detail in Figure 4.2. It is responsible for two functions, to accommodate *timing* differences and to make the necessary *domain conversions* between the domains in which data are encoded.

Timing differences between the computer's internal clock and the time frame of the external device occur at the computer's bus. Interfaces must synchronize data transfer. When an external device is ready to provide data *to* the computer, the latter is likely to be busy with an instruction other than an input instruction at that instant, and would be unable to accept the datum.

101

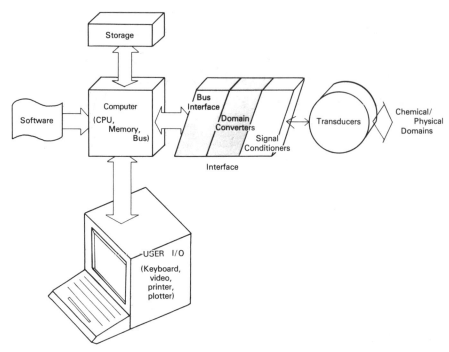

FIGURE 4.1. A laboratory computer system. The small computer's bus interfaces and domain converters move an electrical-domain signal into the computer.

When the computer is attempting to *send* a datum to an external device, that device may be unable to immediately accept the datum.

For example, a printer typically cannot accept and print a character every time the computer's software requests that one be printed because it is already busy printing the last character that was sent. The computer must hold (buffer) that character until the printer is ready for another character, and the interface must inform the computer that the printer is not ready so that the computer will not attempt to send the next character. Then when the printer is ready, the interface will transfer that character to the printer.

Similarly in the input direction, if a device sends digital data from an experiment at a typical rate of 10 kHz, the computer is unlikely to be ready to accept information precisely when the datum becomes available. Each new datum must be held by the interface for a few microseconds while the interface informs the computer of the waiting datum, and the computer responds with the input instruction sequence.

The second function of the interface is *domain conversion*. The chemical or physical information that is encoded by the transducer as an electrical value must be made compatible with the digital domain of the computer. That is, the interface must translate between the electrical domain of the transducer and the digital domain of the computer which, for all small computers, is **transistor-transistor-logic (TTL)**; (see Chapter Six if necessary).

Consider, for example, the conversions involved in one method of acquisition of a datum representing the intensity of a light beam (Fig. 4.3). The intensity information is first encoded as a continuous analog electrical form by the transducer. A common transducer for this purpose is a photodiode, which produces an electrical current proportional to the flux of photons. The digital computer cannot directly accept information encoded as current, so steps must be taken to convert to binary TTL values. This involves an interdomain converter, the **analog-to-digital converter (ADC)**, which accepts analog electrical input and produces a binary result. However, the electrical input to the ADC lies within in a predefined voltage and frequency range. Consequently, signal modifiers are required, in this case a current-to-voltage converter and a voltage-gain amplifier.

A similar example can be shown for output of information. Again using

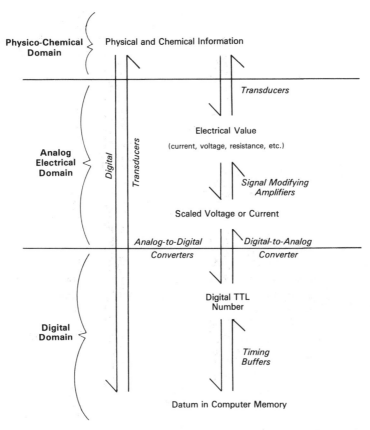

FIGURE 4.2. Interdomain conversions in data acquisition and control.

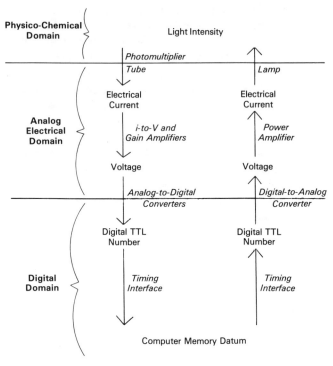

FIGURE 4.3. Interdomain conversions involved in acquisition of light internsity information and control of intensity.

light intensity as the physical data domain, a lamp can be used as an output transducer. An interdomain converter, a DAC, is required to translate the information to the electrical domain. Suitable signal modifiers (such as a power amplifier) are required to match the lamp's current and voltage requirements.

That sequence—transducer, signal modifier, interdomain converter, and timing buffer—moves from the specific phenomenon to the general computer. Of the many types of information that are to be acquired by the computer, light intensity is only one representative. However, many transducers of physical phenomena yield current, and most transducers produce an electrical signal (current, charge, or resistance) which can be converted to voltage and finally conditioned for an ADC.

The designer of a specialized interface confronts a decision: Which part should be built? Which should be purchased? In the interface sequence, as the signal moves closer to the bus, the required component is common to more types of interfaces. Consequently, these interface components are more likely to be found on generally available interface subsystems. If the entire interface is constructed on a custom basis, from transducer to bus, (a *bus-level interface*,) it can be optimized for the application, and higher performance is possible. However, if a general application subsystem is obtained, the user is spared many design problems, and time is saved. The trade-offs resemble those described in Chapter Three regarding machine language and high-level language programming: in most cases, it is to the user's advantage to purchase the subsystem and design only the signal modifying and conditioning interface.

While recognizing that cases do occur in which the interfacing problem is so unique that no suitable module exists, in this chapter we will deal with higher level interfaces such as are commonly available for small computers: parallel input–output (PIO) interfaces, ADCs, DACs, and counter-timer-clocks (CTCs). In many cases, as when the input transducer is an already existent scientific instrument, these building blocks will usually suffice to form the active part of the interface. However, a few basic characteristics of the timing buffer must first be understood.

4.1. INPUT–OUTPUT REGISTERS

If considered from the vantage of the programmer, the interface level of the timing buffer consists of one or more **registers** (Fig. 4.4). A register is an electronic unit capable of holding one or more bits of data. A random access memory location is one example of a register, but I/O registers differ, accepting a datum from one device and providing it to another (peripheral to CPU or the reverse). One of those registers, the **input data register** or **output data register**, holds the datum to be received or transmitted.

In most cases, the computer must be able also to direct commands to the interface in its various functions, and so an interface often includes a **control register**. Additionally the computer may need to obtain information about the status of the data transfer process, thereby requiring a **status register**. Often these are combined into a control–status register (CSR) which is viewed by the computer as having an address separate from the data register. In the CSR, each bit typically has a separate function; for status registers, one bit may signal the presence of input data while others might indicate various error conditions. Similarly, the control registers use single bits or small groups of bits to determine operating parameters for the next interface level, enabling conversions, interrupts, direct memory access, etc.

4.1.1. Polling and Interrupts

During a typical input operation, a datum is sent by some external device and stored in the input data register, directly available to the computer. Upon this occurrence, a single bit in the status register, the Data AVailable (DAV) bit, changes state. The CPU then can be notified in one of two ways: by interrogating the status register (polling) or by allowing the interface to interrupt the program and request service.

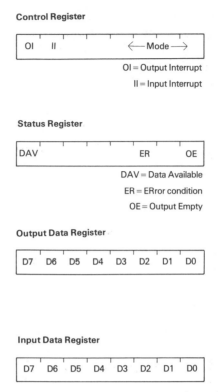

FIGURE 4.4. Control, status, and data registers for an interface port.

Commonly in small computers the status port is **polled**; that is, at regular intervals as represented in Figure 4.5, a software routine is executed which directs the computer to read the byte in the CSR to determine if the DAV bit is set: if it is, the computer proceeds to read the input data register. In many cases, the act of reading the data register automatically clears the DAV bit.

The alternative, **interrupt-driven I/O**, is somewhat more complicated in both hardware and software, but can be more efficient in program execution. The subroutine to acquire the datum is called by the hardware instead of by a program statement. An interrupt, generated in response to an electrical pulse sent to the CPU (Fig. 4.6), causes the CPU to immediately suspend execution of the current program sequence and execute a subroutine called an **interrupt completion routine**. When that routine is completed, the CPU is restored to the position in the program at the moment it was interrupted. The entire handling of the interrupt may only require a few tens of microseconds.

During the acquisition of a datum, the change in the state of the DAV bit can be fed to the CPU's interrupt input to generate an interrupt. The interrupt

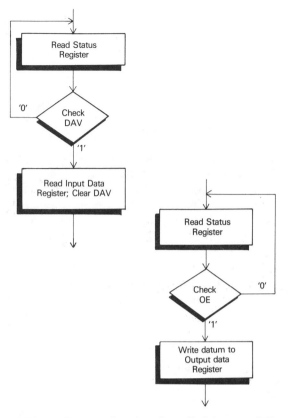

FIGURE 4.5. Program flowchart for polled data acquisition.

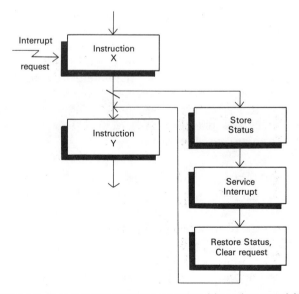

FIGURE 4.6. Program flowchart for interrupt-driven data acquisition.

107

completion routine then reads the data register and stores the datum for later use in the main program. An advantage is obtained since, if data arrives while the computer program is busy (concurrently processing previous data, for example), that datum can be placed in temporary storage by the interrupt completion routine until the main program is ready to handle it. The primary data-processing routine is slowed only slightly while the computer maintains prompt response to incoming data.

A typical computer system with interrupt capability such as the IBM PC or most S-100 bus computers has as series of lines on the bus for accepting interrupt request pulses. An interrupt controller circuit in the system responds to the pulse by signaling the CPU Interrupt Input that an interrupt is pending; when the CPU responds back with an interrupt acknowledgment, the controller replies with information that directs the CPU to the correct interrupt completion routine.

The process for output is similar although the sequence is reversed. When the computer is to output a datum, it may or may not need to determine if the previous datum has been accepted by the next level of the interface. If it does, the computer must first check a bit of the status register, Output Empty (OE). That bit is cleared when the external device sends a signal to acknowledge that it has accepted the datum. Again the term, polled I/O, is used to refer to the programming method in which the status register is regularly checked to see if the output register is empty. Only when that register is empty does the computer send the next datum. The alternative technique uses the external device's acknowledgment signal to generate an interrupt; the interrupt completion routine writes the next datum to the output register before returning to the main computer program.

The interface at the level of the bus needs to be described only briefly. An I/O register is specified by its unique *address* which must be used by the CPU in order to communicate with the register. During an I/O operation, the CPU places that address on the bus's output lines along with control signals which indicate I/O and control the timing. The interface subsystem must respond to the address and control signals using its address/control logic; that logic produces a **device select pulse (DSP)** which occurs *only* when the correct address *and* control bits are specified. The DSP is used to control transfer of data to or from a register.

4.1.2. Direct Memory Access

When blocks of data are moved between interface and memory, another style of I/O can be used for much higher performance at the cost of greater complexity. **Direct Memory Access (DMA)** is an alternative method of data transfer which bypasses the time-consuming processes of moving the data through the CPU. Additional DMA controller hardware added to the interface allows the interface to become the "temporary master" of the bus; temporarily the DMA unit rather than the CPU controls the computer.

Before DMA transfers can begin, the DMA controller hardware must be set up; the address register and data count register are initialized by a series of programmed data output sequences to the DMA unit. When a datum is ready at the input port, the DMA logic sends a signal to the CPU which forces the latter to enter a "hold" condition; the CPU responds by isolating itself from the bus after it finishes its current instruction. The DMA controller can then take over the bus as the temporary master; it must generate the memory address of the datum's destination and transfer the datum directly to memory. Then control of the bus is returned to the CPU, the DMA address register is incremented, the data count register is decremented, and the DMA system prepares to transfer the next available datum. When the data count reaches zero, an interrupt is sent to the CPU to inform it that the task has been completed.

Transferring data from memory to the output port under DMA is similar: when the port is ready for a datum, it sends the hold signal to the CPU and transfers the datum.

A dramatic improvement in data throughput is obtained by side-stepping the CPU, up to two orders of magnitude. At most, the CPU only loses a clock cycle for each datum transferred. If the transfer rate is not excessively high, a DMA transfer can be made without affecting the CPU by "cycle-stealing," transferring data during cycles when the CPU does not use the bus. Pipelined CPUs can accommodate DMA transfers without significantly affecting processor efficiency. However, for the very highest DMA rates, the CPU must be brought to a halt with a software instruction; the interrupt automatically restarts the processor.

DMA requires complex logic circuitry to handle bus timing, accept instructions from the processor, maintain the memory addresses, count the number of transfers, and generate interrupt signals. Fortunately, DMA peripheral-integrated circuits are produced for the major microprocessors which greatly simplify DMA design.

4.2. PARALLEL INTERFACES

The parallel interface is both the foundation of most other I/O interface units and a common module on its own; it in fact consists simply of a timing buffer with control and status register logic. Commonly, the parallel interface module contains output ports, and/or input ports, plus a control–status register. In many cases, a single control–status register may control more than one port and even the *direction* of the data transfer for the ports under its control. We will begin this discussion by considering very simple I/O parallel interfaces and finish with a detailed look at a typical integrated-circuit parallel interface module.

4.2.1. The Simple Output Port

The simplest form of output register is a simple latch for digital data. The data input to the latch is tied to the data bus, and the output is available to any peripheral devices. When the appropriate DSP activates the latch (as in Fig. 4.7), the datum is captured by the latch and is held at the output of the latch. There is no attempt to communicate with the peripheral through a separate control–status register to see if the peripheral is ready for a new datum; the register is simply updated with new data. In Fig. 4.7, an LED display is interfaced which displays the current contents of the register at any time.

4.2.2. The Transparent Input Port

In some cases, one wishes simply to monitor the condition of some external device. The input port hardware then consists of only a register which, when activated by a control signal, acts as an electrical buffer and a gate, open between the external device and the bus. When that control signal is inactive, the gate is closed, separating the bus from the external device.

For example, a binary counter constructed from integrated circuits might be used to count the pulses coming from a turnstyle (Fig. 4.8) in the entrance

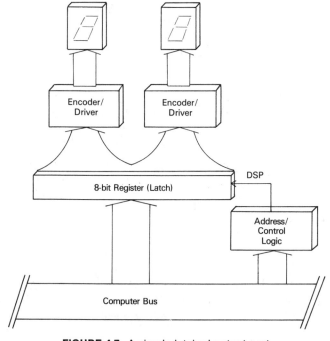

FIGURE 4.7. A simple latched output port.

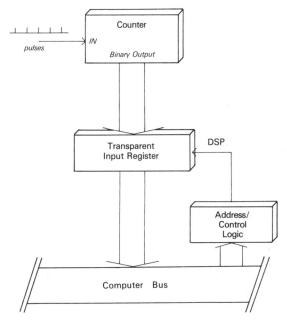

FIGURE 4.8. A simple transparent input port.

to a closed system (for example, a theater); the current count of objects inside the system can be obtained at the output of the counter. Whenever the current count is required in the program, the transparent input port is read. As with the simple output, there is no need for handshaking, no attempt by the counter to inform the input port that it has changed its sum, and no attempt by the input port to tell the counter when it will gather a new datum.

4.2.3. Parallel Output Interfaces with Handshaking

The device to which the computer sends data may not always be ready for the data. Consequently, a technique often is necessary that, in addition to loading the output register, will inform the external device that a new datum is available and will check whether the datum has been accepted. **Handshaking** is required. An output signal line notifies the external device that a new datum has been latched by the output register, and an input signal line obtains either an acknowledgment or an indication of the external device's READY status. A CSR that can be read by the computer provides the status of those lines. The bit in the status register which signals the condition of the external device is called a **flag**.

Figure 4.9 shows the timing diagram for such an exchange which would be suitable for the orderly transfer of information to an external device such as a printer. Using the $\overline{\text{READY}}$ line, a printer that is interfaced through a

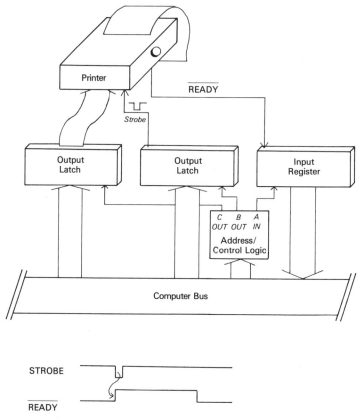

FIGURE 4.9. Parallel output port with handshaking. The printer signals readiness to accept data with READY, and the computer signals that it is sending a datum with the strobe.

parallel port can send a signal to the computer indicating that it is ready to accept a character. The computer first checks the bit of the register that contains $\overline{\text{READY}}$; if the flag is HI, the computer must wait, but if it is LO, the computer can follow by sending a new datum to the data port. (Recall that the bar over READY to form $\overline{\text{READY}}$ causes the line to be called "not ready"; that is, the printer *is* ready when the logic level is LO, and is *not* ready or $\overline{\text{READY}}$ when the logic level is HI.) An additional output action on a separate line is needed to generate a pulse on the STROBE line which goes to the printer. The STROBE is used by the printer to latch the datum and set $\overline{\text{READY}}$ to HI until the printer is prepared to accept another datum.

The sequence just described requires an output port to transmit the character byte plus one or two additional ports that can be configured to have the capacity for enough bits to transmit the STROBE and receive the $\overline{\text{READY}}$. Some ports can be configured so that in the same port, some bits are for input and some are for output. The programming sequence shown in BASIC (Fig. 4.10) uses the following sequence:

Test $\overline{\text{READY}}$ by reading the port, performing a logical AND with the corresponding bit, and looping back until the result is **0**.

Output the datum through the data output register.

Generate a strobe pulse by sending a **0** to the appropriate bit of the control register to set STROBE to LO followed by a **1** to that same bit to set the STROBE to HI.

If an interrupt system is configured, the $\overline{\text{READY}}$ line can be used also to trigger an interrupt, following which the interrupt completion routine will output the next datum. In order to ensure that spurious interrupts are not triggered when the main program is unprepared, a bit of the control register is often used to inhibit the interrupt.

A variant of that handshake procedure uses an $\overline{\text{ACK}}$ signal (acknowledgment) from the external device. Instead of reporting whether it is ready or not ready, the external device acknowledges every character byte that it receives with a pulse.

4.2.4. Parallel Input Interfaces with Handshaking

The input interface performs a function which is the inverse of the output port, both in data transfer and in the CSR. The input port is a latch, rather than a transparent buffer which must be activated by the external device (Fig. 4.11). That latching pulse also sets a Data AVailable (DAV) bit of the

```
2200 REM    A Subroutine written in BASIC for output with handshaking.
2210 REM        The subroutine must check until the external device is ready,
2220 REM    then output the byte, and then send a strobe pulse.
2230 REM        The port numbers for the status and data are PORTSTAT and PORTDATA.
2240 REM        The READY line is bit 0 of input PORTSTAT; the STROBE line is
2250 REM    bit 7 of the output port number PORTSTAT.
2240 REM        Check if the external device is READY.
2250        ISTAT = INP ( PORTSTAT ) AND 1:          REM Read the status port and
2260        IF ISTAT = 0 THEN GOTO 2250:             REM   check bit 0 for READY.
2270 REM Next output the byte, IDATA.
2280        OUT PORTDATA, IDATA
2290 REM Now send the strobe to the external device.
2300        OUT PORTSTAT, 127:                       REM Bit 7 is LO.
2310        OUT PORTSTAT, 255:                       REM Bit 7 is HI.
2320 REM The data was sent and strobed; routine is done.
2330        RETURN
```

Example Software for Parallel Output

FIGURE 4.10. A program to operate the port in Figure 4.9. The program is written in BASIC. It checks the READY line, and if READY is LO, it sends the datum and creates a strobe pulse.

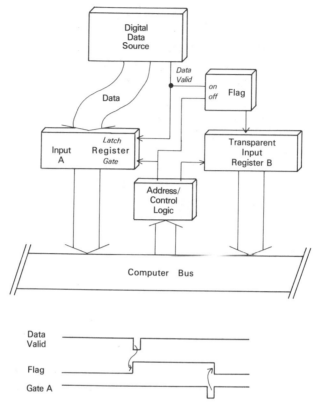

FIGURE 4.11. Parallel input with handshaking. When the Digital Data Source sends a datum, it signals data valid as well, which creates a flag. The flag is cleared when the computer reads the datum.

status register, and, if the interrupt is enabled, triggers an interrupt. When the port is enabled with an input DSP to place the datum on the computer bus, (that is, it is read,) the DAV bit may be automatically cleared.

The programming sequence, illustrated in Figure 4.12, for polled input is simple:

- Poll the status port and test the DAV bit; if not ready, repeat.
- Otherwise, read the datum.
- If necessary, clear the DAV bit.

4.2.5. Typical Integrated Circuit Interface Modules

In order to provide I/O ports which are as versatile as possible, most microprocessor manufacturers also have developed general-purpose parallel interface modules such as the Motorola 6821 PIA (Peripheral Interface

```
2600 REM      A Subroutine written in BASIC for input with handshaking.
2610 REM        The subroutine must check until the external device sends some data, checking
2620 REM   the DAV bit.  Then it reads the byte, automatically clearing the DAV bit.
2630 REM        The port numbers for the status and data are PORTSTAT and PORTDATA.
2640 REM        The DAV line is bit 0 of input PORTSTAT.
2650 REM
2640 REM   Check if the external device is READY.
2650          ISTAT = INP ( PORTSTAT ) AND 1:          REM Read the status port and
2660          IF ISTAT = 0 THEN GOTO 2650:             REM  check bit 0 for Data AVailable.
2670 REM   Now read the datum.
2680          IDATA = INP ( PORTDATA ):                REM The act of reading clears DAV.
2690 REM   That is all that there is to it.
2700          RETURN
```

FIGURE 4.12. A program to operate the port in Figure 4.11. The program is written in BASIC. It checks for Data AVailable, and reads the datum if DAV is true.

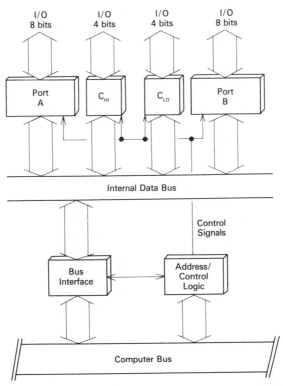

FIGURE 4.13. A typical integrated circuit interface circuit. An Intel 8255 in Mode 0 is represented. Each of the four port units can be set for input or for output.

Adapter), the Intel 8255 PPI (Programmable Peripheral Interface), and the Zilog Z-80 PIO (Parallel Input–Output). These units contain *both* the logic for interfacing to the bus and the latches and internal logic for handshaking; consequently, most new general-purpose parallel I/O subsystems incorporate them as foundation modules.

The 8255 PPI, whose block diagram is shown in Figure 4.13, is typical. This single integrated circuit, with up to 24 I/O lines, was designed with the flexibility to operate in any of the modes discussed above or even in a combination. Many personal computer systems use this chip to build the printer port. The operational modes are programmed by entering appropriate codes into a control register. The type of handshaking is set by the **mode**.

In *mode 0*, shown in Figure 4.13, ports A, B, and C can be set independently to become *either* input or output ports. Port C is divided into a pair of nybbles, each of which is configured independently; one can be for input while the other is for output if that is convenient.

If port A and/or port B is set to *mode 1*, it can be configured for input *or* output, but some associated bits in port C are used for handshaking (Fig. 4.14). For example, if A is an input port, the input data can be latched using an input signal entering a bit of port C; another bit of port C will automatically

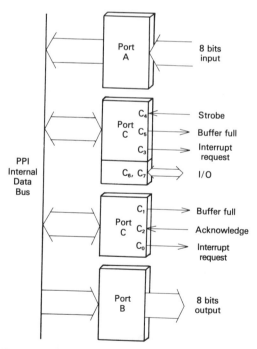

FIGURE 4.14. The 8255 interface circuit in mode 1. Port A is set for input; a datum can be latched by a STROBE pulse, and C_5 indicates its presence. Port B is set for output; bit C_1 indicates the presence of a datum to be moved to an external device.

signal to the external device that the datum has been accepted. If A or B are output ports, port C bits are allocated to the STROBE and $\overline{\text{ACK}}$ functions described earlier. Another bit of port C can be used to request an interrupt if that is desired. Two bits are still left over, and these can be turned ON and OFF, independently of all other bits.

Mode 2 is available in port A creating a *bidirectional* port; under the control of port C bits, the same port can be used to both send and receive data.

Full descriptions of this device can be found in data sheets from the manufacturer, Intel Corporation.

4.3. ANALOG INTERFACES

At the beginning of this chapter, the process of encoding physical- or chemical-domain information in a form suitable for the internal use of the computer was divided into a series of steps. Typically, the transducer yields information in the analog electrical domain; this information can be modified by amplifiers, as will be described in Chapter Five, but it remains continuous. Consequently, the next step in building an interface for the acquisition of that datum after the parallel port stage is to add an ADC. The ADC accepts the electrical information which is **analog** (continuous, not quantized) and encodes it in a **digital** form suitable for the parallel port. The symmetric step for output, to encode digital information as an analog electrical value, employs a DAC interfaced through a parallel output port. These steps are sufficiently ubiquitous in real-time computing that general-purpose add-on modules are readily available to add to most bus-oriented computers.

In this chapter, ADC and DAC subsystems will be examined from only two vantage points: that of the designer who specifies the part, and that of the user–programmer.

Several specifications are significant in the evaluation of the ADC and DAC circuits:

1. The first is **resolution**, the number of bits which are converted. Typically, the ADCs and DACs used in computer subsystems have 12-bit (1 part in 4096) resolution; 8-bit modules are quite inexpensive but have insufficient resolution for many laboratory applications, and the other extreme, 16-bit modules are very expensive as a result of the need to manipulate analog signals with a precision of 0.003%.

2. The **analog range** is the difference between the maximum and minimum ADC inputs and DAC outputs. It may be either a voltage or current range.

3. The **digital coding** format is binary in most computer subsystems, but for some other applications, BCD may be useful.

4. The **accuracy** is characterized by the relative error in the output, comparing actual with expected.

5. The **linearity** is described in terms of the maximum error in the difference between the outputs for consecutive inputs. The minimum tolerable standard is that ADCs and DACs must be **monotonic**; that is, when the input changes, the output must change in the correct direction. Lack of monotonicity can be devastating when the ADC or DAC is used in experimental control-loop applications.

4.3.1. Analog-to-Digital Conversion Subsystems

The ways in which the ADC subsystem is used can be compared with the ways in which data might be manually acquired from an instrument using an analog meter readout. The meter is continuous, moving continuously from the position at which one reading might be taken to the position at which the next is taken. Typically the operator determines when to acquire a reading, reads it, and writes it in a notebook. The data listed in the book are discrete; there is no information about the movements of the meter needle between readings. However, if the signal is "noisy" (the needle oscillates or vibrates), the operator must "condition the signal": that is, average the readings. By visually estimating the average position of the needle, the noise is filtered or subjected to a type of integration. These same functions of quantizing and filtering are assigned to the ADC subsystem.

The type of configuration that is suitable for an ADC *subsystem* depends on several factors:

- The required precision and accuracy.
- The number of separate analog input signals.
- The sampling rate, both overall and per channel.
- The signal conditioning requirements.

To meet as many requirements as possible, the ADC, designed and constructed for general application, must have more components than just the foundation of the parallel port and the ADC itself. Other components can be included, either required by the ADC technology or added as extra-cost options: a rapidly changing signal is momentarily captured by a **sampling-and-hold amplifier**; signal-conditioning flexibility is provided by a **programmable gain amplifier**; when there are more than one signal source, a **multiplexer** is used; other signal conditioners may be required. Figure 4.15 illustrates their interrelationships.

Both the requirements of the information and the requirements of the ADC determine which of the components in the figure are required. In the following section, before considering the design of the entire ADC subsystem, each of the components will be described. We will begin with the ADC

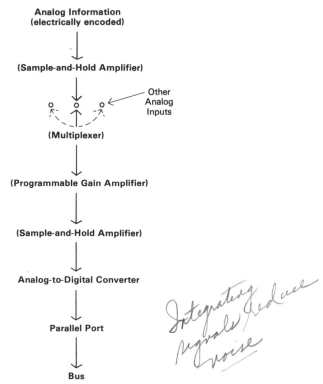

FIGURE 4.15. An ADC subsystem. Optional components are shown in parentheses.

itself, and work backwards to determine how the signal must be prepared for conversion.

4.3.2. The Analog-to-Digital Converter (ADC)

The heart of the subsystem is the ADC; its performance typically is the limiting factor in the speed, accuracy, and resolution of the entire subsystem. At least a dozen different techniques have been proposed and implemented for producing the conversion; those which are useful can be grouped into three categories: integrating, successive approximation, and parallel. The choice usually depends primarily on speed, but also on resolution requirements and cost.

Integrating ADCs. As is well known from the world of analog electronics, the precision of the measurement of a constant signal that has some electrical noise added to it can be improved by *integrating* the signal for a set period of time; if no higher frequency information is present within the integration period, the electrical noise will be reduced. Two common types of integrating converters integrate the signal in the process of making a conversion, and consequently they are useful for relatively slow data-acquisition rates, less than about 50 conversions per second.

The more common of the two uses the **dual-slope** method (Fig. 4.16). The converter is a single integrated circuit containing an integrating amplifier, a reference voltage source, a comparator amplifier, and a timer. A digital logic unit coordinates the conversion in the following sequence.

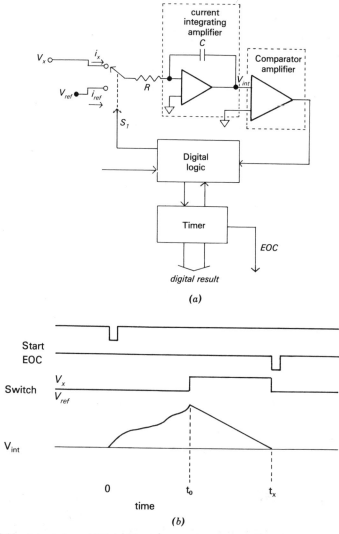

FIGURE 4.16. A dual-slope ADC. (*a*) Block diagram showing the integrator and comparator which compares current from V_x with current from the reference, V_{ref}. (*b*) Timing diagram showing the output of the integrating amplifier.

First, the integrator integrates a current, i_x, which is derived using Ohm's Law from the unknown voltage and a resistor, V_x and R. The signal is integrated by the integrating amplifier for a set period of time, t_0. During this integration, the amplifier output, V_{int}, increases; V_{int} is proportional to current and the integration time, t_0, but inversely proportional to the size of the integrating capacitor, C:

$$V_{int} = \int_0^{t_0} (V_x/RC) \, dt$$

The integrator input then switches to the reference current, i_{ref}, whose polarity is opposite to that of the input voltage; i_{ref} is derived from the reference voltage, V_{ref}, and resistor R. This reference current is integrated until the integrator output returns to zero; that is, the two integration periods are summed:

$$V_{int} = 0 = \int_0^{t_0} (V_x/RC) \, dt + \int_{t_0}^{t_x} (V_{ref}/RC) \, dt$$

If for now we assume the unknown voltage to be a constant, we can solve directly for the time that the reference voltage was integrated since V_{ref} and t_0 are constants:

$$V_{int} = 0 = V_x t_0/RC + V_{ref}(t_x - t_0)/RC$$

then

$$t_x - t_0 = -(t_0/V_0) \times V_x$$

Therefore the time that the reference is integrated is related to the input voltage by a constant. (Integrating and comparator amplifiers are described in Chapter Five.)

Since the dual-slope converter is *ratiometric*, integrating both input and reference voltages with the same electronic circuitry, the drift in the amplifiers, especially that due to temperature, can be minimized. The accuracy of the device depends primarily on the precision of the time measurement, which can be very good, and the linearity of the integrator, which can be made sufficiently linear that 14–16-bit converters (with a resolution less than 0.002% of full scale) are readily available.

The dual-slope method dominates in digital panel meters where stability is very important, but high conversion rates are unnecessary and even undesirable. Often, the conversion cycles are synchronized to the line frequency (60 Hz) so that this relatively low-frequency noise source is correlated and thereby eliminated. The time for a single conversion is typically 30 ms.

The second integrating ADC method is conceptually simpler, using the **voltage-to-frequency converter (VFC)** and a counter/timer. The VFC is simply a circuit that converts an input voltage to a train of pulses whose rate is directly proportional to the input; the pulses are fed to a counter which is gated by a precision timer so that a high-precision frequency measurement can be made. Commercially, they are available both as a complete ADC unit and also as a VFC alone; with the VFC the user can add as large a counter as desired to digitally integrate the signal for longer periods.

Although there are several techniques for making voltage-to-frequency conversions, the best uses the **charge-balance** technique, illustrated in Figure 4.17. An integrating amplifier balances the charge due to current generated from the input voltage by adding precise charge pulses of opposite polarity. When the output of the integrating amplifier exceeds a threshold, as detected by the comparator, a gate is opened which allows the reference pulses to reach the integrator. The reference pulses are derived from a precision reference voltage. These reference pulses also are electrically buffered and converted to TTL to become the output.

This ADC integrates the input on two levels. First, on the time scale of individual signal pulses of electrical charge, the inputs are integrated by the analog integration amplifier. Second, on the time scale of the entire conversion, the pulses are integrated by counting them.

The VFC method has several important advantages over the dual-slope method. First, the signal is integrated over the entire conversion period rather than less than half the period as is the case for the dual-slope converter. Second, the input range can be large; like the dual-slope converter, the VFC integrates current, but the voltage-to-current resistor which determines full-scale input range is often external to the VFC. Consequently, the user sets the range with a resistor appropriate to the input signal. Third, since the full-scale frequency can be designed to be as high as 10 MHz, the measurement period for a 12-bit conversion can be as short as 2.5 ms; how-

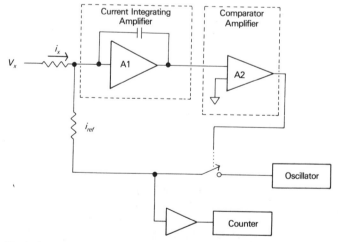

FIGURE 4.17. A charge-balance integrating ADC. Charge from V_x is balanced against i_{ref}; the pulses of charge then are counted.

ever, for commercial units, 10–100 kHz is typically the maximum frequency. Fourth, units have been built to measure very small currents, and it has been shown that, within limits, such a device can integrate short input pulses with large amplitudes. Since electrical current is the charge per unit time, a charge balancing VFC used for pulses acts as a charge-to-count (*q*-to-*n*) converter. Finally, the method has a logistical advantage. When the analog source must be located remote to the computer, the VFC can be located adjacent to the transducer; the digital pulses can be transmitted with very high noise immunity over long distances by a pair of wires, by laser light pulses, or by radio to the site of the computer where they are counted.

Successive Approximation (SA) ADCs provide flexibility which make them the most common in ADC subsystems for computers. The integrating ADCs, although inexpensive and quite accurate, have a very limited conversion rate, much less than 1 kHz, but most laboratory computers can, without DMA, accept data by programmed input at rates up to 20–80 kHz; using DMA, the rates may reach a few megahertz. The SA ADC fits that time scale.

In the conversion process, the input voltage is compared with test voltages produced by a DAC under the control of a logic unit called an **SA register (SAR)**, represented in Figure 4.18. A trial-and-error procedure is followed, very much like adding weights on an analytical balance: each weight is tested starting with the largest. A voltage from a DAC is compared with the input.

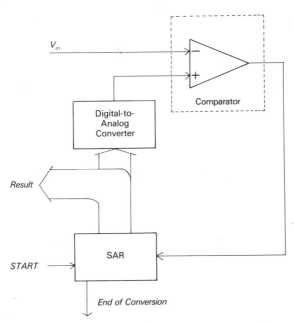

FIGURE 4.18. A successive approximation ADC. The voltage generated by the Successive Approximation Register (SAR) is compared with the unknown voltage, V_{in}.

The SAR tests 1 bit at a time as diagrammed in Figure 4.19. First the MSB is turned on; if the comparator determines that the input voltage is exceeded by the DAC voltage, the bit is turned off but otherwise it is left on, and the SAR moves to the next bit.

The conversion time of the SA ADC is directly proportional to the number of bits being tested, not to the resolution. For example, the conversion time of a 12-bit converter is $1\frac{1}{2}$ times that of an 8-bit converter although the resolution is 16 times greater. The rate at which the converter can step through the sequence (the conversion rate) is limited by the speed with which the DAC can settle to its new value after being changed by the SAR. In the moderate price range, 12-bit ADCs convert in 1–100 μs.

The primary advantage of the SA ADC is speed; when compared with those ADCs which are even faster, a second advantage is price although most SA ADCs costs considerably more than typical integrating ADCs. The cost results from the very tight specifications for the DACs.

One disadvantage of the SA technique is that, since the input voltage must remain constant during the conversion, a sample-and-hold amplifier is usually required (discussed below) which samples the signal for only a short period compared to its hold period, and then maintains that voltage until the conversion is complete. Thus no integration of slower signals is inherent to the technique, although pseudo-integration methods will be considered later.

The **computer-controlled SA converter** may be an inexpensive variant if one has only a DAC component available in the computer system, as shown in Figure 4.20. A test value is sent from the computer, which replaces the SAR, to the DAC and that output is compared, by a comparator, with the unknown voltage. Although it may be slower than using an SAR, it is equally effective.

Parallel (flash) converters are the fastest of all, and as might be anticipated, the most expensive. Since one of the largest applications for ultrafast ADCs is the digitization of video signals, they are also termed **video** ADCs.

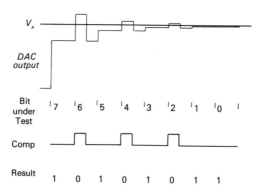

FIGURE 4.19. The DAC output for an 8-bit SA ADC. Beginning with bit 7, each output is set to **1**. If the result is less than the unknown voltage, it remains **1**; otherwise it returns to **0**. Then the next bit is tested.

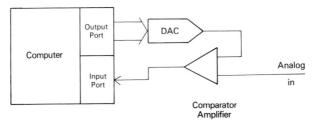

FIGURE 4.20. Using the computer as an SAR. If a DAC is available, the computer can be used as an SAR for analog data acquisition.

For an n-bit conversion, there are 2^n levels of resolution or **resolution elements**. In order for a parallel ADC to achieve a resolution of 2^n elements, the input voltage is simultaneously compared with $2^n - 1$ reference voltages by $2^n - 1$ comparator amplifiers; since the converter must also detect an over-range error, one additional reference and comparator is required. The reference voltages are generated with a linear voltage divider; for an 8-bit converter, there are 256 resolution elements. The reference voltages for the 256 comparators are derived from a series divider chain of 256 resistors with resistance R plus $\frac{1}{2} R$ at each end. As suggested by the block diagram of Figure 4.21, manufacturers have succeeded in integrating this entire unit into one integrated circuit.

This converter's name results from the fact that all of the bits are determined simultaneously; all comparators with reference voltages below that of the input voltage will register a logical **1**, and those above will be **0**. A priority encoder, consisting of a network of TTL gates, accepts the $2^n - 1$ comparator outputs, determines which was the highest to produce a **1**, and converts the comparator levels to straight binary code. An input "trigger" signal may be required, not to start a conversion, but to transfer the data into the priority encoder and output latch.

The speed of the conversion is dependent almost entirely on the settling time of the comparator amplifiers; 1–70-MHz conversion rates are typical. The delay induced by the priority encoder is usually minimal although the designer of the converter must take care that all bits of the output change simultaneously. For example, if bit 0 were to change more rapidly than bit 1, the change in two bits from a 01_B (decimal 1) to a 10_B (decimal 2) would momentarily produce a 00_B (decimal 0) or a 11_B (decimal 3) which are not intermediate values. Errors due to this lack of synchronization are commonly termed "glitches" and are eliminated by the encoder–latch circuitry.

Since the number of required comparator amplifiers increases exponentially with the number of bits, parallel converters tend to have low resolution. (However, they can be partially cascaded in some cases.) Some high-speed converters place a parallel converter and an SA converter in tandem. The effects of large-scale integration also have been found in the realm of ADCs; 6-, 8-, and even 9-bit flash converters, integrated on a single chip requiring 64, 256, and even 512 comparator amplifiers respectively, are available!

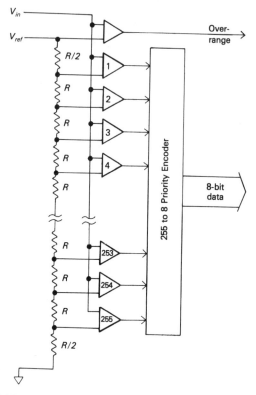

FIGURE 4.21. An 8-bit parallel ADC. The 8-bit output is the number of the comparator that triggers which is highest up the ladder.

The comparator amplifier itself can be considered a **1-bit flash ADC**. It can be used simply to indicate digitally whether the input voltage is above or below a reference level; for example, a photodiode transducer and a comparator can determine whether a light is on or off. If there is any noise in the signal, a comparator with hysteresis, the **Schmitt trigger** should be used; the hysteresis produces a lower effective reference voltage when the input goes from high to low than when it goes from low to high. For details, the section on operational amplifiers (Chapter Five) may be consulted.

Game ports are built-in to a number of small computers and have limited ADC applicability. Computers including the Apple II series, the Commodores, and the Ataris have provision for a "paddle" which is simple a variable resistor (potentiometer). The game port supplies 5 V to the potentiometer, and the potentiometer modulates the current according to Ohm's Law (Fig. 4.22). The ADC, which converts that analog signal to a digital signal, consists of a circuit that generates a pulse whose length is proportional to the resistance (a monostable; see Chapter 6 if necessary), and the length of that pulse is timed by the computer's internal clock. The result is an inexpensive and reliable ADC.

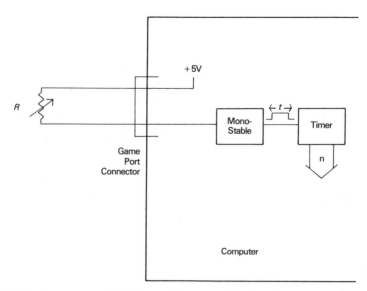

FIGURE 4.22. A game port ADC. The result is proportional to the resistance, *R*. This port is found in computers that use paddles for game playing.

The limitations of this approach are twofold: first, most computers that have game ports designed them with only 8-bit resolution which may or may not be sufficient; second, the result is proportional to resistance or to the *reciprocal* of either the current or the voltage.

Practically, one is limited to measurements involving transducers which (1) are resistive and (2) experience a resistance range that extends down to near zero. Study of Chapters Seven and Eight may suggest several good applications. For example, position can be measured using rotary or linear potentiometers; wrapping the pen-drive cord of an analog recorder around a 10-turn potentiometer, which is connected to the game port, yields a very quick interface. Temperature can be measured with thermistors, and light can be monitored with photoresistors. Additional measurements can be made of a variable voltage using a fixed resistor, but they are difficult to linearize with either accuracy or precision.

Increasing the Resolution of the ADC. The addition of random noise to an otherwise static signal actually can be used to increase resolution. The noise must exceed one quantization level and be random. This practice of adding noise to obtain higher resolution is called **dithering**, and the effect can be seen in the example in Figure 4.23. Suppose that a 2.427-V signal is to be quantized by an 8-bit ADC with an input range of 0.00–2.56 V. If there is no noise in the signal, the average of an infinite number of measurements will still be 242. However, if noise that appears random with respect to the conversion rate is intentionally added to the signal, the converter will return results of both 242 and 243, and the average of those measurements will be

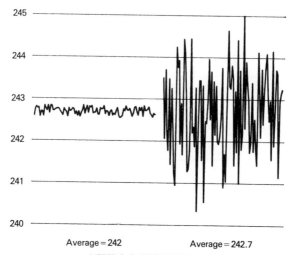

FIGURE 4.23. Increasing resolution by dithering. The average of a series of acquisitions in the first part is 242.0. The average of the integer results in the second part is 242.7. Noise was required to improve resolution.

242.2; the result is one-half unit lower since the ADC actually truncates rather than rounds the result. Dithering has effectively increased the resolution of the measurement.

4.3.3. The Sample-and-Hold (S&H) Amplifier

The S&H amplifier has the capacity to accurately track a signal while its switch is closed (Fig. 4.24), and when the control input commands the switch to open, it can hold at its output the voltage which was at the input at the time of opening. The term **Track-and-Hold** also can be applied, although it suggests a slightly different mode of operation in which the amplifier spends most of the time in the tracking state.

A S&H amplifier provides two potential benefits to the system: first, the analog information to be converted is representative of the signal at the precise instant that the hold state was entered even though there might be a delay from that moment until the ADC operates on it; second, that analog level can be retained until the ADC is ready to convert and for as long as the ADC requires to complete the conversion.

The S&H amplifier is a requirement for subsystems with SA ADCs since the input must not change within the conversion period; the effect of neglecting the S&H amplifier when the signal is changing faster than the conversion time is to dramatically decrease the effective resolution.

However, the S&H amplifier is only required with integrating ADCs when it is necessary to sample the signal so that the result is representative of the signal at a very precise instant with an uncertainty of less than a microsec-

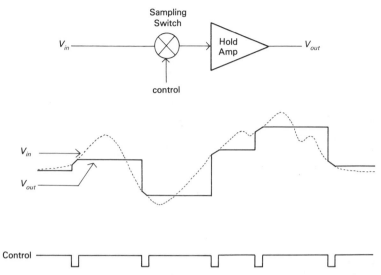

FIGURE 4.24. A sample-and-hold amplifier. When the control input is LO, the hold amplifier follows the input; otherwise the output is fixed.

ond. These occasions are rare since the acquisition rate of these converters is still relatively low.

Many parameters are used by the manufacturers to specify S&H amplifiers, and some are illustrated in Figure 4.25. Several are related to the amplifying characteristics, those which would apply even in the absence of the sample-and-hold capability. Ideally, the output when the input is zero

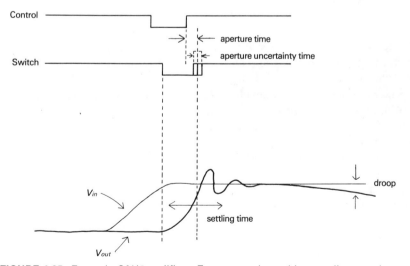

FIGURE 4.25. Errors in S&H amplifiers. Errors experienced in sampling are shown.

also should be zero; the **offset** is a measure of the actual output at zero input. Most S&H amplifiers have *exactly* unity gain; the **gain error** is the ratio of the actual output to the ideal output, corrected for the offset. Ideal amplifiers follow the input with no time lag; the **settling time** is the time required for the output to reach the ideal value to within a specified error after an instantaneous jump in the input.

A second set of parameters are related to the sample-and-hold capability: How fast can it accurately sample an input? How long can it hold that value? The ideal S&H amplifier opens its switch instantaneously; the actual delay is an **aperture time**, and the uncertainty in determination of when the switch is open is the jitter or **aperture uncertainty time**. Together, the settling time and the aperture time define the overall delay from the time that sampling was initiated until the output is stable, the **acquisition time**. A final important characteristic is the decay rate or **droop**; the voltage is stored on a capacitor, and all real capacitors leak charge with time which results in droop.

Typically, the ADC follows immediately after sampling so that the droop is less than a resolution element of the ADC. Acquisition times under 100 ns are common. The primary error introduced by the S&H amplifier is usually in gain and offset; however those are remedied in the subsystem calibration procedure.

4.3.4. The Programmable Gain Amplifier (PGA)

Obtaining the maximum precision from the digitization of the input requires that the signal be amplified so that the maximum analog signal produces a maximum digital value. However, the signal must not generate error by exceeding the input range of the ADC. One way of adjusting the ADC subsystem so that full-scale changes in the signal conditioning stage produce full-scale changes in the ADC result is to provide a gain amplifier that can be controlled by the computer (Fig. 4.26). If an over-range error is detected through an error bit in the status register, the PGA can be used to decrease the range of the input to the ADC; if all of the input signals are too low, the gain can be increased.

The PGA also *may* be considered to be a method of increasing the overall resolution of the conversion. In principle, if the PGA provides gains of $\times 1$ and $\times 2$ before an ADC with an input range of 0–10 V, one extra bit of precision is added; the higher gain is used when the signal is in the 0--5-V range, the lower gain for the 5--10-V range. However, the existence of a PGA does *not* guarantee increased resolution; that would require the amplifier gain ranges to have accuracy greater than that of the ADC (0.025% for a 12-bit ADC), and that performance is by no means guaranteed.

Gain changes require switching an appropriate resistance into the circuit of an amplifier using solid-state semiconductor switches. An additional 2–4 bits of an output register are required for the computer to output the code for the desired gain. Typically, binary gain values are used: 1, 2, 4, 8, 16.

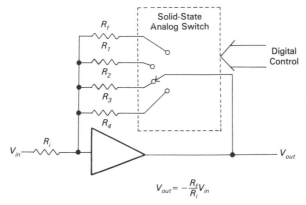

$$V_{out} = -\frac{R_f}{R_i} V_{in}$$

FIGURE 4.26. A programmable gain amplifier. A solid-state switch under computer control can change the gain of the amplifier.

Autoranging PGAs can detect the range of the input signal and automatically shift input ranges to accommodate it. However, this process, being a trial-and-error sequence, can be relatively slow since the amplifier and the over-range circuitry must settle after each trial.

4.3.5. Multiplexers

The multiplexer is actually nothing more than a multiposition solid-state switch for analog signals. An ADC can be shared by as many analog inputs as desired with the help of a multiplexer that, under digital control, selects a single signal for conversion. It is employed because, in many cases, the most expensive parts of an ADC subsystem are the ADC and the fiberglass board; a separate ADC for each input signal would be too expensive to implement. An output port is used to control the multiplexer in order to direct the correct signal to the ADC (Fig. 4.27).

Often an option exists that automatically increments the channel after each conversion. This is useful primarily for the routine monitoring of many transducers such as in a process control system. In most current systems, the multiplexer is made from solid-state field-effect transistor switches which can be controlled from TTL-compatible inputs. The "cross-talk" between channels is negligible, and the speed of the switch is not a factor when conversion rates are slower than about 1 MHz.

4.3.6. Signal Conditioning and Modifying

In most cases, the ADC subsystem as described so far does not provide a suitable input for the experimental signal. Signal conditioning systems, if present, may be required for several possible purposes, illustrated in Figure 4.28.

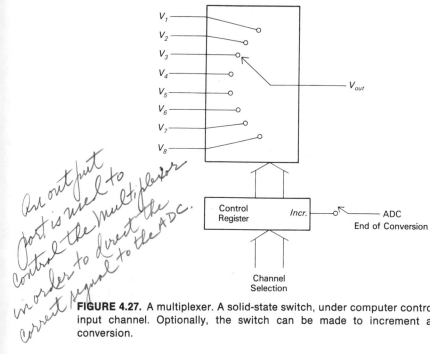

V_1
V_2
V_3
V_4
V_5
V_6
V_7
V_8

V_{out}

Control
Register

Incr.

ADC
End of Conversion

Channel
Selection

An output port is used to control the multiplexer in order to direct the correct signal to the ADC.

FIGURE 4.27. A multiplexer. A solid-state switch, under computer control, can select the input channel. Optionally, the switch can be made to increment after every ADC conversion.

The first stage may provide a **high-impedance buffer** with no other effect on the signal. The input to the buffer amplifier draws a negligible amount of current from the source, typically 1–100 nA, but the output can give as much current as is needed to get to the next stage. Such a buffer is often required to avoid "loading" the transducer by extracting more current than the transducer can accurately provide. It may be located on the board with the ADC, or it may be placed in a module with the transducer so as to amplify the signal above electrical noise that might be acquired between transducer and ADC.

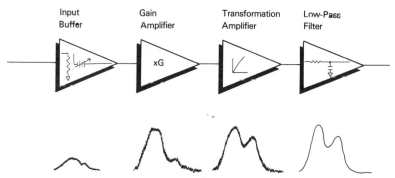

Input
Buffer

Gain
Amplifier

Transformation
Amplifier

Low-Pass
Filter

xG

FIGURE 4.28. Components of signal conditioning and modifying subsystems.

A second conditioning function is **fixed gain**; the gain of the amplifier can often be adjusted, usually by the installation of resistors, to increase or decrease the range of the input signal to match that of following components. In order to retain maximum precision, the full range of the analog signal must correspond to the full range of the ADC without ever exceeding the ADC input.

A third function is **transformation** of the input data; for example, in order to handle a very wide dynamic range, an analog logarithmic converter can be used so that good precision can be maintained at low input levels without amplifying the high input levels to the degree that they cannot be handled by the remaining electronics.

Finally, **filtering** can be added. Filtering removes frequencies in the electrical signal which have no useful information while passing unchanged those that contain useful information. **High-pass** and **low-pass** filters remove low and high frequencies, respectively. **Bandpass filters** remove frequencies both above and below the desired frequency band. **Notch filters** remove only a narrow frequency band of offending noise. These filters are further described in Chapter Five.

Correct filtering may be necessary to avoid **aliasing**. If the signal contains frequencies that are more than half the acquisition rate, they may appear in the result at a frequency which is some fraction of their actual frequency. Avoidance of this source of error frequently requires analog filters, since these errant components of the digitized signal are not easily removed afterwards by software. Aliasing is discussed in detail in Section 4.5.

Satisfactory filtering often requires only the simple addition of a capacitor to the amplifier to filter high-frequency noise. Notch or bandpass filters of greater complexity combat specialized interference problems. Further information on circuits used for signal conditioning and modifying is found in Chapter Five.

4.3.7. ADC Subsystem Integration

Figure 4.15, shows the ADC subsystem and all the options previously discussed. Programmable gain amplifiers are generally extra-cost options, and signal conditioning and modifying amplifiers must be designed for the specific application, and are seldom mounted on the subsystem board. (An exception is a signal conditioner for thermocouples which can be obtained as an option on some ADC subsystems.)

The position of the S&H amplifier and the software component of the interface also bear discussion.

The S&H amplifier is a *required* component only when employing the successive approximation ADC. It could be placed anywhere between the transducer and the ADC but is generally placed before or after the multiplexer.

In most applications and in nearly all commercial ADC subsystems, when data from several sources are to be acquired, the S&H amplifier is placed between the multiplexer and the ADC so that the S&H amplifier can be shared by all input channels (Fig. 4.29*a*). This requires the computer to acquire multichannel information *sequentially*; the conversion of each datum must be preceded by setting the multiplexer and then switching the S&H amplifier to the hold state. Often the time difference between reading successive channels is small enough to be negligible, typically around 50 ms. However, in some cases, data from all channels must be accurately correlated in time, and the various inputs must be sampled *simultaneously*; this can be accomplished by allocating each channel a separate S&H amplifier (Fig. 4.29*b*). All of the data can be acquired simultaneously by pulsing all of the S&H amplifiers at the same time and then successively multiplexing them to the ADC for sequential conversion.

That software is a vital component of any interface bears emphasis. However, before writing sample programs for the acquisition of data through an ADC, the architecture of a simple ADC subsystem must be outlined. This example will assume a 12-bit ADC that is triggered by the program. An 8255 interface adapter, as described in Section 4.2.5., can be used to interface the ADC to the computer bus; the interconnections between the ADC and the 8255 are shown in Figure 4.30.

Mode 1 operation can be invoked to latch data into two input ports where the availability of new data can be detected from a *Buffer Full* bit, such as

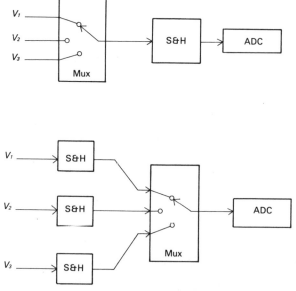

FIGURE 4.29. Positioning the S&H amplifier. Conventionally, the S&H Amplifier is placed between the Multiplexer and the ADC as in the top diagram. If all signals are to be acquired simultaneously, S&H amplifiers would have to acquire signals before the multiplexer.

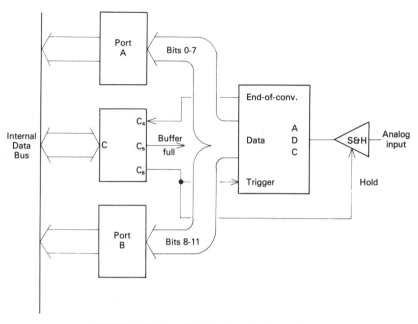

FIGURE 4.30. An ADC interfaced with an 8255.

bit C_5 of port C. To trigger the conversion from the 8255 PPI, bit 6 of port C must be forced low and then high, creating a pulse.

The design of the supporting software, listed in Figure 4.31, assumes that the modes of the port have already been set in an earlier subroutine. Therefore, only the following steps are needed:

- Create the trigger pulse.
- Check for completion of the conversion.
- Acquire the 12 bits of data.
- Store the datum in the correct member of a data array.

The subroutine is written first in BASIC and second in assembly language for the Intel 8088/8086 as examples of both a simple higher level language implementation and a typical assembly language implementation.

There are a number of simple variations of the arrangement in Figure 4.30. For example, use of a programmable clock to trigger the ADC might be more versatile. Second, the Interrupt Request line of the 8255 could be led to a vectored interrupt controller in which case the acquisition subroutine would need to be an interrupt completion routine. Third, alternative configurations of the 8255 can free additional bits for such purposes as control of the multiplexer or programmable gain amplifier.

Direct Memory Access. DMA is a useful option for an ADC subsystem when

••

BASIC version

```
4000 REM Trigger an ADC conversion and return datum in array IDATA, element J.
4010 REM The addresses of the 8255 ports A, B, and C are PORTA, PORTB, and PORTC, respectively.
4020 REM
4030 REM First trigger the conversion with a pulse at port C, bit 6.
4040      OUT PORTC, 191:                    REM  191 = 255 - 2⁶.
4050      OUT PORTC, 255:                    REM  Raise bit 6 again; conversion starts.
4060 REM Now check to see if the conversion is complete.
4070      JSTATUS = INP ( PORTC ) AND 32:    REM  Read the status of bit 5 of port C.
4080      IF JSTATUS  = 0 THEN GOTO 4070:    REM  If the buffer is not full, then wait.
4090 REM Now get the datum as two bytes.
4100      DATALO = INP ( PORTA )
4110      DATAHI = INP ( PORTB ) AND 15  :   REM  For high byte, read only four bits.
4120 REM Pack the datum into the array, combining into an integer variable.
4130      IDATA ( J ) = DATALO + 256 * DATAHI
4140 RETURN
```

Equivalent subroutine written in assembler for an 8086/8088 CPU.

Comment %

This subroutine, GET-DATUM, triggers an ADC conversion and places the datum into array IDATA, the number of the element being contained in register SI. The 8255 ports are PORTA, PORTB, and PORTC.

%

```
GET_DATUM:
    MOV     AL, 255-64      ; Load the AL register with bit 6 LO.
    OUT     PORTC, AL       ; Output to port C.
TEST_EOC:
    IN      AL, PORTC       ; Get the contents of Port C, the status port.
    AND     AL, 32          ; Isolate the Buffer Full bit, and
    JZ      TEST_EOC        ; repeat if the bit is zero.
    IN      AX, PORTA       ; AX is a 16-bit register, so PORTA and PORTB
                            ; are acquired in a single operation.
    AND     AX, 0FFFh       ;Mask off 12 bits.
    MOV     IDATA[SI], AX   ; Store the datum in the array that begins at address
                            ; IDATA plus an offset of the number in SI.
    ADD     SI, 2           ; Increment the pointer by two bytes for the next datum.
    RET                     ; Mission accomplished; return.
```

••

FIGURE 4.31. Programs to drive the configuration of Figure 4.30. The driver is written both in BASIC and in 8086/8088 assembly language.

high-speed acquisition is desirable. This technique was described earlier as part of Section 4.1. ADC throughput can be increased almost an order of magnitude by using DMA and an ADC with an adequate conversion rate. Assuming a sufficiently fast ADC in the absence of DMA, the throughput is limited by the rate at which the subroutine for data acquisition can execute. The processor must check for the completion of the conversion, acquire the data, store it in memory, adjust the memory address, and check to see if enough data have been acquired. Depending on the processor and CPU clock speed, the upper limit might be 10–40 kHz. In order to achieve higher rates, DMA is required.

DMA techniques can provide about an additional order of magnitude of throughput, but that matches the limits of conventional SA ADCs. Conversion at higher rates requires waveform digitizers which are usually independent instruments (see Section 4.6.2.).

4.3.8. Digital-to-Analog Conversion Subsystems

The DAC subsystem is a much simpler module than the ADC. As has been described, the DAC output is an analog reflection of the digital input. In the following section, the techniques for making this conversion will be described, and, at the end, we will return to a typically integrated DAC subsystem.

The Converter. The DAC has been mentioned on numerous occasions thus far, and has been defined in terms of its function only: producing an electrical signal which is proportional to a reference voltage and linearly related to the digital input (Fig. 4.32). (Nonlinear DACs are available also, called "companders" since the near-logarithmic transfer function can be used to compress or expand the range.)

Conceptually, the simplest form of a DAC is the **weighted current source** of Figure 4.33. The DAC is simply a reference voltage, V_o, with a series of resistors acting as calibrated current sources, each having a solid-state switch. The incoming digital word simply closes the switches corresponding to the data bits thereby enabling the appropriate current sources. The resistance corresponding to the nth bit is precisely calibrated to be proportional to 2^{-n}, so the current is proportional to 2^n. The sum of the currents for all the bits that are turned on is proportional to the input datum.

A quick calculation demonstrates the fundamental limitation of this approach. For a 12-bit converter, the ratio of the resistance of the LSB to that of the MSB is 2048:1. In order to keep the current for the MSB within a reasonable range, the resistance should be at least 1 K Ω; then the resistance of the LSB will be 2.048 MΩ which is about the upper practical limit. If the reference voltage, V_0, is 10 V, the currents passing through individual resistors range from 10 mA to about 2.4 μA; the power that must be dissipated from a single resistor ranges from 100 to 0.024 mW. This presents two problems: the difficult calibration of the resistors over such a wide range, and

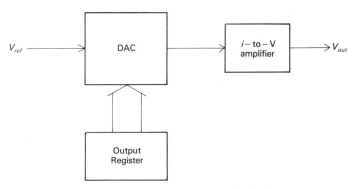

FIGURE 4.32. Block diagram of a DAC.

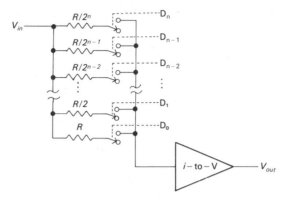

FIGURE 4.33. Weighted current source DAC. Each data bit switches a resistor that can provide current proportional to the value of the bit.

the variation in both the temperature coefficient and heat produced by the resistors.

The **R-2R ladder network** provides the solution to these problems. Figure 4.34 shows the resistor network, the derivation of which is somewhat tedious and will be omitted. The ratio of the largest to the smallest resistor is only 2:1 making the DAC simple to build and calibrate. The switches and resistors can be integrated into a single chip, and the resistors can be trimmed by the manufacturer using an automated laser. The result is lower cost, better accuracy, and a smaller temperature coefficient of the error in both the absolute accuracy and the linearity.

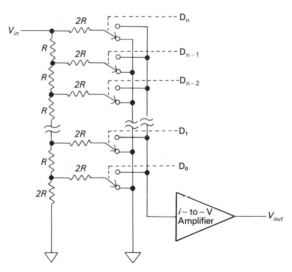

FIGURE 4.34. R-2R DAC. With the network shown, the DAC can be constructed with only two sizes of resistors.

Since the desired form of analog output is frequently voltage, a current-to-voltage amplifier is usually required as an output stage. Operational amplifiers (OA) configured for this task (Chapter Five) require an external resistance to set the gain, and this resistor is typically integrated into the DAC chip. Where a high-power output is required, OA power stages can be added.

The previous examples have depicted a unipolar output; the output current is proportional to the product of the reference voltage and the digital input. If as in Figure 4.35 the digital input is a signed integer, that is, the MSB is used as a sign bit, then the MSB can be inverted, and an offset current is added to the DAC current output. For an 8-bit DAC, a binary $1000\ 0000_2$ will give a 0.0 current output with the MSB inverted, but the offset current will make the total current and therefore the output voltage negative. When the number is $0000\ 0000_2$, the DAC current plus the offset current should be zero.

DACs also can serve the function of multiplication, applying a digital multiplication factor to an analog input signal which replaces the reference voltage; as such, the circuit can be used as a **programmable attenuator**. If the input analog signal, V_i, is directed to the reference input and the digital input is n, the output will be proportional to $V_i \times n$. This arrangement is termed a **multiplying DAC**.

Where both the analog input and the digital input can be bipolar, a **four-quadrant multiplying DAC** results. However, if the analog switches of the DAC cannot switch bipolar currents or the DAC is not equipped to handle a digital sign bit, a **two-quadrant multiplying DAC** results. If neither input can be bipolar, the DAC is, of course, single quadrant.

The DAC Subsystem Integration. The DAC differs from its reciprocal, the

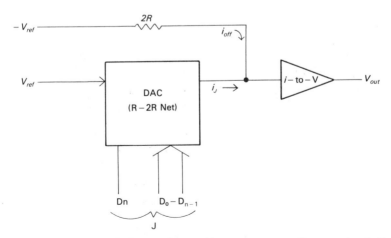

FIGURE 4.35. A bipolar DAC. Current from $-V_{ref}$ generates an offset equal to half of full scale. By inverting the most significant bit, D_n, it becomes a sign bit.

ADC, in that it does *not* require a sequence of events to be completed; consequently, no handshaking is required. In its simplest form, the subsystem then requires only a simple digital latch, a current DAC, and usually an amplifier.

In a typical experiment, several devices might be controlled by this programmable voltage source: many types of power supplies, oscilloscopes, and recorders. Since multiplexing the output is not practical, the inclusion of 2–4 DACs on a single board is common. Each DAC is independent, being associated with a separate output register.

A pair of DACs can display a waveform on an X–Y oscilloscope or X–Y recorder (Chapter Ten includes further details). On the oscilloscope, the data may appear as dots in the X–Y field. However, the members of a data pair are entered into the DACs sequentially, and a spurious dot will appear between the update of the X and Y DACs. If a pair of data at coordinates, X_1, Y_1 and X_2, Y_2 are to be displayed in sequence, and the Y DAC is updated before the X DAC, three dots will appear: X_1, Y_1; X_1, Y_2; and X_2, Y_2. The problem is further aggravated when the DAC has greater than 1-word resolution; if a pair of 12-bit DACs are loaded from an 8-bit bus, *four* output operations are required.

There are several possible solutions. The common procedure is to provide a circuit that produces a pulse on the *trailing edge* of the pulse that loads the last word. That pulse, as in Figure 4.36, can be used to control either

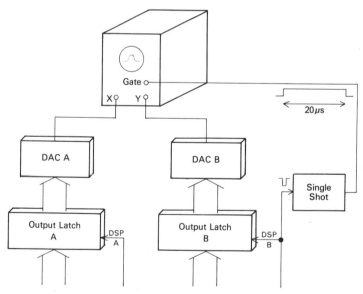

FIGURE 4.36. Using a pair of DACs for graphics generation. Each X–Y location is positioned on the oscilloscope by a pair of DACs. When both are stable, a pulse turns on the beam.

the pen-lift input of a plotter or the gate input of an oscilloscope; the latter can be employed to turn the oscilloscope beam off.

4.4. COUNTER/TIMER SUBSYSTEMS

There are several useful counting interface functions which have hardware components in common, but perform different functions according to the configuration. The first is a **programmable clock** whose function is to generate pulses at precise intervals which can be set by the program. In view of previous discussion, an obvious application is the use of these pulses to trigger an ADC. The second function is a **timer**, a device that can count clock pulses during an externally controlled period. The third is an **event counter**, a device that could count external events such as scintillation counts or animal responses.

The minimum components for these subsystems include (1) a precision oscillator that generates the reference time-base pulses, (2) a gate for turning the counters on and off, and (3) a counter that can be preset to a constant for counting pulses. Normally, the counter counts down from a preset number to zero.

The precision oscillator is easily constructed using a crystal in the range of 1 to 10 MHz; these inexpensive devices provide a frequency which is typically accurate to five or six significant digits, more if the temperature is stabilized. Many computer systems provide a frequency reference on the bus: for example, 2 MHz in S-100 computers and 4.77 MHz in the IBM PC.

The remaining components are either assembled from standard TTL devices, or are contained in LSI circuits from the major microprocessor manufacturers. The Intel 8253 is an oft-used example. A block diagram is shown in Figure 4.37. Three 16-bit presettable counters with gate control are integrated onto a single chip. Where 16 bits are insufficient, the counters can be used in tandem.

4.4.1. Programmable Clock

The objective of this system is to generate pulses with precision timing. The configuration is shown in Figure 4.38. The precision oscillator, at typically 1–5 MHz, is passed through a gate which feeds the next stage, a presettable down counter.

The counter is configured as a **modulo-n** down counter: after n is loaded into this stage, the counter decrements on each pulse that it receives. When it reaches 0, an underflow pulse is generated at the output, and n is automatically loaded into the counter again. The process repeats, generating underflow pulses at precise intervals. The resultant frequency, f_n, is f_0/n where f_0 is the frequency of the precision oscillator.

The presettable counter typically has a 16-bit input, so the input frequency

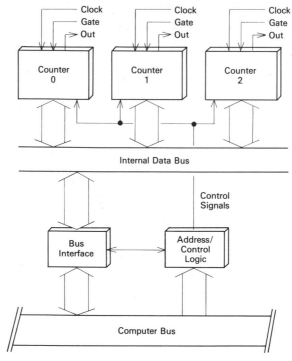

FIGURE 4.37. An integrated circuit counter/timer. An Intel 8253 contains three counters that can be configured for various operations.

can be divided by up to 2^{16}-1. If the resultant frequency is still too high, two counters can be used in tandem.

The output underflow pulses can be used by any external device (ADCs, stepping motors, audio tone generators, etc.) or they can trigger interrupts for timing within a program. The time-of-day clock for many computers is based on a precision oscillator divided down to 100 Hz or less, and those pulses generate interrupts; the interrupt completion routine then updates the time-of-day count.

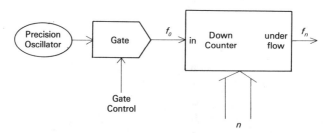

FIGURE 4.38. Programmable clock configuration. The oscillator, turned off and on with the gate control, is divided by the programmable down counter. The output, f_n, is f_0/n.

By setting the mode in the control/status register of the CTC, the user can choose how to control the gate. In some cases it may be sufficient to open and close with the CSR. In others, the capacity to open the gate with an external pulse is useful; this allows the gate to be turned off while n is loaded, and the clock can be started *synchronously* with the experiment. For example, if the clock is used to trigger a burst of ADC conversions, the gate control assumes the function of the sweep trigger on an oscilloscope; conversions can be made at precisely separated intervals after being triggered by a feature of the data source.

4.4.2. Period Timer

To measure the length of a pulse, the components are reconfigured to count the number of oscillator pulses during the period being measured. As can be seen from Figure 4.39, the differences between it and the programmable clock are only that the gate is controlled directly by the external device whose period is being measured, and the down counter is loaded only once with a constant (n_1). Opening the gate causes clock pulses to reach the counter which counts down until the gate is closed; the closure of the gate also might be detected by an interrupt if it is simultaneously fed to the interrupt circuitry. At this point the contents of the down counter are stored as n_2. The elapsed time, t, is simply

$$t = (n_1 - n_2)/f_0$$

where f_0 is the input oscillator frequency. The period timer is useful for timing external events; for example, a race could be timed by using a pulse derived from the starter's gun to open the gate, and a signal from an interrupted light beam at the end of the race to close the gate.

4.4.3. Event Counter

To count events, the oscillator of the clock circuit is here exchanged for an external input, coming from the experiment (Fig. 4.40). A preset number, n_1, is loaded at the beginning of the counting period. The event pulses will

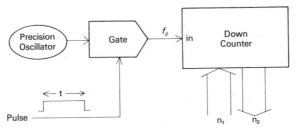

FIGURE 4.39. Period timer configuration. The down counter counts the number of clock ticks that the pulse allows through the gate.

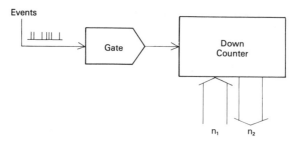

FIGURE 4.40. Event counter. The down counter counts the number of pulses until it is reset by the computer.

cause the counter to count down, and at the end of the counting period the counter is read again. The number of events is found by difference.

A frequency meter is not much different from an event counter except that the count (now given in Hertz instead of events) is made for a precise period. If the pulses are counted directly by the down counter, the output of a second counter in the programmable clock mode can be used to gate the input signal and time it for a precise interval.

An alternative counter method might be mentioned: the interrupt-driven counter. This is an excellent alternative if (1) the programming language supports interrupts, and (2) the hardware for vectored interrupts is already included in the computer system (the IBM PC has some; the Zenith Z-100 has many). The pulse to be counted triggers an interrupt, and, in the interrupt completion routine, a location in memory is incremented or decremented as required. A few higher level languages support the installation of the interrupt-completion routine software; unfortunately most do not.

4.5. THE TIMING OF DATA ACQUISITION

We have seen that the conversion of analog information to digital data is a discrete process, usually requiring that the process be *triggered* and that the result be *acquired*. The problem of when the acquisition should take place requires some attention, because the temporal precision is often as important as the precision of the conversion. If data are acquired to characterize the exponential decay of some process, uncertainty in the timing of the data is as damaging as uncertainty in the data themselves.

4.5.1. Triggering the Acquisition

Data acquisition from devices such as an ADC may be triggered from several sources, illustrated in Figure 4.41, and often the choice of the source is under the control of the computer. Four categories might be considered that can

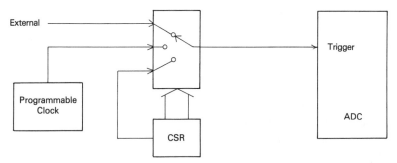

FIGURE 4.41. Sources of trigger pulses for the ADC. The ADC may be triggered by the CSR of the computer, by pulses from the programmable clock, or by pulses from an external device.

be used to describe the timing of an acquisition and therefore the ADC trigger source:

1. **Time-driven** acquisition. Models for data processing that treat the data as the function of time require an accurate and precise time clock, as in the experiment in Figure 4.42. The pulse for initiating the acquisition would probably come from the programmable real-time clock such as described in Section 4.4. Should none be available, the computer could keep reading a time-of-day clock and initiate the acquisition when the time reaches a predetermined point. Most experiments in which a time-related response to a stimulus is under study are time driven.

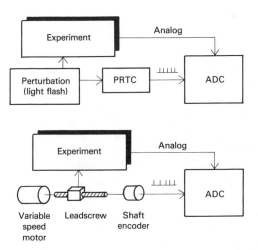

FIGURE 4.42. Time and event-driven data acquisition. Above, the light flash triggers the experiment, and the ADC is repeatedly triggered by a precision clock. Below, the leadscrew turns an encoder, so the pulse rate is proportional to the rate of screw movement, but synchronized with position.

2. **Event-driven** acquisition. One might envision an animal behavior experiment in which each time that the animal pushes a bar, the computer records the temperature, time, or light intensity. Some infrared spectrometers provide a pulse at every wavenumber; use of those pulses to trigger acquisitions synchronizes the data to wavenumber, independent of changes in scan-speed or pauses in the scan (Fig. 4.42). In these cases, the experiment provides the timing; the acquisition is precisely synchronized to the experiment.

3. **Program-driven** acquisition. At a certain place in a program, information from the experiment is required, and that datum is acquired. One assumes in this case that the data is not changing in a meaningful way with time. Such an occasion exists when the computer is simply to monitor one or more data sources; the displays might be updated whenever the computer can acquire data.

4. **Data-driven** acquisition. Consider the previous case where data sources are monitored. If one source should be found out of line, the following data from that channel might be acquired more rapidly at the expense of the others. Another application could be to chromatography when the retention times are very long but the peaks must be recorded precisely. When only the uninteresting baseline is recorded, few acquisitions are made, but as the signal increases on a peak, the rate of acquisition also increases giving higher reliability to important data than to baseline data.

Some of these methods may be used together; for example, an external event might be used to turn on the programmable real-time clock; data then could be acquired at regular intervals. The beginning of the data block is then precisely synchronized to the external event.

A special case of using the timing functions to acquire a repetitive waveform with very high time resolution is the **boxcar averager**. This can be built from even a slow ADC using the block diagram of Figure 4.43. As illustrated in Figure 4.44, the timer is triggered on at a reproducible point in the waveform or from another synchronous event. When the timer counts down to zero, the resultant pulse closes the sample-and-hold amplifier and triggers a conversion. The process is repeated on the next sweep, but with a slightly longer timer period. Slowly, by taking 1 datum per repetition of the signal, the waveform is acquired with a time resolution determined by the increase in the timer period for each sweep. If the input frequency of the timer is 2 MHz, the time resolution can be as low as 0.5 μs.

Although this technique may sometimes be necessary, it has several drawbacks. First, it cannot be used with a single-shot event. Second, the preponderance of the acquisition time is spent waiting to acquire the data. For example, if the waveform being acquired has a period of 100 μs, the time to acquire the 200 data needed for one 100-μs period of the waveform will be 20 ms.

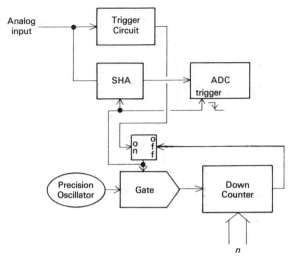

FIGURE 4.43. Block diagram of a boxcar averager ADC. A repetitive signal is accumulated by acquiring one datum on each repetition. The signal triggers a delay, after which the ADC is triggered.

4.5.2. Aliasing

The *rate* of acquisition of time-dependent data is also important. It must be matched to that signal under two constraints. First, we must prevent meaningful high-frequency signals from appearing as an incorrect signal frequency. Second, we must prevent higher frequency noise signals from creating an incorrect data component. Both of the problems are **aliasing**.

According to the Nyquist theorem, the rate of data acquisition must exceed twice the highest frequency present in the signal. The reason is understood from illustrations of aliased signals. In Figure 4.45, the solid line is a representative signal. If, because the signal is not well understood, the ADC

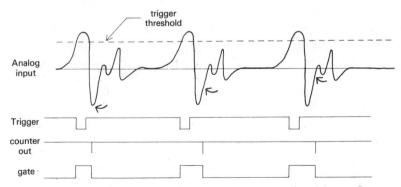

FIGURE 4.44. The waveforms in a boxcar averager ADC. Arrows indicate the time at which the S&H amplifier is triggered.

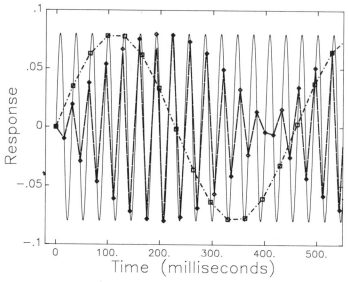

FIGURE 4.45. Aliased acquisition of a sine wave signal. The light line is the actual signal. In one case, acquisitions were made at a point denoted by squares, and the acquired signal apparently has a very low frequency. In a second case, acquisitions were made at points noted by the diamonds.

is triggered at too slow a rate, that is, less than twice the highest frequency component of the signal, the image that is acquired (generated by connecting the data points) is that of a single broad peak. In order to ensure that the signal does not so disguise itself, the acquisition rate must be more than twice that of the highest signal frequency.

The highest frequency component in a signal is readily determined when the signal is known to be a sine wave, but the problem is more complicated for the typical exponential decay or peak-shaped signal. At a minimum, the rate must be sufficient to provide a half-dozen or more points across the half-life of the decay or the peak width for accurate determination of peak position or area. (Further discussion is found in Chapter Five.)

In the second example (Fig. 4.46), the high-frequency component is noise. An exponential decay curve, which can easily be sampled at a sufficiently high rate, is somewhat obscured by the presence of noise. For simplicity in the illustration, the noise is sinusoidal, such as would be produced by interference from line power. The distortion, which is visually observed in the sampled curve, could easily lead to fundamental misinterpretation.

To avoid aliasing, there are two options. One is to sample at a sufficient rate that the noise is not aliased; the noise can then be removed by digital filtering techniques. The second method is to use analog filters to remove the noise component before conversion to digital. The latter method is typically the practical alternative.

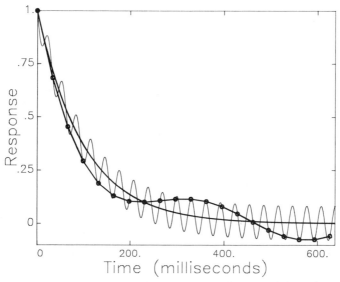

FIGURE 4.46. Aliased acquisition of an exponential decay curve. The curve has been corrupted by sinusoidal noise. Acquisition at a rate which aliases the noise generates a deceptive and incorrect signal.

The type of ADC will have an effect on the type of filter. If an integrating ADC is employed and the duty cycle (fraction of the total conversion time spent in integration) is high, that filtering is performed by the ADC. On the other hand, an SA ADC with a fast S&H amplifier converting at the same rate can easily give an aliased result.

4.6. REMOTE INTERFACE CONTROLLERS

Up to now, we have assumed that the computer has an open-ended architecture; that is, it has a bus capable of accepting interface boards which can be addressed directly by the computer. There are cases in which this suggestion cannot be accommodated. It may be that the only available computer or the computer chosen because of other unique features is a large mini-computer or a microcomputer without an available bus. In other cases, the available lab computer may be shared with colleagues and consequently, it might be inconvenient to change boards mounted internally each time the computer is used.

A solution to this problem is a Remote Interface Controller. The entire interface unit is placed outside the computer; the computer communicates with the unit via a general-purpose interface, typically a serial interface. The remote interface operates as a slave computer for which the lab computer is the master.

There are several advantages to this approach over the usual bus-level approach:

1. The remote interface controller gives access to computers not otherwise suitable for data acquisition and control operations, that is, those which cannot accept an interface card on the bus. One example is an Apple Macintosh which offers a unique operator interface advantageous for developers of real-time software; a second is a time-shared DEC VAX which provides substantial on-line computing power. Neither of these computers can easily accept interface boards which can be directly addressed.

2. Interface portability is possible. We have previously referred to the problems of moving software from one computer to another. Interfacing carries this problem into three dimensions: the experimental systems vary between users; the computers used to automate those systems also vary; even when those dimensions are constant among several researchers, the design of the interface boards can vary. An interface that can be used in the same way in different systems makes it easier for more users to share techniques.

3. It brings more concurrency to the operation of the experiment. While a remote unit is acquiring data, the master computer can be asynchronously processing data.

4. It makes programming easier. In most cases, drivers for a serial I/O port are already written into the OS. Therefore, communication with the interface can be accommodated through simple calls which send and receive character strings.

There also are disadvantages to using an external interface unit:

1. The user gives up some control. The remote computer is restricted to a relatively small number of commands which are fixed in ROM. While a good set of commands will cover most experiments, there may be control applications in which the sequence of events is too complex for the remote computer's command set. (Some interface units may allow the master computer to down-load routines into the slave for increased flexibility; typically these routines must be passed to the slave as BASIC code or binary code.)

2. The external interface is probably more expensive to manufacture since the circuitry will usually require its own cabinet and power supply.

3. There are restrictions on the quantity and/or rate of data acquisition. Data can be acquired faster than they can be transmitted over the serial line; if higher rates are necessary, the slave computer must store data

locally before it transfers them to the master, and this storage space may be quite limited.

We now consider the general concepts of using two types of controllers, a General Purpose Remote Interface Controller and a more specialized Waveform Digitizer.

4.6.1. General-Purpose Remote Interface Controller

This unit is a self-contained microcomputer (a microcontroller). We cannot exactly define it, but typical characteristics can be considered.

The controller is a microcomputer with its own memory and interfaces, as shown in Figure 4.47. At a minimum, it must contain the following items:

- An 8-bit CPU, and often an IC designed as a controller rather than the type used for general-purpose small computers.
- RAM and ROM—the ROM contains the operating program, so the RAM can be quite small unless the unit has the capability to buffer acquired data before transmitting to the master.
- A general-purpose communication interface; typically this is an RS-232 serial interface, but it also may be an IEEE-488 instrument bus (both are described in Chapter Nine.)
- Control and data-acquisition functions including one or more of the following: ADC, DAC, analog multiplexer, general-purpose amplifiers, timers, and TTL parallel interfaces.

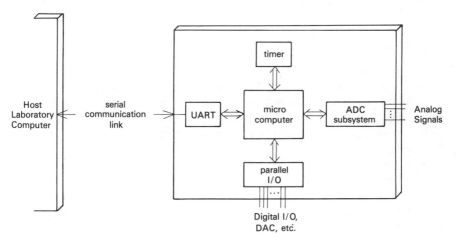

FIGURE 4.47. A remote interface controller. A slave microcomputer is used to acquire data and transmit them by a common serial link.

With this array of components and the appropriate software, a typical sequence for acquisition of a single datum could be to send a "C" to the remote unit, and it responds with the results of the conversion. From there, many additional features could be added using commands sent from the master computer:

- Change multiplexer channel.
- Increment multiplexer after each conversion.
- Sum together a number of conversions before returning the result.
- Make conversions at precisely defined intervals.
- Wait for an external event before acquiring data.
- Enable and disable the controller; that function could allow the controller and another peripheral to share a port or the controller and a terminal could share a port on a large computer.
- Set the direction of a parallel port; read from the port; write to the port.
- Count events or generate a square wave of a desired frequency.
- Output an analog value *via* the DAC.
- Accept subroutines downloaded from the master computer to accommodate sequences for specialized functions as pulse polarography or stepping motor control.

The reader should not expect that any existing commercial unit can handle exactly the functions described above. However, there are units that do accommodate most of the functions and should be viewed as a viable alternative to a bus-oriented interface.

4.6.2. Waveform Digitizers

One data-acquisition problem *requires* an external interface. When data-acquisition rates significantly exceed 1 MHz, the data cannot be acquired directly by computer, even with direct memory access. The Waveform Digitizer is conceptually similar to the Remote Interface Controller described in Section 4.6.1. in that it will acquire data at a desired rate and buffer them for later communication to the master computer. However, its special ability is high speed; commercial digitizers can acquire data at rates up to 70 MHz with 9-bit precision. An example is illustrated in Figure 4.48.

An instrument with that capability costs many times more than the typical small laboratory computer. Consequently, Waveform Digitizers are designed as stand-alone instruments; the operating parameters can be set on the front panel. The ability to transfer data to the master computer and accept parameters from the master computer are special options. Most commercial units communicate over an IEEE-488 interface, described in Chapter Nine.

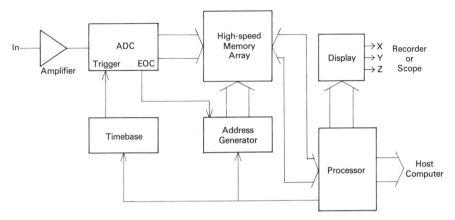

FIGURE 4.48. A waveform digitizer. This unit is used to acquire data at very high rates and transfer them to the computer at lower speed.

One particularly important feature is the **pretrigger**. Whether or not an experiment is proceeding, the Waveform Digitizer continuously makes conversions and places the results in its internal memory which is used as a ring-buffer. Therefore, a trigger pulse, instead of commanding it to acquire data over the next few microseconds, also might command it to hold the data just acquired. In contrast to the usual analog oscilloscope, the trigger can be at the beginning, middle, or end of a single sweep.

4.7. INSTRUMENT INTERFACING

Using the building blocks described in this chapter, many laboratory interfacing applications can be accomplished. Typically, a scientific instrument contains or is connected to some *transducer* which produces an electrical signal related to the physical or chemical phenomenon of interest. The remainder of the instrument *conditions the signal* and provides a *readout*; the type of readout determines the type of interfacing.

Most scientific instruments can be grouped into three categories plus a category for counters, as suggested in Figure 4.49. Older instruments are entirely analog; the readout has no digital components, but usually contains an analog panel meter and a voltage output for a recorder. The next group dates from the period when digital panel meters first became available; instruments were upgraded by the simple replacement of the analog meter with a digital panel meter, a self-contained module that accepts a voltage and displays it digitally. Instruments that are counters of the same generation are handled similarly. Finally, as microprocessors were incorporated into instruments with increasing sophistication, more functions for communication with other computers were added.

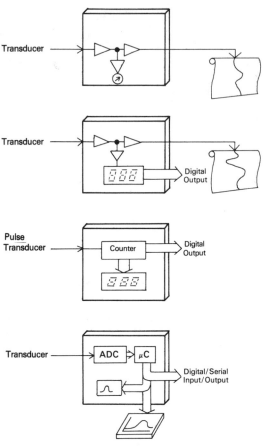

FIGURE 4.49. Scientific instruments and their electronic outputs. Top, an instrument with only a recorder output. Next, a similar instrument upgraded with a digital panel meter. Third, a counting instrument which has a direct digital output. Bottom, a microprocessor-controlled instrument.

4.7.1. Analog Instruments

Before 1970, most instruments employed entirely analog signal conditioning and fed either or both of two types of readouts: analog meters and strip-chart (or sometimes X–Y) recorders. The interface with a computer needs only to read the instrument output values.

Many measurements by these instruments are made of stable signals, involving no independent variable that changes during the course of the measurement. However, if the signal is measured as a function of some independent variable other than time (for example, wavelength in a spectrophotometer), that variable is transduced to time by linearly varying (scanning) the independent parameter. The constant speed of the recorder allows accurate recording of the dependent variable as a function of the independent variable.

The computer must acquire the analog output, substituting for the recorder in two functions. First, the recorder signal must be converted to a digital value. Second, if the signal varies as function of time, the time base must be accurately established.

Acquisition of the signal requires an ADC subsystem. The ADC must often be preceded by a gain amplifier. The instrument's recorder output typically covers a range of between 1 and 100 mV full scale; ADC inputs typically accept signals in a 10-V range. Consequently, a gain amplifier is usually required.

If a time base must be established, either the programmable real-time clock, described earlier, can generate trigger pulses for the ADC or, in most 16-bit computers, the time-of-day function of the OS can be monitored to determine when to generate trigger pulses.

4.7.2. Counters and Analog Instruments with Digital Readouts

These instruments have a digital readout and a connector which gives the digital values, but no other digital functions. The instrument's signal processing is entirely analog.

The simplest "digital readout" for an analog instrument requires only that a modular digital panel meter (Fig. 4.50) be mounted in the instrument and connected to the recorder output. These meters usually have a resolution of $3\frac{1}{2}$ or $4\frac{1}{2}$ digits meaning that the output can read ± 1.999 in the former case and ± 1.9999 in the latter. Nearly all of these meters contain dual-slope ADCs in which the internal DAC and counter follow a BCD sequence.

Panel meters have several types of computer-compatible readouts. The first type is none; many panel meters are entirely self-contained and have no digital signal output at all. For interfacing, one must revert to analog interfacing.

The second type of panel meter does have a digital signal output. A connector is included which simply provides the BCD codes for each digit at the rear of the meter. For a $3\frac{1}{2}$-digit meter, each full digit requires four pins plus one pin for the half-digit, and one for the sign. Another output signal may be available to inform the computer of the process of a conversion. Usually the signal levels are TTL-compatible so that the signals can be connected directly to the parallel port, two BCD digits per byte, at least 2 bytes for a $3\frac{1}{2}$-digit meter.

The timing of the acquisition may require some extra care. If an acquisition is attempted while a conversion is in process, the result could yield half of the digits from the previous conversion and the other half from the conversion in process. For example, if the previous conversion had a result

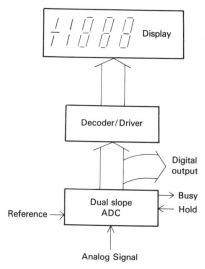

FIGURE 4.50. Block diagram of a panel meter.

of +1.001 and the next had a result of +0.998, the steps of reading two bytes might first return the "+1.0" part of the first value followed by the "98" part of the second; the apparent result is "+1.098," which is not close to either. The best solution to this synchronization process entails either monitoring the signals to determine when a conversion takes place, and acquiring the datum immediately thereafter, or controlling the conversion timing.

Consider the example shown in Figure 4.51. The BUSY line is HI *during* a conversion. In BASIC, we would monitor that line to determine when it went HI and then LO (lines 20–30); then the 3½ digits would be acquired immediately and organized into a value.

Using advanced features of various interface adapters such as the 8255, it also may be possible to use a BUSY or End-of-Conversion signal from the meter to latch the datum into the parallel ports, thereby simplifying the timing process.

Occasionally in a panel meter, not only are the BCD values present, but greater control is possible so that the panel meter can be triggered on command from the computer. Although there are several ways that this may be done, in some cases the panel meter contains an input line marked HOLD; while it is held LO, no conversions take place, but if it is briefly raised, the meter is allowed to make a single conversion. An output port from the computer will then be necessary. If the HOLD line as described were connected to port C, the additional line shown in Figure 4.51 would need to be added to the program in order to trigger the conversion.

A third type of panel meter, not often encountered, *multiplexes* the digits at the output. A computer *output* port must be used to select the data to be

Least Significant Digit,	1's Place	Port A,	Bit 0
	2's Place		Bit 1
	4's Place		Bit 2
	8's Place		Bit 3
Second Digit,	1's Place		Bit 4
	2's Place		Bit 5
	4's Place		Bit 6
	8's Place		Bit 7
Third Digit,	1's Place	Port B,	Bit 0
	2's Place		Bit 1
	4's Place		Bit 2
	8's Place		Bit 3
Most Significant Digit	1's Place		Bit 4
Sign Bit			Bit 5
BUSY Bit			Bit 6

```
1000 REM  Program to acquire a datum from the panel meter.
1010 REM    PORTA is the address of Port A, PORTB is the address of Port B (POR)
1020 REM    The conversion is triggered automatically by the meter.
1030    N = INP(PORTB) AND 64: IF N = 0 THEN GOTO 1030      'Wait for the panel meter to go BUSY.
1040    N = INP(PORTB) AND 64: IF N = 1 THEN GOTO 1040      'Wait for conversion process to end.
1050    N = INP(PORTA):  DIGIT1 = N AND 15                  'Get Least Significant digit from Port A
1060         DIGIT2 = (N AND 240) / 16                       'Mask the second digit and shift it 4 bit
1070    N = INP(PORTB): DIGIT3 = N AND 15                   'Get the third digit from Port B.
1080    DIGIT4 = ( N AND 16 )/16                            'Get the half-digit.
1090    SIGNBIT = N AND 32                                  'If Bit 5, SIGNBIT = 0, sign is +.
1100 TOTAL = DIGIT1 + 10*DIGIT2 + 100*DIGIT3 + 1000*DIGIT4
1110    IF SIGNBIT<>0 THEN TOTAL = -TOTAL                   'If SIGN is negative, ...
1120 RETURN
```

     If the HOLD feature is present in the panel meter, and is connected to Port C, Bit 0, line 1020 would be replaced by the following lines:

```
1020 REM   The Conversion is triggered by the following statement.
1030     OUT PORTC, 1: OUT PORTC, 0                         'Create a pulse on the HOLD line.
```

FIGURE 4.51. Computer program for acquisition from a panel meter with fully parallel output.

acquired. A 3-bit value selects the digit which will be made available at the BCD output of the connector. To acquire the value, the status must be requested first, and when the conversion is complete, the digits are selected and acquired one at a time; a sample acquisition subroutine is found in Figure 4.52. The circuitry of the digital multiplexers are discussed in Chapter Six.

4.7.3. Microcomputer-Controlled Instruments

When the instrument includes microcomputer-based processing, there are many ways in which the instrument could be designed to allow data transfer and control. Increasingly, instruments are designed to take full advantage

••

BCD Output,	1's Place	Port A,	Bit 0	
	2's Place		Bit 1	
	4's Place		Bit 2	*Single Digit*
	8's Place		Bit 3	*Input*
BUSY			Bit 4	
Control Output	1's Place	Port B,	Bit 0	
	2's Place		Bit 1	
	4's Place		Bit 2	*Digit Selector*
	8's Place		Bit 3	

```
1000 REM  Program to acquire a datum from a panel meter with a multiplexed output.
1010 REM        PORTA is the address of the input port, PORTB is the address of the output port.
1020 REM        The conversion is triggered automatically by the meter.
1030    N = INP(PORTA) AND 16: IF N=0 THEN GOTO 1030    'Wait for the panel meter to go BUSY.
1040    N = INP(PORTA) AND 16: IF N=1 THEN GOTO 1040    'Wait for conversion to end.
1050    OUT PORTB,0                                     'Select the Least Significant digit,
1060        DIGIT1 = INP(PORTA) AND 15                  ' & get the 4-bit BCD digit.
1070    OUT PORTB,1                                     'Select the second digit,
1080        DIGIT2 = INP(PORTA) AND 15                  ' & get it.
1090    OUT PORTB,2                                     'Select the third digit,
1100        DIGIT3 = INP(PORTA) AND 15                  ' & get it.
1110    OUT PORTB,4                                     'Select the fourth digit, which is only
1120        DIGIT4 = INP(PORTA) AND 1                   ' 1 bit & get it.
1130    SIGNBIT = INP(PORTA) AND 2                      'SIGNBIT is 2nd bit with the half-digit.
1100    TOTAL = DIGIT1 + 10*DIGIT2 + 100*DIGIT3 + 1000*DIGIT4
1110        IF SIGNBIT THEN TOTAL = -TOTAL
1120    RETURN
```

••

FIGURE 4.52. Computer program for acquisition from a panel meter with a multiplexed output.

of their architecture. They communicate as would any other small computer: through an RS-232 serial interface or through the General-Purpose Interface Bus (defined by the IEEE-488 standard). Both are discussed further in Chapter Nine.

Typically the small computer must send command character strings to request a datum or to command the instrument. Usually the data are both sent and received as strings of ASCII characters.

In other cases, the interfaces are fully parallel and the small computer must set up the communication in a manner similar to that described for a multiplexed panel meter, the difference being that a much wider range of information can be passed.

The advantages of this mode of communication are very compelling. Nearly every small computer has a standard serial port, and the hardware interface consists *only* of the cable between computer and instrument. The software for communication through the serial port is frequently a part of the OS, so no low-level commands (IN or OUT) are required.

However, the disadvantages of the approach are very difficult to overcome indeed.

1. If the operating software of the instrument's microcomputer failed to implement an important function, a fix-up is nearly impossible. A separate interface may be required anyway. For example, several analytical instruments are on the market with options to cycle through a number of sample cells. However, the operating software allows neither the number of the current sample nor the signal indicating that a change of sample is in progress to be transmitted to the computer. All of the hardware is available for providing the information required for the experiment, but a simple oversight by the programming team, fixed into ROM, makes the information unavailable.

2. The information must be passed on the instrument's terms which are fixed. Again, in a well-known analytical instrument, capability is provided to produce data at precise, regularly timed intervals. This function makes experiments which must monitor a changing phenomenon very simple. However, after the computer issues a command to the instrument to begin the experiment, there is an indeterminate delay of several seconds. Again, the instrument manufacturer's oversight cannot be rectified by modifying the instrument since the operating program is proprietary and is fixed in ROM.

CHAPTER FIVE

Analog Electronics

Often, in designing the electronic portion of an interface, a special-purpose analog module is required to tailor the signal form and range, or to filter the signal derived from a transducer. This is the **Signal Conditioning and Modifying** stage illustrated in Figure 5.1. Although the computer and its general-purpose peripherals are flexible, adapting to a wide range of applications, the specific chemical and physical phenomena under study present unique problems. Because of the variability in the phenomena being measured and the transducers required, this stage often must be designed specifically for the particular application. The output of the transducer must be converted to a compatible signal form, signal range, and frequency range in order to mate transducer and ADC input. Similarly, the DAC output must be appropriate for the output transducer.

In this chapter, the components required for many of these operations will be introduced. Without apology to readers with a rigorous electronics training, this discussion will emphasize the practical aspects of design. The design of this portion of the interface should be understood by those with a minimal background in electronics. Consequently, a "black box" approach will be used: this will be achieved by presenting components as simple blocks with inputs and outputs; the internal components are ignored. In most cases, integrated circuit (IC) solutions to problems will be sought, thereby taking advantage of the expertise of the company designing the IC. The difficulties in component-matching are reduced to connecting pins.

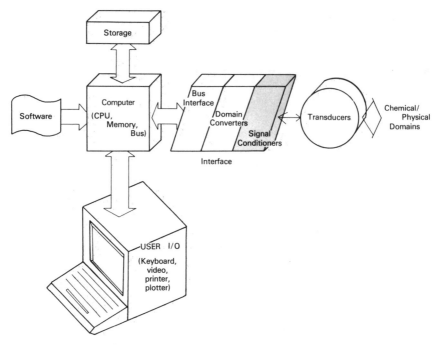

FIGURE 5.1. A laboratory computer system. The signal conditioning stage is highlighted for study in this and the following chapter.

5.1. ANALOG SWITCHES

Interfacing analog signals to the computer may require the capability of switching signals while in the analog domain. One example was described earlier: the multiplexer in an ADC subsystem. The analog switch is a solid-state device with the favorable characteristics of an ordinary mechanical switch—very low resistance when ON and very high resistance when OFF. However, the solid-state characteristics also are present: high speed and freedom from the errors that result from the contact bounce generated when metal contacts close.

Analog switches are constructed from various types of field-effect transistors (FETs). These transistors have three significant pins, shown in Figure 5.2: the source is the input connection, the drain is the signal output, and the gate controls conductance between source and drain. When an FET is used as a switch, potentials are applied to the gate which cause it either to have no effect on the signal or to "pinch off" the signal. That is, the source–drain resistance is either low, typically less than 100 Ω, or high, much greater than 10^6 Ω. (Although from the diagram, there is no apparent difference between source and drain, in practice the internal gate–drain and gate–source capacitances make the source input preferable.)

a)

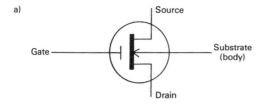

b)

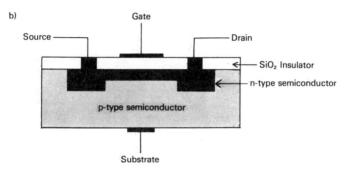

FIGURE 5.2. The FET transistor. (*a*) The symbol shows the gate which controls a signal passing from source-to-drain. (*b*) A diagram of an FET. The gate, though fully insulated from the conducting channel, can stop current flow with an electric field.

The specifications of the FET switch will often be important. First, the *ON resistance* (when the switch is conducting) could lead to a significant ohmic voltage drop ($V = iR$) when the signal takes the form of a current. Second, the *OFF resistance*, which fails to reach the ideal of infinity, may play a part when the signal is a voltage fed to an amplifier with high-input resistance. The consequence of nonideality in these parameters may be **crosstalk** in a multiplexer: When a group of switches act as a multiposition switch, signals meant to be *OFF* may still be partially transmitted. Finally, the switch does not act instantaneously; the speed of some applications, such as in a successive approximation ADC, is limited by the switching time of these switches.

A number of FET technologies have been developed, each with its own advantages and special design requirements. Fortunately, most designers need not be concerned with these differences because of the availability of integrated circuits, which include on one chip both the FET and the electronics to convert from TTL levels to the appropriate levels for the FET gate.

Two integrated circuit packages are illustrated in Figure 5.3. Simple single-pole/single-throw switches, illustrated at the top, are used for steering signals in some applications shown later in the chapter; the dotted line rep-

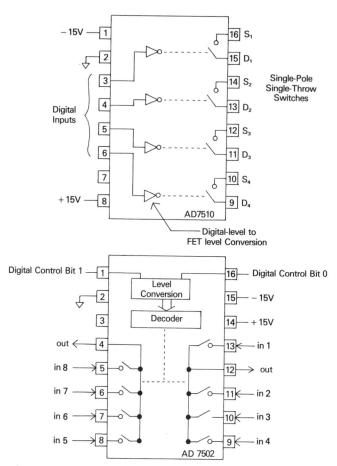

FIGURE 5.3. Two configurations of analog switches. Above, an integrated circuit with four switches. Below, an integrated circuit with two switches that can be used as a dual 4-in-1-out multiplexers (Analog Devices, Norwood, MA).

resents control of the switch by the gate, which has a digital logic input. Multiplexers, as the one in the lower part of the figure, may be packaged as single-pole or double-pole switches; the latter are useful when differential signals are to be measured, that is, the *difference* between two signals is measured rather than the difference between a single signal and ground. Demultiplexers are the simple reverse of multiplexers; in most cases, the same IC can be used in either direction with minimal degradation in performance.

5.2. OPERATIONAL AMPLIFIERS

When conditioning and modifying analog signals with discrete components (transistors, capacitors, resistors), the designer must pay careful attention to the subtleties of electronics which affect the linearity of the signals or the noise content of the signal. Fortunately for the person who is not an electronics expert, there are modules that may be used almost as a series of black boxes. By adding only resistors and capacitors, the characteristics can be controlled in a predictable fashion.

These devices are operational amplifiers (OAs). Because of their remarkable versatility and predictable behavior, they were used as the active components of analog computers until the 1960s; input voltages were entered with potentiometers and could be added, subtracted, multiplied, divided, and transformed by simply plugging wires; the answer was read on a meter. In the 1960s, the value of OAs in scientific instrumentation became apparent to several analytical chemists, and new areas of instrument design were opened to scientists with minimal electronics training.

From a simple description of the operation of an OA, we can go on to derive a host of applications. The OA is simply an amplifier that acts on the *difference* between two inputs, V_+ and V_- (Fig. 5.4). The operation of the OA is controlled by a **feedback element**, a circuit component between input and output that determines the form of the OA's action. A few calculations, presented in the following section, will help understand those negative feedback circuits.

5.2.1. Basic Characteristics of the Ideal OA

The ideal OA, Figure 5.4a, is represented by a triangle with two input terminals, the **inverting input**, V_-, the **noninverting input**, V_+, and a terminal for **output**, V_0. The basic operation of the OA is to amplify the *difference* between the V_- and V_+ inputs. That is

$$(V_+ - V_-) \times A_o = V_0 \qquad (5.1)$$

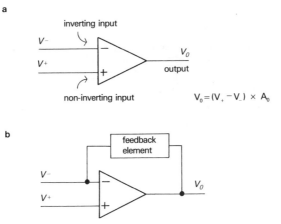

FIGURE 5.4. Operational amplifiers. (*a*) With no feedback element, it is an infinite-gain amplifier. (*b*) A feedback element controls the operation.

where the amplification factor, A_o, is the gain of the amplifier; since no external components have been added, A_o is termed the **open-loop gain**. In the ideal OA, A_o is infinite, and this ideality is usually a good approximation unless the frequency of the signal is high (see Section 5.4.).

An infinite-gain amplifier has limited utility. However, by adding a **feedback loop** between the inverting input and the output, negative feedback from the output continuously "adjusts" the input, and the performance of the OA can be controlled and customized according to the characteristics of the elements in that loop.

The effect of this adjustment is to stabilize the amplifier. Rearrangement of (5.1) implies that, in the absence of other constraints, $V_0/A_o = 0$, and V_+ must equal V_-: whenever the feedback loop is configured so that (5.1) can be satisfied, the two inputs will be at the same potential. The gain of the amplifier circuit *with* the feedback loop is termed the **closed-loop gain**, A_c.

A second ideal characteristic is **high-input impedance**. Simply stated, this means that the inputs, V_- and V_+, require essentially no current from the input signal in order to operate on those inputs. Consequently, the amount of current flowing in the feedback loop can be accurately determined. Conversely, the **output impedance** should be near zero; that is, no matter what the resistance of the device that the OA is driving (the load) and the amount of current it must draw from the OA, V_o should remain invariant.

Of course, as with any real device, the ideal characteristics can never be achieved. In the past, it was always necessary to evaluate a trade-off between these characteristics, other characteristics, and cost; modern OAs come remarkably close. The common CA3140 (see Table 5.1) has an open-loop gain of 10^6 and an input impedance of 10 GΩ (10^{13} Ω) which is sufficient for most applications.

There are many other specifications which describe *real* OAs, and these will be considered later in this chapter together with deviations from ideality

TABLE 5.1. Some Typical OA Specifications

	OP-07 [1]	CA3140 [2]	741 [3]	Units
Input Resistance	6.0×10^6	1.5×10^{12}	2.0×10^6	Ω
Open Loop Gain	500	100	200	V/mV
Unity Gain Bandwidth	0.6	4.5	1.5	MHz
Output Resistance	60	60	75	Ω
Input Offset Voltage	0.06	5	15	mV
Input Noise Voltage	0.35	12	1	μV

[1] Precision Monolithics, Santa Clara, California
[2] RCA Solid-State, Somerville, New Jersey
[3] Signetics, Sunnyvale, California

for all characteristics. First, some of the useful OA circuits will be explored, most of which result from the design of a feedback loop.

5.2.2. The Voltage Follower

The voltage follower has a gain of one produced by the configuration of Figure 5.5, and is employed to take advantage of the OA's I/O impedance characteristics. The operation of the circuit in this figure can be derived by adding to (5.1) the external connection between the output and the inverting input: $V_0 = V_-$. Therefore,

$$(V_{in} - V_0) \times A_o = V_0 \tag{5.2}$$

Rearranging,

$$V_0 \times (1 + A_o) = V_{in} \times A_o \tag{5.3}$$

and since $A_o \gg 1$, and $1 + A_o = A_o$,

$$V_0 = V_{in} \tag{5.4}$$

$$A_c = 1 \tag{5.5}$$

As long as A_o is large (recall that ideally it is infinite), the output will be equal to the input. This configuration's usefulness lies in its ability to act as a buffer; the input can be connected directly to high-resistance transducers including, for example, a glass pH electrode, and the output provides the current required by the following stage of the interface.

5.2.3. The Follower with Gain

The impedance characteristics of the simple follower can be maintained while the OA provides a degree of linear amplification with the circuit of Figure 5.6. Again starting with (5.1) and setting V_+ equal to V_{in}, we must write an equation for V_-. Since the currents at any point must sum to zero, and the current into the OA at V_- is essentially zero, the current, i_b, through the resistors R_1 and R_2 must be equal.

By Ohm's Law, $i = V/R$, so that by setting the currents through R_1 and R_2 equal, we get

$$V_-/R_2 = (V_0 - V_-)/R_1 \tag{5.6}$$

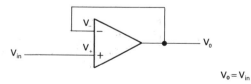

FIGURE 5.5. The voltage follower configuration.

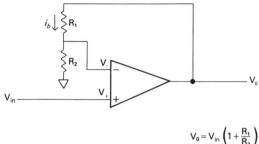

$$V_0 = V_{in}\left(1 + \frac{R_1}{R_2}\right)$$

FIGURE 5.6. The follower-with-gain configuration.

Noting that $V_+ = V_{in}$, and solving (5.1) and (5.6) for V_0

$$V_0 = \frac{V_{in}A_o(1 + R_1/R_2)}{A_o + 1 + R_1/R_2} \tag{5.7}$$

As long as $A_o \gg 1 + R_1/R_2$, that equation reduces to

$$V_0 = V_{in}(1 + R_1/R_2) \tag{5.8}$$

and,

$$A_c = 1 + R_1/R_2 \tag{5.9}$$

For any reasonable value of A_o, that condition will hold as long as A_c is not extremely large. Equation (5.9) shows another restriction on the choice of A_c that can be implemented; it must exceed unity.

5.2.4. The Current-to-Voltage Converter

By changing the configuration so that an input current is applied to the circuit at the inverting input, the output will follow the input current according to a simple equation. Again, the configuration is analyzed beginning with (5.1). Using Figure 5.7, we can write an equation for the current since the sum of i_{in} and i_b must be 0. Therefore,

$$i_{in} = -(V_0 - V_-)/R_b = -i_b \tag{5.10}$$

Substituting into (5.1),

$$V_0(A_o + 1) = -i_{in}R_bA_o + V_+A_o \tag{5.11}$$

This unwieldy relationship is reduced by noting that since no current flows into V_+, the voltage drop across R_+ is 0, and V_+ is 0. (R_+ is present only to balance offsets due to nonideality in the OA; the description comes later in the chapter.) If $A_0 \gg 1$

$$V_0 = -i_{in}R_b \tag{5.12}$$

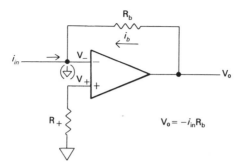

$$V_0 = -i_{in}R_b$$

FIGURE 5.7. The current-to-voltage converter. The ground symbol in parentheses, indicates virtual ground.

The conversion factor from current to voltage is fixed by R_b alone and is negative.

This simple relationship is the basis of the conversion of the current output of many transducers to voltage; examples include photomultiplier tubes and temperature transducers. Before OAs were available, it was common to simply pass the current through a calibrated resistor and measure the voltage drop according to Ohm's Law; however, for large signals, significant error can result (see Chapter Seven).

The advantage of this OA circuit is that the transducer terminal, which produces the current, is tied to a terminal which is always at the same potential as ground. Since it is at the same potential but not wired to ground, the inverting input is said to be at **virtual ground**, and this is symbolized with a ground symbol in parentheses.

5.2.5. The Inverting Amplifier with Gain

By adding an input resistor, R_1, as shown in Figure 5.8, the input current, i_1, is generated from an input voltage, V_1. Equation (5.12) then becomes

$$V_0 = V_1(R_b/R_1) \qquad (5.13)$$

The circuit has a closed-loop gain of

$$A_c = -R_b/R_1 \qquad (5.14)$$

There is no limit to the gain for an *ideal* amplifier: it may be greater than, less than, or equal to one.

The inverting amplifier merits wide application since it can provide gain over a wide range, and the gain is *linearly* related to the feedback resistance, R_b. In some applications, a ten-turn potentiometer can be used for R_b, and the amplifier gain can be read directly from the dial. It is also possible to control the gain from digital logic by using the resistor/solid-state switch network of a DAC as programmable resistance.

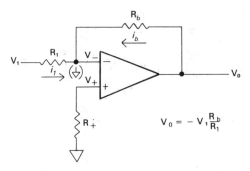

FIGURE 5.8. The inverting amplifier with gain.

The inverting amplifier, besides providing a fixed gain, also can be used as a **summing amplifier**. The total current entering at V_- can originate in several sources as illustrated in Figure 5.9. The output is

$$V_0 = -R_b (V_1/R_1 + V_2/R_2 + \cdots V_n/R_n) \tag{5.15}$$

If the input resistors are equal, i.e., $R_1 = R_2 = \cdots = R_n = R_b$, the output is simply the negative linear sum of the inputs. Thereby two or more signals may be added. Besides adding signals, this circuit also is useful for including an offset or "zero adjust". The combination of an inverter with a gain of one and this summing amplifier also is used to compute difference.

5.2.6. The Difference Amplifier

A small modification in the simple inverting amplifier produces a difference amplifier, that is, an amplifier whose output is *linearly* related to the difference between two inputs, V_1 and V_2. In Figure 5.10, the potential at V_+

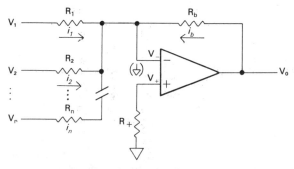

FIGURE 5.9. The summing amplifier. $-V_0/R_b = V_1/R_1 + V_2/R_2 + \cdots V_n/R_n$. If all the resistors have the same value, the output is a simple negative sum of the input voltages.

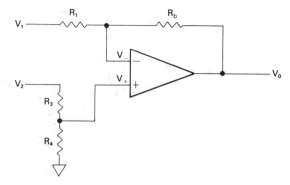

FIGURE 5.10. The difference amplifier. The output, V_o, is equal to the difference of the inputs, $V_2 - V_1$, if $R_1 = R_b$ and $R_3 = R_4$.

is the output of a simple voltage divider

$$V_+ = V_2(R_4/(R_3 + R_4)) \qquad (5.16)$$

When (5.16) is substituted into (5.11), and the approximation repeated to eliminate A_o, the output is

$$V_0 = -V_1 R_1/R_b + (1 + R_1/R_b)V_2 R_4/(R_3 + R_4)$$

That equation is not particularly useful unless $R_1 = R_b$ and $R_3 = R_4$. Then a simple differential amplifier results:

$$V_0 = V_2 - V_1 \qquad (5.17)$$

5.2.7. The Instrumentation Amplifier

The difference amplifier of the previous section suffers from two limitations: Its relatively low-input impedance, since the signal must be accepted as *current*, and because the resistors must be carefully matched in order to obtain an accurate unit gain. A solution to the input impedance problem is to add a pair of voltage follower amplifiers, as illustrated in Figure 5.11*a*. Calibration requires that each input be grounded while the gain of the other is checked and adjusted with trimming potentiometers. This arrangement is less than ideal since the external components require careful adjustment and often are responsible for temperature-dependent error.

Where a differential input with accurate gain is required, an **instrumentation amplifier module** rather than a configured OA may provide an answer. Normally in such a device the gain is determined either by choosing from a small fixed number of alternatives by connecting terminals or by adding a single external resistor. The performance is improved since the module is optimized for differential input and it has fewer external components than the OA instrumentation amplifier.

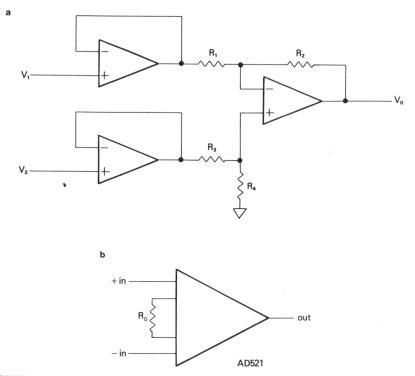

FIGURE 5.11. Instrumentation amplifiers. (*a*) A difference amplifier with high-input impedance. (*b*) An integrated circuit version (Analog Devices).

5.2.8 The Integrating Amplifier

A capacitor can store electrical charge; its voltage is q/C where q is the
charge and C is capacitance. Since electrical current is the rate at which
charge is transmitted, the capacitor is used to integrate a signal. As was the
case for the simple inverter, the input current, i_1, and the feedback current,
i_b (Fig. 5.12) must sum to zero. When the capacitor is the feedback element,
the current through it is proportional to the change in voltage across it, dV_0/dt. Consequently when V_- is at virtual ground

$$i_1 + i_b = 0 = V_1 R_1 + C \, dV_0/dt \tag{5.18}$$

Integrating that equation over time,

$$V_0 = \frac{-1}{R_1 C} \int V_1 dt \tag{5.19}$$

A shorting switch in the feedback loop, S1, is required to establish the starting time, $t = 0$, when the capacitor is discharged and V_0 is 0. The switch
may be a solid-state analog switch or a mechanical switch depending on

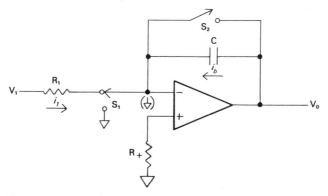

FIGURE 5.12. The integrating amplifier. The output is proportional to the integral of the input over the period that the switch S_2 is open and S_1 is closed.

timing constraints. The integration can be terminated by opening switch S2, interrupting the input current.

Where a signal must be integrated, there is often a choice between hardware and software techniques: By summing the results of a series of ADC conversions or using more sophisticated numerical methods, the integration is completed in software; the integrating amplifier places integration in hardware. Where both are feasible so that the choice actually exists, the software approach is preferable because (1) hardware integrators can be difficult to calibrate due to internal offsets in the OA that are also integrated, and (2) adding the results of a series of conversions can help to overcome some of the limitations of ADC resolution which would still be present if the analog result of an integration were converted only once. However, when the signal to be integrated is short, less than the ADC conversion time, the integrator is very valuable; for example, if the energy of a subnanosecond laser pulse is to be measured, the transducer current for the entire peak can be integrated and held before conversion.

Some practical notes about integrators should be added. The current in the feedback loop is actually the *sum* of currents through various paths in the feedback loop; only a single path is shown in Figure 5.12, but leakage in switch S2 (usually a FET) and leakage in the capacitor itself may decrease V_0 and/or make the integrator nonlinear. Care must be taken to choose high-quality analog switches and low-leakage capacitors (polystyrene or polycarbonate).

5.2.9. The Differentiating Amplifier

By reversing the positions of the capacitor and the resistor from the configuration of the integrator, the differentiator (Fig. 5.13) results. Following

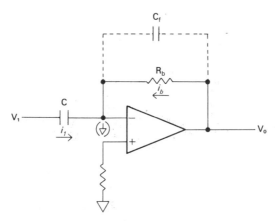

FIGURE 5.13. The differentiating amplifier. The output is proportional to dV_1/dt, where t = time. C_f provides some low-pass filtering.

the method of (5.16),

$$i_1 + i_b = 0 = C \, dV_1/dt + \dot{V}_0/R_b \tag{5.20}$$

The output is then the differential of the input:

$$V_0 = -R_b C \, dV_1/dt \tag{5.21}$$

Any differentiator must be used with great care since the noise in a signal, even when its of much lower magnitude than the signal, usually changes more rapidly than the signal itself; differentiators such as this one tend to amplify that noise. A software technique, such as the method introduced in Chapter Eleven, is often preferable; a digital differentiating and filtering algorithm can be used without irreversibly distorting the data. If a hardware approach still seems necessary, a filtering capacitor, such as C_f shown in the figure, is mandatory.

5.2.10. The Logarithmic Amplifier

A precise logarithmic current–voltage relationship is found in the world of electronics at the semiconductor junction. The current across the junction is proportional to $\exp(k_1 V_j)$ where V_j is the potential across the junction and k_1 is a constant which includes the inverse of the absolute temperature. Again substituting into the equation for an inverting amplifier and referring to Figure 5.14,

$$i_1 + i_b = 0 = V_1/R_1 + k_2 \exp(k_1 V_0) \tag{5.22}$$

where k_1 and k_2 are both constants. Then the output voltage is

$$V_0 = -(1/k_1)(\log V_1 + \log R_1 + \log k_2) \tag{5.23}$$

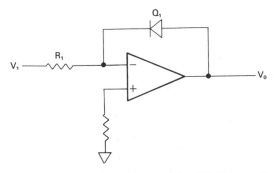

FIGURE 5.14. The logarithmic amplifier. $V_o = -k \log(V_1)$ This configuration is operational, but somewhat thermally unstable; a unit from an OA manufacturer should normally be used.

The simple circuit in the illustration has several significant drawbacks led by the temperature dependence: The output is proportional to absolute temperature. Practical circuits using matched transistor pairs or feedback elements in thermostated ovens can be found in OA design references. The best practical solution in most cases is not to design with diodes but to use a logarithmic circuit module available from one of the major analog component manufacturers.

The advice given for integrators and differentiators again applies: the conversion is frequently performed better by software. There are two possible reasons for using hardware, however. The first is speed: In software the logarithm is somewhat slowly computed, and where large numbers of data points are to be transformed, the possible improvement in precision may be less significant than the improvement in speed by use of hardware. The second reason is based on precision: in some applications a measurement must be made on signals that span several orders of magnitude, but greater resolution is required at the low end than at the high end. The compression produced by analog logarithmic conversion may lead to a statistically more reliable result.

5.2.11. The Comparator

The comparator was introduced earlier in Chapter Four as a single-bit ADC. Ideally, it is simply an OA following the basic operation of (5.1); the output goes to either infinity depending on the relative magnitudes of the inputs. As a practical matter, such extremes are, of course, not possible, and the actual "open loop" limits are bounded by the power supply voltages for the amplifier; the power supply voltages often are ± 15 V, and the output limits are generally 0.1–5 V less than the power supply depending on the semiconductor technology of the OA.

Many OA comparators should not be operated in the simple open-loop configuration of Figure 5.1 since some OAs tend to briefly "latch up" when driven to their limit; that is, if the input polarity changes, there will be a delay for recovery before the output follows. However, among integrated circuit OAs, some are already configured and optimized as comparators, including outputs which are compatible with logic levels. The use of these will ensure high-speed performance. When a general-purpose OA is employed, the output will often be bipolar. The circuit in Figure 5.15 illustrates a simplified method for conversion to TTL.

Perfect comparison is not achieved with real amplifiers. There is a "dead zone" of ± 1 to ± 10 mV in which the amplifier fails to react. Sometimes, this zone is advantageous, eliminating extra oscillations (see the following section).

5.2.12. The Schmitt Trigger

This variant of the comparator is used when the input has two discrete states; instead of varying above and below a single precise threshold, the signal has definite, separated ON and OFF conditions. For example, if we need to test whether a light beam is blocked or not blocked, the light detector should give us either a large signal or no signal. However, an input voltage cannot instantaneously change from one state to the other, and while it is changing, the noise on the signal could cause it to cross any absolute threshold more than once, thereby creating spurious transitions. The Schmitt Trigger, by using *positive* feedback, creates **hysteresis** region or "dead zone" between the comparator's states.

This amplifier is best understood with reference to an example. Consider the amplifier of Figure 5.16 in which the open-loop limits are ± 10 V. If the input voltage to the inverting input is -10 V, then the output will be $+10$ V; the voltage at the noninverting input is $R_1/(R_1 + R_2) \times 10$ V which for $R_1 = R_2$ is $+5$ V. Therefore as the input increases, the output will not change until the input exceeds $+5$ V.

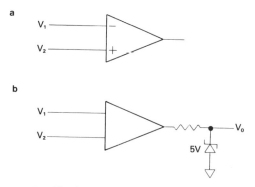

FIGURE 5.15. A comparator. The lower circuit can drive many TTL circuits. ICs already configured as comparators are preferable.

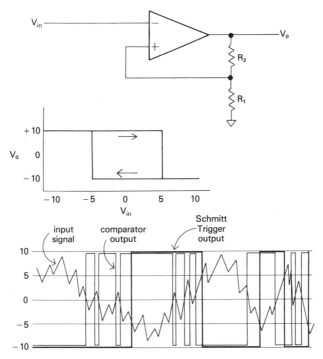

FIGURE 5.16. Schmitt trigger. The positive feedback in the circuit diagram gives rise to the hysteresis shown in the center. The noise in the waveform (bottom) will generate transitions of the output which will be suppressed by the Schmitt trigger.

However, when the signal goes above $+5$ V, the output will become -10 V, and V_+ will be -5 V. The level at which the change in state will occur when the input returns negative is then -5 V. A noisy curve, in the lower portion of Figure 5.16, makes only a few low-frequency oscillations, but the curve crosses zero many times; the output of a simple comparator makes many state changes, but the dead zone of the Schmitt trigger immunizes the output from the effects of the noise on the input which would result if a simple comparator were used.

5.2.13. Sampling Amplifiers

The need for an amplifier which can sample a signal at a precise position in time for later measurement was presented in the previous chapter; such amplifiers fall into two categories. The first is an amplifier which will sample a signal when commanded by some external signal and hold that signal at its output until ready for a new sample; this is the **sample-and-hold (S&H) amplifier**. The second uses a characteristic of the input signal to determine when it should sample the signal; that characteristic is the peak value, and the amplifier is called a **peak detector amplifier**. The latter measures the maximum (or minimum) of a transient signal, and is useful for a signal whose lifetime is too short for the entire peak to be digitized or otherwise measured.

A simplified sampling amplifier is shown in Figure 5.17; it consists simply of a follower amplifier with a capacitor at the input and a switch to effectively connect and disconnect the signal. When the switch is closed, the capacitor is charged and discharged at a rate related to the time constant, $R \cdot C$. Since R is small, the amplifier simply acts as a follower which removes high frequencies (low-pass filter). However, when the switch is opened there is no path for the capacitor to be further charged or discharged (remember that the resistance of the OA input is large), so the capacitor maintains its voltage and the amplifier stays in a hold state.

The difference between the peak detector and the S&H amplifier is the nature of the switch. For the S&H amplifier (Fig. 5.18), the switch is a solid-

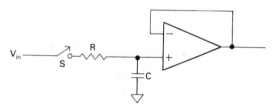

FIGURE 5.17. A generic sampling amplifier. When the switch is closed, the amplifier is a simple low-pass filter. When the switch is opened, the signal level is stored on the capacitor and buffered at the output.

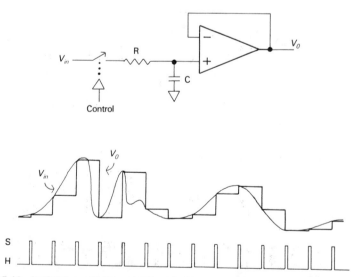

FIGURE 5.18. An S&H circuit. The switch of Figure 5.17 is a solid-state switch under digital control. At the bottom, the waveform illustrates the operation.

state analog switch; these switches can be opened and closed very rapidly and come close to possessing the ideal characteristics of zero resistance when closed and infinite resistance when open. The peak detector (Fig. 5.19) uses a diode as a switch; the diode is effectively closed when current flows in the forward direction (it is "forward-biased") but is open when the polarities change ("reverse-biased"). As long as the input signal is greater than the voltage on the capacitor, the switch remains closed, and the amplifier follows the signal; after the signal has reached a maximum, the input is less than the stored value, and the switch is open. In order to measure negative peaks, the polarity of the diode is simply reversed. After the measurement has been made, the peak detector must be reset by discharging the capacitor with a solid-state or mechanical relay.

As a practical matter, the amplifier circuit shown suffers from its low-input impedance; the current required for charging and discharging the capacitor must come from the input. Furthermore, the switch itself may have significant resistance when it is conducting which adds a voltage drop according to Ohm's Law; this is particularly the case for a silicon diode which has a characteristic voltage drop of 0.6 V. The circuit of Figure 5.20 solves these difficulties easily. The follower (A1) input provides high-input impedance and a second follower (A2) at the output acts as a buffer. When switch S1 is closed, the capacitor voltage is equal to the input voltage, S1's voltage drop being compensated by the feedback loop through A2 and R; when the switch is opened and S2 is closed, two OAs are effectively isolated from each other as independent followers. If an S&H amplifier is used, S1 and S2 are solid-state switches, but if the circuit is a peak detector, they are signal diodes.

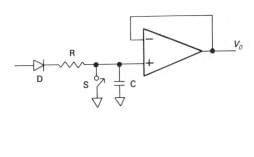

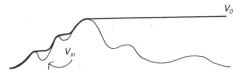

FIGURE 5.19 A peak-detector circuit. A diode is used as a switch. If the signal is greater than the voltage on the capacitor, the switch conducts. Switch S is necessary to reset the peak value.

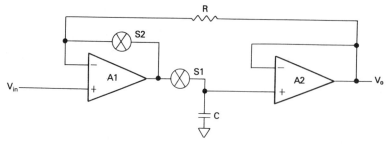

FIGURE 5.20. A real sampling amplifier. This configuration, using solid-state switches or diodes as switches, provides much higher input impedances, faster response, and correction for voltage drop across the switch.

5.2.14. Isolation Amplifiers

In some applications, chiefly those dealing with live subjects, the transducer must be electrically isolated from the instrumentation for safety. Complete electrical isolation also is needed to avoid noise in certain situations, and to extract a small difference between two high voltages. The isolation amplifier is represented schematically in Figure 5.21. Two methods of achieving isolation have been developed by OA manufacturers.

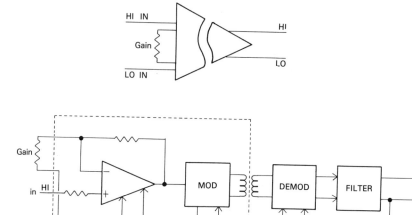

FIGURE 5.21. An isolation amplifier. The input voltage is converted to AC by the modulator so that it can be transmitted across a transformer. The demodulator converts back to a DC voltage. The transformer provides several thousands of volts of isolation.

The first uses a transformer: The dc signal controls the amplitude of an oscillator so that a modulated signal feeds the primary winding of the transformer. Then the transformer output is rectified and filtered. If the primary and secondary windings are the same size and the frequency of the signal is much lower than the modulation frequency, the signal should be accurately transmitted.

The second method is optical: the dc signal feeds an OA which drives an LED; a matched pair of photodiodes both feed back and feed forward, the feedback linearizing the optical link and the feed forward providing the isolation. The latter method is capable of higher frequency response, and both have isolation of up to 6500 V.

Isolation amplifiers also require separate power supplies for the two stages. This need is satisfied by separate transformers providing power to the input stage from the output stage. Typically, as in the figure, an additional ± 15 V is available for powering the transducer.

5.3. FILTERS

We nearly always need to filter a signal while it is still in the analog domain; the examples of aliasing in Chapter Four made that clear. The concept of filtering is not foreign to anyone who has worked with analog instruments before. The signal, when plotted on a strip-chart recorder, is often "noisy." That is, the signal entering the recorder is the sum of the useful information and additional signals at higher frequencies which have no information content; the latter is usually observed as rapid oscillations by the recorder pen. Many recorders have a knob marked "DAMPING" which removes much of this noise. The damping function applies a *low-pass* filter, a filter which lets through the information which is present at low frequencies while "damping out" the high frequencies which are assumed to be noise.

In other cases, we observe that the baseline does not stay constant on a strip-chart recorder; it "drifts." The drift also is noise, but it possesses a *lower* frequency than the desired information. We cannot always do it electronically, but we then must find a way to eliminate the effect of the low-frequency noise; doing that requires a *high-pass* filter.

In this section, we will consider the various types of filters, how they are designed, and their effect on both the signal and the noise.

Four general categories of active filters are used in analog circuitry. **Low-pass** filters remove high-frequency signals and pass the low; conversely **high-pass** filters remove the low-frequency signals and pass high-frequency information. **Bandpass** filters remove frequencies both higher and lower than the chosen frequency range, and **notch** filters pass everything *but* the frequencies in a narrow range.

The rigorous and highly mathematical discussion of signal filters is out of the scope of this book. However, this does not relieve the computer user of the need to use filters nor does it mean that filters cannot be used effectively by persons other than electrical engineers to eliminate unwanted noise.

The significant components of simple filters are resistors and capacitors. (Inductors also can be used in developing filters, but they are less practical and will be ignored here.) The resistance of a resistor, defined by Ohm's Law, is ideally a constant and should be invariant with frequency; the resistor's **impedance** (frequency-dependent resistance) is equal to its resistance.

A capacitor also has the effect of resisting or impeding the flow of electrical current, but the extent depends on the frequency of the signal. Since the effect must be defined over a range of frequencies, the impedance must be used, and the impedance of a capacitor (also the capacitive reactance), X_c, is

$$X_c = \frac{1}{2\pi f C}$$

where f is the frequency of the signal. In contrast to a resistor, the impedance of a capacitor to current flow is dependent on frequency. A simple filter includes only a resistance and capacitance in parallel or in series, but the total impedance of a capacitor and a resistor in series must be computed as a vector—the square root of the sum of the squares.

A filter circuit has two effects on a signal:

- It **diminishes the amplitude** of a particular range of frequencies; for example, a low-pass filter diminishes the amplitude of higher frequencies, increasing the attenuation with increasing frequency.
- It causes **phase-shift**; for example, if a simple cosine wave is subjected to a low-pass filter, the output signal lags behind the input; signal peaks lag, and square waves become rounded.

The desired effect would be for a low-pass filter to remove 100% of the amplitude of all frequencies above a chosen cutoff, and to transmit the others with unity gain. The filters, however, are not ideal. We wish not to have phase shift, and we will see that the relationship between the attenuation and the frequency is not exactly ideal. An understanding of the effects of a filter on both the desired signal and the noise will require that we look at these two properties.

5.3.1. Frequency Components of Signals

In the laboratory, we encounter signals which take a variety of shapes. Occasionally we encounter a true cosine wave signal that we want to acquire: a sound wave, a cam position, and so on. However, the signals are usually more complicated: repeating square waves and exponential decay curves are examples. Most experiments do not yield a repeating function at all; a complicated sequence of peaks is typical. Probably the most representative is chromatographic output, a series of unevenly spaced "peaks," Gaussian-shaped signals which appear at apparently random intervals.

Upon analysis, every signal can be broken down into a combination of a number of cosine waves; the magnitudes and frequencies of these cosine waves determine the shape of the signal that is observed. Examples will be shown later. We want a filter to remove the offending frequencies (noise) without affecting the useful signal. Consequently, we must get a feel for whether some of the desired signal also will be removed by a filter. Therefore, the total effect of a filter can be understood only by finding out what frequencies are present in the signal.

Earlier, two effects of a filter on the signal were stated: the filter changes both the amplitude and the phase or lag of any frequency. After finding out how these two effects affect the signal, we can apply them to each separate frequency component of the signal; then we will know the overall effect of the filter.

The technique used to analyze a signal to determine its cosine wave content is called *Fourier analysis*. We will leave the mathematics and simplify by showing results in graphical form. (The detail of Fourier analysis can best be found in the references; we will take a nonrigorous approach.) Three examples of signals illustrate how the cosine wave components add up to form the signal.

First, if the signal is a pure cosine wave, the analysis is simple. A measure of the periodic time gives the frequency. In Figure 5.22 we find a cosine wave drawn as a function of time. A second graph shows its frequency domain spectrum. This plot shows the amplitude of each cosine wave component that contributes to the signal wave. As we would expect, in the frequency domain of the cosine signal, there is only one spike representing the single frequency of the cosine wave signal itself.

The second example, a square wave, is much more complicated. Intuitively, we might guess that the square wave is similar to the cosine wave. However, a circuit that transmits a square wave must be able to transmit signals that contribute much higher frequencies than the fundamental frequency of the square wave; if the elements of the circuit are to follow the "sharp corner" of the square wave, they must be able to respond rapidly. That becomes clear from a Fourier analysis of the square wave.

The square wave can be built up from a series of cosine waves. This is illustrated in Figure 5.23. The series begins in curve A with a base cosine wave that has the same frequency, f, and the same amplitude, y, as the square wave: $\cos(2\pi ft)$. To that fundamental cosine, a negative cosine wave with three times the frequency and one-third the amplitude of the original is added: $-(\cos(6\pi ft)/3$. The sum (the dashed curve) is "squarer" than a cosine wave, and looks somewhat like a square wave.

In curve B, additional cosine waves are added to the sum in A. The frequencies are higher odd harmonics, up to 15, of the base frequency, f, whose

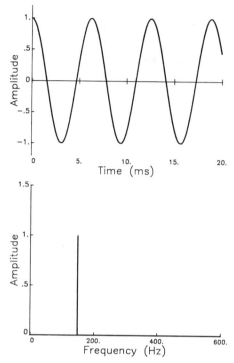

FIGURE 5.22. A cosine wave in the time domain (top) and in the frequency domain (bottom).

amplitude is that of the base cosine wave divided by the multiple; the signs are alternated. The sum of these first eight components looks even more like a square wave.

When fully analyzed, the square wave is proportional to

$$\cos(\omega t) - \frac{\cos(3\omega t)}{3} + \frac{\cos(5\omega t)}{5} - \frac{\cos(7\omega t)}{7} + \frac{\cos(9\omega t)}{9} - \cdots \quad (5.28)$$

where $\omega = 2\pi f$, the angular velocity of the signal, and t is the time. The series goes to infinity but is represented in curve C of Figure 5.23 with moderate accuracy by the first 127 components. From the example, it is clear that even a low-frequency square wave signal requires that the amplifier be capable of passing higher frequencies.

A gaussian signal curve also can be constructed from a series, but in this case the series is the integral of another Gaussian. A Gaussian signal peak is proportional to

$$\exp(-\alpha t^2) \quad (5.28)$$

where α is a measure of the reciprocal of the width of the peak.

The cosine components of the Gaussian peak are not a series of discrete frequencies as was true for the square wave, but are continuous; each frequency component has an amplitude determined by another Gaussian:

$$\exp \frac{-\pi^2 f^2}{\alpha} \qquad (5.29)$$

If the cosine waves of all frequencies in the frequency Gaussian are summed, we will obtain a Gaussian-shaped signal in the time domain.

An approximation will illustrate. Curve *A* in Figure 5.24 shows three cosine waves—one at zero frequency (a constant) and two at increasing frequencies but decreasing amplitude; although we cannot graphically show all component frequencies, this figure shows that as more are added in proper proportion, a peak shape develops. If the envelope of the amplitudes of these individual curves was plotted against frequency, the shape would be Gaussian.

The sum of the first four components in curve *A* begins to look somewhat Gaussian (the dashed line). In *B*, four additional frequencies are added to the sum in *A*, and the resultant curve is closer to Gaussian. By continuing to add component frequencies out to high frequencies in curve *C* (127 total

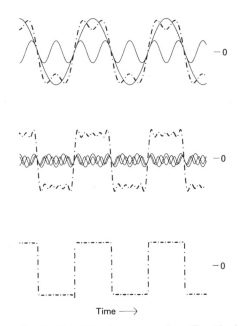

Time $\longrightarrow$

FIGURE 5.23. Component frequencies of a square wave. Top, the broken line is the sum of the first two cosine components. Center, the next three components are shown; the broken line is the sum of the first five components. Bottom, the sum of the first 127 cosine components of a square wave; a small overshoot is still present.

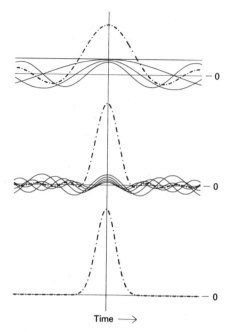

FIGURE 5.24. Component frequencies of a Gaussian curve. Top, the broken line is the sum of four low-frequency cosine curves which are components of a Gaussian curve. Center, five additional cosine curves at higher frequencies are added. Bottom, 118 additional cosine curves are added.

components which are not shown), we have improved the Gaussian still further so that it looks exact.

From the descriptions of the time and frequency domain Gaussians (5.28) and (5.29), one other feature can be noted. The widths of the pair are reciprocals: if the Gaussian-shaped envelope in the frequency domain is made only half as wide, the resulting time-domain signal is twice as wide. That means that a narrow signal peak has larger amplitude components at higher frequencies than does a wide peak.

With these examples in hand, we can go on to study filters. To obtain the effect of an analog filter on the composite signal, we must determine the effect on both the amplitude and the phase shift (lag) at each frequency.

5.3.2. First-Order Low-Pass Filters

A **low-pass filter** is included in one form or other in nearly every analog circuit in the computer interface. This is for two reasons: (1) the amplifier itself cannot respond with infinite speed, and (2) a filter is intentionally included to remove some noise.

One can nearly always assume that a signal from an experiment contains high-frequency components that have no information value; in electronic as well as in audio signals, any unwanted component of a signal is **noise**. The noise may arise from discrete external sources such as 60 Hz from the power lines or radio frequency interference (rfi) from electric motors, pulsed lasers, and radio stations. Other noise is inherent to electronic components and sometimes can be characterized statistically; for example, Johnson noise arises from temperature-dependent effects in various conductors. Removal of these higher frequencies is nearly always required to avoid aliasing the signal.

Two simple low-pass filters based on follower and inverter amplifiers are shown in Figure 5.25. If in the follower the capacitor and resistor are treated as a voltage divider, an equation can be written to relate the output, V_o, to the input. The denominator of a voltage divider equation is the sum of the two resistances or impedances; however when these impedances are summed, they must be treated as vectors at right angles so that the sum is the square root of the sum of their squares:

$$V_o = \frac{X_c}{(X_c^2 + R^2)^{1/2}} V_i = \frac{1}{(1 + (2\pi fRC)^2)^{1/2}} V_i \qquad (5.30)$$

From this equation, we conclude that if f is small, $V_o \to V_i$, but where f is large, $V_o = 0$.

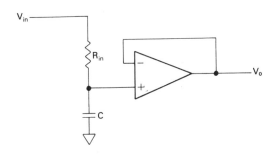

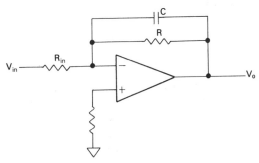

FIGURE 5.25. Two operational amplifier configurations for low-pass filters.

A similar result can be demonstrated for the inverter circuit in which the two impedance elements, R and C, are taken in parallel; the total feedback impedance, X_t, can be calculated as in the usual parallel resistance problem except that the terms are squared:

$$\frac{1}{X_t^2} = \frac{1}{R^2} + \frac{1}{X_c^2}$$

The result is the equation for the inverter with a filter term added:

$$V_o = -V_i \frac{R}{R_{in}} \frac{1}{(1 + (2\pi fRC)^2)^{1/2}} \qquad (5.31)$$

Where the filter elements consist only of a single RC circuit, the circuit is a **single stage filter**. The response of this circuit is customarily plotted as in Figure 5.26, which is known as a **Bode diagram**; the frequency response, V_o/V_i, as a function of frequency is shown on a log–log plot. The gain remains relatively constant at low frequencies and falls off nearly linearly at high frequencies with a slope of -1 on the log–log plot. (The frequency response is customarily measured by engineers in **decibels**. Where the linear gain is

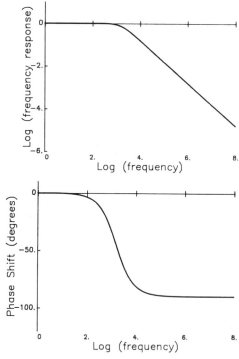

FIGURE 5.26. Response curves for a single-stage low-pass filter. Top, the attenuation with frequency is shown. Bottom, the phase shift, or lag, with frequency is shown.

A, the decibel scale provides a concise representation where

$$dB = 20 \log_{10} A$$

Therefore in this case, the gain drops by a factor of 10, which is 20 dB, for each tenfold increase in frequency.)

The **upper cutoff frequency**, f_2, is defined as the frequency at which the filter attenuates the amplitude by a factor of 0.707 (3 dB). It can be calculated from the relationship $f_2 = 1/(2\pi RC)$. Since single-stage low-pass filters are easily implemented, it is good practice, whenever an inverter amplifier configuration is used, to establish a reasonable cutoff frequency and add the appropriate filter capacitor to the feedback loop. However, f_2 must be chosen with considerable care.

The other feature of low-pass filters is the **phase shift**. The cosine wave subjected to a first-order low-pass filter lags behind the input by a phase angle, θ, where

$$\theta = \arctan (X_c/R) - 90^0 = \arctan (2\pi fRC) - 90^0 \qquad (5.32)$$

As the frequency increases, the phase angle also increases. At f_2, this lag is 45^0. The phase shift for a first-order low-pass filter is also shown as a function of log (f/f_2) in Figure 5.26.

These functions, attenuation and phase shift, are illustrated in Figure 5.27. The solid output cosine wave both is attenuated and subjected to phase shift compared with the dashed input wave. For this case, there is no change in the waveshape because it is a pure single frequency.

Returning to the remaining examples, the square wave and the Gaussian-shaped peak, a first-order filter can be applied to each cosine component using (5.30) and (5.32). The components are then summed as in Figures 5.23 and 5.24. Without belaboring the mathematics, the results are illustrated in Figures 5.28 and 5.29. Notice that the frequencies of the component curves are the same with and without filtering, but attenuation and particularly the

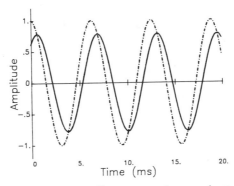

FIGURE 5.27. The effect of a low-pass filter on a cosine waveform. The output (solid) is both attenuated and lags the input.

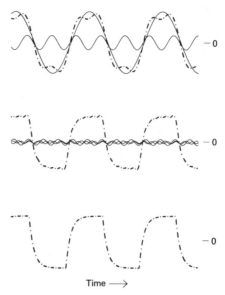

FIGURE 5.28. Square wave subjected to a low-pass filter. The contributions to the synthesis of the square wave are the same as for Figure 5.23 except that the attenuation and phase shift of a low-pass filter was applied.

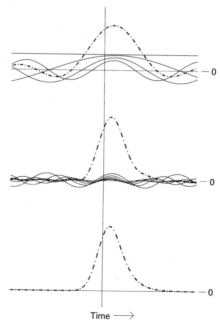

FIGURE 5.29. Gaussian curve subjected to a low-pass filter. The contributions to the synthesis of the Gaussian curve are the same as for Figure 5.24 except that the attenuation and phase shift of a low-pass filter was applied.

phase shift are clearly visible and are responsible for the distortions in the output.

5.3.3. Higher Order Low-Pass Filters

An ideal filter for removing high-frequency noise would have a much sharper cutoff than provided by a first-order filter, and it would produce no phase shift. **Higher order low-pass filters** are characterized by a much steeper slope in the Bode diagram, generally $-n$ in the log–log plot where n is the order of the low-pass filter. The relationship is shown in Figure 5.30.

A second-order low-pass filter can be implemented with single operational amplifiers. A **Sallen–Key** second-order low-pass filter is shown in Figure 5.31); $f_2 = 1/(2\pi R_1 \sqrt{C_1 C_2})$ when $R_1 = R_2$. Another route to higher order filters is to cascade first- and second-order filters. The references should be consulted to find appropriate circuits.

We also should note that other filters based on more complicated R–C networks have been devised whose features are closer to the ideal: the frequency response below f_2 is flatter, the frequency response above f_2 is much

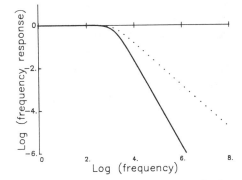

FIGURE 5.30. Response curve of a second-order low-pass filter. The dotted line is a first-order curve. The steeper cutoff is clearly visible.

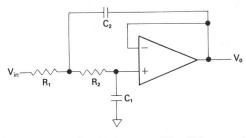

FIGURE 5.31. A Sallen–Key second-order low-pass filter. This second-order filter requires only a single amplifier.

lower, the lag as a function of frequency is flatter. The **Butterworth filter** maximizes flat response; the **Chebyshev filter** has the sharpest cutoff. None of these circuits nor their equations will be shown here, but it is interesting to compare responses for some sixth-order filters, shown in Figure 5.32.

Electronic modules may be obtained in which the filter characteristics have been optimized; a single module contains the entire circuit. In a few cases, such as the Frequency Devices 745PB-3 Butterworth fourth-order filter module, the time constant is digitally programmable and can be placed under computer control.

5.3.4. High-Pass Filters

The **high-pass filter** is used to remove a DC voltage level from a high-frequency signal; this is typically required when the signal is modulated by use of a chopper or another modulator. The principles are the same as those for the follower low-pass filter except the positions of R and X_c are reversed (Figure 5.33).

$$V_o = \frac{R}{(R^2 + X_c^2)^{1/2}} V_i = \frac{2\pi fRC}{((2\pi fRC)^2 + 1)^{1/2}} V_i \qquad (5.33)$$

This equation leads to the conclusion that at high frequency, the gain is 1, and at low frequency, the gain is 0. Once again, a Bode diagram (Figure 5.34) can be constructed and a **lower cutoff frequency**, f_1, can be found at the point where the gain is attenuated by 0.707 (3 dB). The value of f_1, $\frac{1}{2}\pi RC$, the phase shift at f_1, 45^0, and the slope of the Bode diagram, $+1$ on the log–log scale (20 db/decade), are the same as the characteristics of the low-pass filter except for sign. The slopes of the Bode diagrams can be increased by using higher order filters just as was done for low-pass filters.

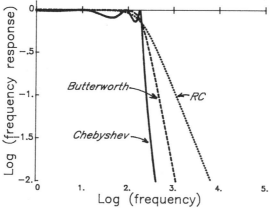

FIGURE 5.32. Response curves of other filters. The Chebyshev filter, known for its sharp cutoff, and the Butterworth filter, having a flatter response, are compared with an *RC* filter. All are sixth-order filters.

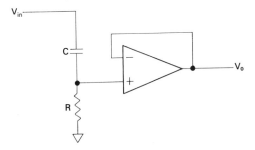

FIGURE 5.33. A first-order high-pass filter.

5.3.5. Bandpass and Notch Filters

These filters are both produced by the appropriate overlap of low-pass and high-pass filter networks. A bandpass filter, or tuned filter, passes only frequencies centered about f in a narrow range and could be implemented by cascading a low-pass filter with $f_2 \geqq f$ and a high-pass filter with $f_1 \leqq f$. Of course, use of higher order filters increases the selectivity.

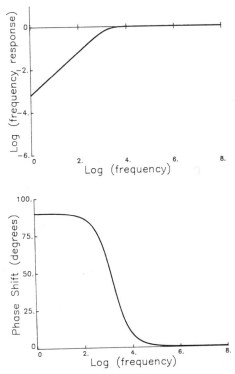

FIGURE 5.34. Response curves for first-order high-pass filters. Attenuation and phase-shift occur at low frequencies, while higher frequencies are unaffected.

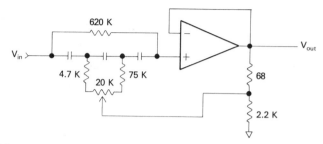

FIGURE 5.35. A notch filter. This filter may be tuned with the potentiometer to remove a narrow range of frequencies at about 60 Hz.

Where the offending noise in a system lies in a narrow frequency range centered about f (for example 60 Hz from the power lines), a notch filter can be used to pass all frequencies but f. In concept, this is a high-pass filter with $f_1 \geq f$ is placed in parallel with a low-pass filter with $f_2 \leq f$. In actuality, that combination can be built with a single OA. A notch filter is shown in Figure 5.35. for which a narrow band of frequencies centered at about 60 Hz is removed.

5.4. DEVIATIONS FROM OPERATIONAL AMPLIFIER IDEALITY

Thus far, the OA has been presented as a nearly ideal device having infinite open-loop gain, infinite internal input impedance, and zero output impedance. Although no real device is perfect, OA performance must be nearly ideal in an application. Consequently, the limitations of OAs must be understood.

In order to simplify the study of the errors inherent in OAs, we can look at models in which the **equivalent circuit** of a *real OA* will be drawn as if it was an *ideal OA* with additional components. As they arise, the nonideal resistances, voltages, and currents can be shown as if they are external components at the inputs and outputs of the ideal OA.

5.4.1. Errors Due to the Nonideal Gain of the Amplifier

Derivation of OA circuits assumed infinite open-loop gain; in practice, this is not so, even at zero frequency. Furthermore, the variation in the gain of an amplifier with frequency became apparent; even if no filter elements are added, an OA has limited bandwidth, an internal low-pass filter which limits the gain of the amplifier.

Consider the *open-loop gain, A_o,* in relation to the frequency. How might it affect the OA circuits which were discussed? In each of those configurations, the development of the equation for the OA's transfer function, A_c, depended on making an approximation; in general, these approximations rested on an assumption that A_o is much greater than unity. When filters were considered, the Bode diagram illustrated the fact that the gain of a circuit with a limited frequency response is *not* constant with frequency. If very high closed-loop gain is required over a wide frequency range, the open-loop gain must be extremely high; the amplification should be made in stages using OAs in tandem, each configured for moderate gain.

The **finite internal input resistance**, although ideally infinite, also can be a source of gain error. In the case of the voltage follower, this error is no different from that produced by attaching other measurement devices (voltmeters, oscilloscopes, etc.) with insufficient input resistance to a voltage source; the internal resistance of the measurement device forms a voltage divider with the source resistance. Usually, the purpose of using a follower configuration is to minimize the current drawn from the voltage source.

For example, a glass electrode used for pH measurements has an internal resistance of about 10 MΩ; it is represented as in Figure 5.36 with its equivalent circuit, a voltage source and its resistance, R_t. The internal input resistance of the OA, r_{inp}, and the output resistance of the electrode, R_t, form a voltage divider so that the voltage at V_+ is attenuated by the factor $r_{inp}/(r_{inp} + R_t)$. If r_{inp} is large, there is no attenuation, but if it is not large, serious error can occur. (Not considered here is the fact that some transducers including the glass electrode itself may behave nonlinearly and respond slowly when the signal does not meet a high-input impedance input.)

5.4.2. Offset Errors

The ideal OA produces an output of exactly zero when V_+ equals V_-; several sources of error produce a constant offset at the inputs. Since the

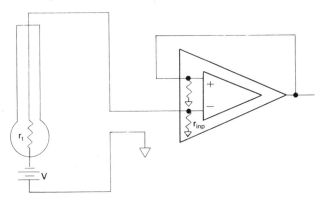

FIGURE 5.36. The input impedance of an OA. If r_{inp} is not extremely high, a large error in the reading of the potential will be produced when the OA is used with this transducer. The glass electrode has an extraordinarily high internal resistance, r_t.

offset may be amplified next by the closed-loop gain, the magnitude of the error is considered in absolute terms by referring to the inputs.

The **input offset voltage**, V_{off}, results from less than perfectly matched input components in the OA. In Figure 5.37, the offset is represented as simply a voltage source, either positive or negative, in series with one input. This offset may be nulled by the addition of an external potentiometer, usually a "trimpot", which adds a voltage to compensate the error; specification sheets for OAs show the trimpot resistance and the terminals to which it is connected. The offset is best adjusted by configuring the OA as an inverter with high gain, grounding the inputs, and adjusting the trimpot so that the output is zero.

Related to the offset voltage is **offset voltage drift**. Even after the offset voltage is cancelled, it may return as the temperature changes; where a high degree of stability is required, an OA with a low-drift specification must be chosen.

Frequently an existent offset voltage presents no problem. That is the case when a background level or baseline is acquired by the computer before and/or after the signal of interest is acquired. If this offset is to be subtracted in software, there is no need to null it out at any stage of the signal conditioning unless it is so large as to significantly affect the signal range at the ADC input. Similarly, in the output of analog signals, the offset can usually be accommodated either through feedback from a sensor which monitors the output device or through an offset control on the output device.

For example, if the speed of a motor is controlled through a DAC, a tachometer is usually used to determine the exact speed, and the offset is determined from the tachometer at rest. Besides eliminating a calibration process, an added benefit of avoiding trimpots is to remove yet another source of offset drift—the trimpots themselves. Where possible, the offset is best left uncompensated.

The **input bias current**, i_{bc}, is a similar source of error; even if both inputs are held at ground potential, a finite input current is required for operation

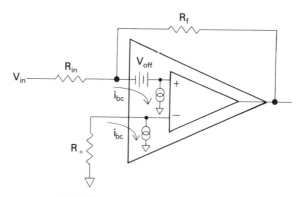

FIGURE 5.37. Offset errors in an OA.

of the OA. The result is additional voltage drop across the external resistances. In the simple inverter configuration, the offset can be largely eliminated by making it equal at both inputs; that is accomplished by adding a compensation resistance, R_+, at V_+ (see Figs. 5.7 to 5.9) which would have no effect if there were no bias current, but will produce a voltage drop of $i_{bc}R_+$. The optimal value of R_+ is the resistance of the input and feedback resistances taken in parallel,

$$\frac{1}{R_+} = \frac{1}{R_{in}} + \frac{1}{R_f} \tag{5.34}$$

5.4.3. Other Limitations

Several other limitations must be considered when choosing or applying an OA.

The **slew rate** is the rate, in volts per second, that the output responds to an instantaneous change in the input. The response is simpler than that of the limited response time of an RC filter. A low-pass filter requires $4RC$ s to reach 98% of the correct value after a step change in the input; this applies whether the step is 10 V or 10 μV. However, the slew rate is the maximum rate of change in absolute terms.

The **maximum output current** is controlled so that a shorted output will not destroy the device; irrespective of the output resistance, the maximum output current cannot be exceeded. A typical value for an integrated circuit OA is about 10–25mA.

When the OA circuit is to drive a device which must draw significant current in order to control, for example, a motor or incandescent lamp, two

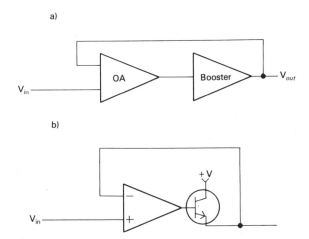

FIGURE 5.38. OAs with current boosters. (*a*) A booster OA can increase the output current capacity up to several amps. (*b*) If the output is always positive, an n-p-n transistor can be used.

methods can be used to provide the additional current. A **current booster** amplifier has a single input and a single output, and is inserted within the feedback loop as shown in Figure 5.38. Less expensive methods use transistors as current amplifiers; by placing them within the feedback loop, the circuit design is simplified, but this design can supply only positive current. More detail can be found in Chapter 7.

The **voltage output range** is limited by the power supply, and for most IC OAs, the power supplies are limited to ± 15 to ± 18 V. There is also a limit to how close to the supply voltages the output may come; with a ± 15-V supply, some output ranges may be close to ± 15 V. For older OAs, the difference between the output range and the power supply may be as much as 5 V. Where that difference is small, the OA can sometimes be conveniently operated from a single low voltage such as a $+5$-V supply used for logic circuits as long as the output is within that range. If high voltages are required, special-purpose OAs can be obtained with an output range exceeding ± 90 V.

5.5. INTEGRATED CIRCUIT OAs

Operational amplifiers suitable for all but the most specialized of applications can be obtained in IC form. Besides low cost, the IC also may offer the advantage of lower offset voltage drift since the internal components are well matched, and the entire unit stays at the same temperature.

Typically, OAs are housed in 8-pin metal cans, 8-pin Dual-In-line-Packages (DIPs), and 14-pin dips, the latter often having two to four OAs in a single package. The OP-07 (Precision Monolithics) OA is a high-quality variant of the most common of all OAs, the 741. Part of the pin definition for the OP-07 in an 8-pin package (Fig. 5.39) is identical to that of many other

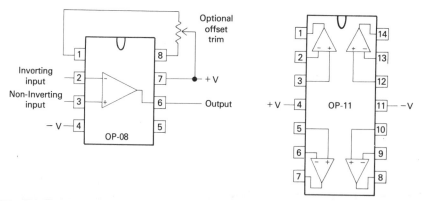

FIGURE 5.39. Integrated circuit OAs. The pinout for the integrated circuit on the left is nearly standard. The pinout of the quad-OA IC on the right is typical.

OAs; the pins shown for the inputs, output, and power are standard for most 8-pin OA packages. OAs differ in how the voltage-offset-null adjustment circuitry is added and in whether and how an external capacitor is used for frequency compensation. Also shown in Figure 5.39 is an IC with four OAs in a single chip.

5.6. BRIDGE CIRCUITS

Many transducers are effective variable resistances, and the changes are small compared with the total resistance. The measurement of small resistance changes is best accomplished with a **Wheatstone bridge**, shown in Figure 5.40. The bridge is composed of four resistance arms with an excitation voltage, V_{ex}, applied across one diagonal, and an output, V_{out}, measured across the other.

The value of V_{out} is the difference between two voltage dividers,

$$V_{out} = \frac{R_1}{R_1 + R_4} V_{ex} - \frac{R_2}{R_2 + R_3} V_{ex} \tag{5.34}$$

V_{out} will be null if $R_1/R_4 = R_2/R_3$.

The transducer resistances may be placed in the bridge in one of four configurations: one, two, or four of the arms may be resistance, R, of a transducer that responds to the phenomenon under study; the fixed resistance in each of the remaining arms is approximately equal to that of the transducer. The arrangement of the transducer resistances determines the overall sensitivity to small changes in the transducer resistance, ΔR, which is defined as the ratio $V_{out}R/V_{ex}\Delta R$.

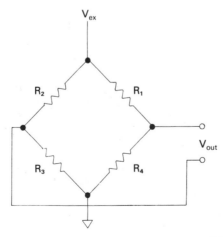

FIGURE 5.40. A simple bridge. The output will be zero if $R_1/R_4 = R_2/R_3$.

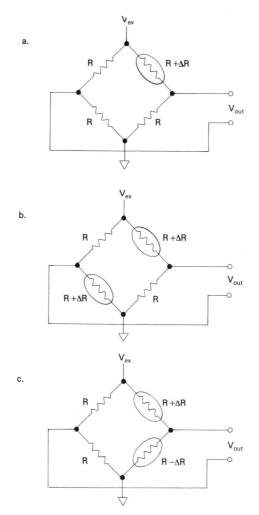

FIGURE 5.41. Configurations for using a bridge to make sensitive measurements. In (*b*), both transducers change in the same direction. In (*c*), they change in opposite directions.

The simplest bridge consists of a single variable element, whose resistance is $R + \Delta R$, and 3 arms whose resistances are R (Fig. 5.41*a*). The ratio of V_{out}/V_{ex} is $\Delta R/(4R + 2\,\Delta R)$. If $\Delta R \ll R$, the sensitivity is $\frac{1}{4}$; a range of ± 10 Ω in a 1 $K\,\Omega$ transducer (1% of full scale) will produce a change of ± 0.025 V in the output when the bridge is excited by 10 V (0.25%). The importance of the bridge, however, is that the output is directly proportional to the *difference* between resistances rather than a small change in a large signal.

The arrangement of two variable elements is shown in Figure 5.41*b*; in this case, the variable elements change identically and are placed in opposite arms. The ratio of V_{out}/V_{ex} is $\Delta R/(2R + \Delta R)$. If $\Delta R \ll 2R$, the sensitivity is about $\frac{1}{2}$.

In some cases, resistive transducers can be arranged so that the resistance of one goes down in the same degree that the other goes up; the resistances are then placed in adjacent arms (Fig. 5.41c). Then the sensitivity, when $\Delta R \ll 2R$, reduces to about $\frac{1}{2}$.

Although the configuration of Figure 5.41c is not more sensitive than others, there is an additional advantage. If the two variable arms change in the same direction, (due to temperature, for example, when pressure is being measured), that change is nearly cancelled.

When active elements are placed in all four arms, with one pair responding in the opposite direction from the other (a combination of Fig. 5.41b and c), the sensitivity is exactly one. The output signal relative to the excitation voltage is equal to the relative *change* in voltage.

The bridge output requires an amplifier with a differential input, since both ends of the output diagonal will be different from ground. The voltage outputs are usually also quite low; therefore instrumentation amplifiers with true differential inputs are often used. When the gain is high, sensitive measurements can be made.

CHAPTER SIX

Digital Electronics

Thus far in the discussion of the computer interface, we have explored the path from the analog signal of a transducer to the digitally encoded datum as well as the return path from the digital form to the analog. Now we will consider information handled electronically in digital form. Some interfaces are inherently digital; in others, digital timing techniques are used in conjunction with the manipulation of analog signals.

In this chapter, sufficient information concerning digital electronics is presented to allow completion of the interface. In the belief that powerful interfaces can be constructed by treating digital components as simple building blocks, this chapter has little detail. The references at the end are outstanding and detailed treatments of electronics at a practical level. We will begin with the common electrical protocol for digital systems, followed by an outline of a few of the basic building blocks of digital electronics.

We might pause to note the importance of the availability of digital electronics to instrumentation. A change in instrumentation development was described in Chapter Five which occurred when the OA made instrumentation design involving analog electronics accessible to the scientist. The introduction of digital electronics was the second revolution for the instrumentation scientist. The precision of digital techniques made designs even better and easier.

6.1. THE TTL SYSTEM, AN ELECTRONIC DEFINITION

A number of different electronic systems for representing digital logic have been developed. Each has had the overall objectives of achieving (1) high

203

speed in carrying out Boolean algebra operations, (2) high density when the components are integrated in a silicon "chip", (3) low power consumption, and (4) high reliability. These electronic systems are called logic families.

Although several others are in use for specialized applications, the dominant families for small computers are Transistor–Transistor Logic (**TTL**) and Metal-Oxide Semiconductor (**MOS**), particularly **Complementary MOS** (**CMOS**). Various MOS techniques are used for the production of most LSI circuits including microprocessors, but interfacing problems and small logic systems usually employ TTL. (LSI-MOS devices are usually adapted to the TTL environment, and include inputs and outputs which are compatible with TTL requirements). TTL remains the most versatile and easiest to use; the examples of digital logic in this book will come from the TTL family.

The output of a standard TTL circuit, being digital, has two valid states representing the logic states of **0** and **1**. If the voltage at an output pin is measured, it would read about 0.6 V when the output represents a logical **0** and 3.6–5 V at a logical **1** (Fig. 6.1). The range between 1.2 and 3.6 V is the **noise margin**. It has no meaning as a logical representation, so even if electrical noise is present, the logic level will not change unless the noise amplitude is large; ambiguity is minimized.

However, the voltage level definition alone does not accurately represent the operation of TTL logic. Although the logical **1** output voltage level is a measurable voltage, the output of a TTL circuit cannot be used as a 5-V source of electrical power. The reason is that the output resistance is quite large when the output is **1**, so very little current can be drawn from the IC. On the other hand, when the output is a logical **0**, the output effectively is connected to ground with small output resistance.

Representations of the equivalent circuits for the outputs of a TTL circuit, representing transistors as switches, are shown in Figure 6.2. From the position of the resistor, we find that the *active* state of TTL is a logical **0**: switch S1 is closed, S2 is open, and the output is effectively grounded. When the output is a logical **1**, switch S1 is open and S2 is closed; consequently, a high-resistance connection to the power supply is formed that could be

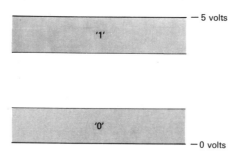

FIGURE 6.1. Transistor-transistor logic (TTL) voltage levels. The range between the **1** and the **0** is the noise margin.

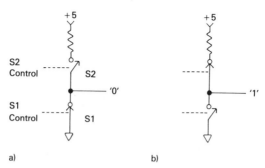

FIGURE 6.2. An equivalent circuit for a TTL output. (*a*) A LO output exists when the output is effectively connected to ground. (*b*) A HI output measures 5 V, but the resistor limits current than can be drawn.

measured by a voltmeter as 5 V, but the current that is drawn from the 5-V supply is limited by the resistor.

In fact, the input to a TTL component is not sensitive to the voltage, but to the ability of the input device to ground the input connection. That is, the TTL system is based on *current-sinking*; the output of a TTL device acts on an input to another stage by sinking current from it to ground. The input to a TTL device is activated by a logical **0** input when it is effectively *grounded*, and it experiences a logical **1** whenever it is *not grounded*.

In part this can be understood from the circuits in Figure 6.3 in which the light bulb represents a TTL input. We might expect that a positive voltage is necessary to activate a TTL input, but, as illustrated from Figure 6.3*a*, when the input to the light bulb is driven from the **HI** output, the current is limited by the resistance; this is a poor method of transmitting the logic information. However, if the input, represented by the light bulb, is driven from the **LO** condition, the current is limited primarily by R_L, the internal resistance of the TTL device's input. The output of a TTL gate is actually

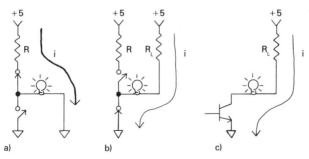

FIGURE 6.3. TTL control of another device. (*a*) An incorrect connection; *R* limits the current. (*b*) Current can flow through the bulb; it is limited primarily by the load resistor, R_L. (*c*) The transistor equivalent of (*b*).

a transistor, shown in Figure 6.3c, which either conducts (in which case it is a logical '**0**' and the light bulb is activated) or does not conduct.

An actual transistor circuit for a typical simple TTL gate is shown in Figure 6.4. It is not necessary to be fully comfortable with the transistor circuit before one can understand TTL, but a description of the circuit is instructive. The gate performs the Boolean **NOT** function: an input level of logical **1** will produce "**NOT 1**" at the output which is **0**. Conversely, an input **0** will yield a **1**.

In the circuit shown in Figure 6.4, if the input is grounded (logical **0**), the base of Q3 will be pulled down to ground, and Q3 will not conduct. Then no current will flow into the base of Q1, and it will not conduct. When Q1 does not conduct, the output is a logical **1**. On the other hand, when Q4 is not grounded, Q3 becomes a conductor, and that turns Q1 on; the result is a logic level **0**.

Transistors Q2 and Q1 form a "totem pole". Q2 is always in the opposite state from Q1, but it actually has no utility in an *ideal* electrical environment. Q2 is included to force the output, as read by a voltmeter, up to about 4 V. This is of some utility in reducing the susceptibility to noise.

Some TTL devices are constructed without transistor Q2, and have no

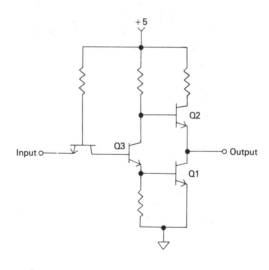

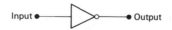

FIGURE 6.4. A TTL inverter circuit. Top, the internal transistor circuit of an inverter. Bottom, the logic symbol.

connection whatever to the collectors of the output transistors, that is, they have **open-collector** outputs. While this device has the same effect as other true TTL logic, an ordinary voltmeter is unable to distinguish a **0** from a **1**.

One application for open-collector outputs is control of power to small loads. Common open-collector TTL ICs can be used to switch the power to loads with voltages up to 35 V and sink up to several hundred milliamps. As suggested by Figure 6.3c, power is applied to the load by making a complete circuit to ground. (Control of loads will be considered further in Section 7.2.2.)

The second function of open-collector TTL logic is in **wired-OR logic**, a simple method of obtaining the logical combination of two or more outputs; an equivalent circuit is found in Figure 6.5. If several open-collector outputs are connected to a single TTL input, that input will experience a **LO** if *any or all* of the open-collector outputs are **LO**. If the "totem-pole" arrangement was used, the outputs would contend with each other if some outputs were in each state; the transistor producing the **0** would need to pull the outputs of all the other devices down to ground.

In order to regain the advantages of having a logical **1** actually produce a 5-V output, a single "pull-up" resistor (R_L in Fig. 6.5) is added between the 5-V power supply and the input of the following stage. If *all* of the open-collector outputs are at a logical **1** level, a voltmeter on the input line will read 5 V, but if any output is at a logical **0** level, the connection to the next circuit is grounded.

The best of both worlds can be obtained with **tristate logic**. As illustrated in Figure 6.6, tristate outputs have the conventional TTL capabilities. Like standard TTL, either transistor switch in the circuit can be closed to create a **LO** or a **HI** condition. Unlike standard TTL, both switches can both be open so that the output is neither **LO** nor **HI**; this is the "third state". As illustrated in the equivalent circuit, the output is effectively disconnected from the rest of the TTL gate.

Several variations of standard TTL are in common usage. The first is Schottky TTL which, because of the addition of special "Schottky diodes",

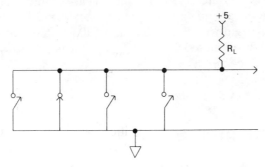

FIGURE 6.5. Equivalent circuit for a wired-OR function. If any switch is closed, the output is LO.

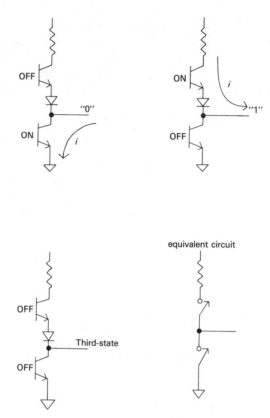

FIGURE 6.6. Tristate TTL outputs. The three combinations, **0**, **1**, and the high-impedance (third) state, are represented. The switch circuit is equivalent to the others.

operates faster but uses more power. The delay from I/O for a typical TTL device is about 15 ns whereas the delay for an equivalent Schottky device is only about 3 ns. Another important variation is low-power Schottky; these devices use only about one-fifth the power of standard TTL, but with the Schottky diodes, they are able to maintain nearly equivalent speed.

The common integrated circuit building blocks for TTL circuitry use ICs whose numbers begin with 74; some will be illustrated in the next section. One IC with the function, **NOT**, is specified by the number 7404. Some "7400 series" chips include an "S" after the "74" (for example, 74S04) to indicate the Schottky varieties; "LS" indicates low-power Schottky. The newer "AL" variety has even greater speed.

6.2. TTL BUILDING BLOCKS

With TTL logic units, design often can be approached as an exercise in connecting simple building blocks. In this section, the fundamental building blocks are described beginning with the logic gates which carry out Boolean algebra in circuitry.

6.2.1. Gates and Buffers

The basic operations of Boolean algebra are the NOT, AND, OR, and EX-CLUSIVE-OR functions. When the AND and OR are combined with the NOT function, the results are the NAND and NOR functions respectively. The symbol for each is shown in Figure 6.7 together with a table of I/O combinations (truth table) for two input versions of each.

The NOT function simply produces the opposite of the input:

"if A is **HI** then the output is **NOT HI** (i.e., **LO**)."

A **HI** or a TRUE is a logical **1**, and a **LO** or FALSE is a logical 0.

The AND and OR functions can be deduced by simply "reading" the inputs. For the AND function (Fig. 6.7),

"if A is **HI** *AND* B is **HI** then the output is **HI**"

Similarly for the OR function,

"if A is **HI** *OR* B is **HI** then the output is **HI**"

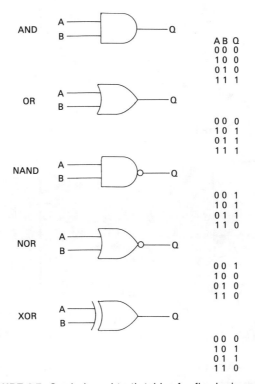

FIGURE 6.7. Symbols and truth tables for five logic gates.

Then if the NOT function is added to the AND function to form a NAND gate, the result can be read

"if A is **HI** *AND* B is **HI** then the output is **NOT HI**"
An analogous statement can be written for a NOR gate.

The EXCLUSIVE-OR gate function can be read

"if A is **HI** *EXCLUSIVELY, OR* B is **HI** *EXCLUSIVELY* then the output is **HI**"

The pin assignments of a few representative integrated circuits which are used to implement some of these functions are shown in Figure 6.8. In the first example, a 7400 IC, four two-input NAND gates are integrated on a single chip. Power is supplied to the chip at $V_{cc} = 5$ V and GND = ground. The 7402 is a quad-NOR gate, and the 7404 is a hex inverter (six inverters on a chip). The 7420 example illustrates the availability of gates with more inputs. Note that the part numbers begin with 74; the basic building blocks

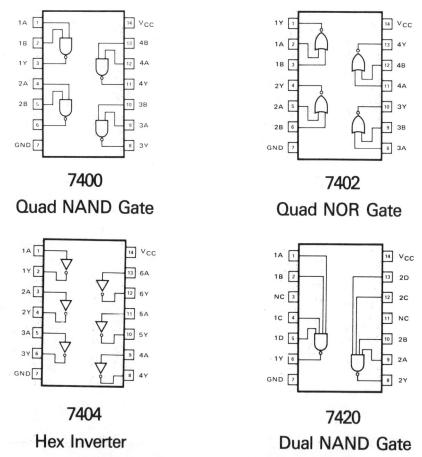

FIGURE 6.8. Pinout for typical logic gate-integrated circuits.

of TTL logic are the 7400 series ICs produced by the major semiconductor houses.

Several diagrammatic conventions are used commonly to denote the addition of the **NOT** function to the other functions. For a gate symbol in circuit diagrams, the addition of a small circle at the input or output serves to indicate that the signal is inverted; hence the symbol for a NAND gate is that of an AND gate with a small circle on the output. To indicate the inversion of a signal having a label such as A or B, a bar is added over that label; hence the inverted A can be written $\overline{A}$ which is read "NOT A".

If the name used as a label indicates a function, that function is in effect when the signal is **HI**, but if a bar is added, that function is in effect if the signal is **LO**. For example, if a signal in a digital circuit is labeled RDY to indicate "ready", the ready condition exists when the signal is **HI**, but if it is labeled $\overline{RDY}$, the ready condition exists when the signal is **LO**. (Because the bar symbol is not available on standard typewriters, some documentation uses an asterisk: A* also is read "NOT A".)

One can show that an OR gate with the *inputs* inverted is equivalent to an AND gate with output inverted (Fig. 6.9). Although the correct use of these representations is all too infrequently followed, the AND representation should be used when the gate is to logically detect that both inputs are **HI**, and the OR representation should be used when the gate is to determine if at least one input is **LO**.

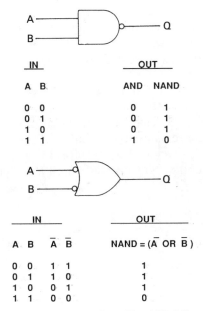

IN		OUT	
A	B	AND	NAND
0	0	0	1
0	1	0	1
1	0	0	1
1	1	1	0

IN				OUT
A	B	$\overline{A}$	$\overline{B}$	NAND = ($\overline{A}$ OR $\overline{B}$)
0	0	1	1	1
0	1	1	0	1
1	0	0	1	1
1	1	0	0	0

FIGURE 6.9. Equivalence of logic combinations. The AND followed by NOT is equivalent to NOT followed by OR.

Often integrated circuits, which have no function in the logic of the system, are required as electrical **buffers**. The symbol for a buffer is simply a triangle. When a subsystem such as a memory board or an I/O port is added to a computer by plugging it into the bus, the data and address lines should be buffered from the bus. Then only the buffer input will place a load on the bus, but without it, the input to each memory IC would be connected to the line on the bus and that would exert too great an electrical load on that line. The buffer also may help protect other parts of the circuitry from noise or dangerous electrical spikes.

These basic 7400 series gates can be combined as building blocks to deliver a wide variety of complex functions, the storage of bits, the decoding of combinations of bits, and the routing of signals. A few of those, that are useful in building interfaces, will now be considered.

6.2.2. Flip-Flops

A **flip-flop** is a circuit which, in the absence of active inputs, is stable in either of two states. The state can be changed by a signal which is only momentarily active. Flip-flops can be sensitive to two types of activity. For most TTL circuits, the input signal is active when at a **LO** level, and has no effect when **HI** or not connected, but for some flip-flops, an input is activated only when experiencing a *change* in state. In that case, it is an **edge-triggered** flip-flop since it responds to a falling or rising edge in the input signal.

The simplest flip-flop is an **RS** (Reset–Set) flip-flop, configured from a pair of cross-coupled NAND gates as shown in Figure 6.10. In digital terminology, **set** refers to the action of setting the output to a **1**, and **reset** refers to the action of resetting the output to **0**. Where there are two outputs, those outputs may have opposite logic and are written **Q** and $\overline{\text{Q}}$; the set and reset action then refer to *Q*.

Both inputs to the RS flip-flop are inactive when **HI**. Whenever the S input goes **LO**, **Q** must go **HI**; since the inputs to the lower NAND gate are then both **HI**, $\overline{\text{Q}}$ goes **LO**. That **LO** is then fed to the top NAND gate and holds **Q HI**. This state remains stable until the R input is activated, and the state of the flip-flop is reversed. The result is the ability to store a single bit of information; the output remains static until a **LO** pulse is applied to the appropriate input.

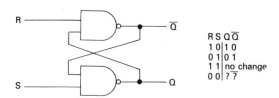

FIGURE 6.10. An RS (reset–set) flip-flop. The state can only change when there is a LO input to *R* or *S*.

Many other types of flip-flops build on the RS flip-flop, each of which can be constructed from NAND gates. Two additional varieties will be presented, but a study of the internal gate structure is not necessary, since each of these flip-flops is available in integrated circuit form with two or more per chip. For example, the 74126 IC contains four RS flip-flops.

A **D flip-flop** is used as a data latch. As shown in Figure 6.11, a *D* flip-flop has a data input (*D*) and complementary outputs, *Q* and $\overline{Q}$. When the clock (*Cl*) input is held **LO**, the level at *D* is transferred to *Q* and its reciprocal is available at $\overline{Q}$. When C1 returns **HI**, the datum is latched and the outputs remain static. *S* and *R* inputs also may be included; they set and reset the output, overriding the latched data.

The **master–slave (MS) flip-flop** (Fig. 6.12) serves a rather different function. It is triggered by an *edge* rather than a level; that is, the outputs respond to the *change* in the clock (*Cl*) input rather than the level. (The name is derived from the internal design which is based on two flip-flops which are activated sequentially as master and slave.)

The MS flip-flop can be understood by first considering only the *Cl* input and the *Q* output. If all of the other inputs (*J*, *K*, Set, and Preset) are **HI** and therefore inactive, the *Cl* input acts as a toggle; after a falling edge at time $n + 1$, $(t_{n + 1})$, the output at *Q* switches to the opposite of the output before the edge (t_n).

J and *K* inputs act to allow the output to store the data represented by either *J* or *K*. If *J* (but not *K*) is **LO**, a falling edge at the *Cl* input will always cause *Q* to be **LO**; if *K* is **LO** (but not *J*), a falling edge at *Cl* will make *Q* **HI**. If *both J* and *K* are **LO**, an edge at *Cl* will have no effect. These results can be seen in the truth table in Figure 6.12.

Set and Reset inputs always override the others, *J*, *K*, and *Cl*. A **LO** at the *S* input will always set *Q* **HI** and a **LO** at the *R* input will always make *Q* **LO**; of course, an exception exists if both are **LO** in which case the output is indeterminate.

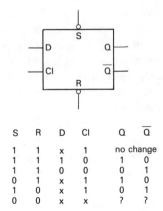

S	R	D	Cl	Q	$\overline{Q}$
1	1	x	1	no change	
1	1	1	0	1	0
1	1	0	0	0	1
0	1	x	1	1	0
1	0	x	1	0	1
0	0	x	x	?	?

FIGURE 6.11. A D-type flip-flop. A LO input to the clock will latch the value of *D*.

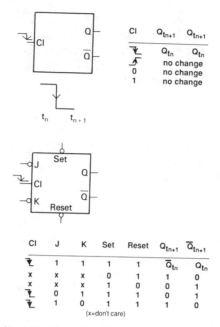

Cl	Q_{tn+1}	$\overline{Q}_{tn+1}$
⌐↘	Q_{tn}	$\overline{Q}_{tn}$
↗	no change	
0	no change	
1	no change	

Cl	J	K	Set	Reset	Q_{tn+1}	$\overline{Q}_{tn+1}$
⌐↘	1	1	1	1	$\overline{Q}_{tn}$	Q_{tn}
x	x	x	0	1	1	0
x	x	x	1	0	0	1
⌐↘	0	1	1	1	0	1
⌐↘	1	0	1	1	1	0

(x=don't care)

FIGURE 6.12. Master–Slave flip-flops. Top, the output, Q, changes state *only* on the falling *edge* of the *Cl* input. Bottom, several additional functions are added. Set and Reset override all other functions. *J* and *K* can be data inputs which are transferred to the outputs on a falling edge of *Cl*.

Two examples of master–slave flip-flops are important in interface electronics, the flag circuit and the counter. Figure 6.13 illustrates one method of providing a flag for a parallel interface circuit. We begin by assuming Q to be **LO**. An edge from a pulse produced by some external device to the *Cl* input will set Q **HI** and $\overline{Q}$ **LO**. Tying $\overline{Q}$ back to the K input will prevent any later pulse at *Cl* from setting Q **LO** again, and the flag is set. The computer can read that flag and act upon the information that it represents; for example, the pulse may result from the end of an ADC conversion, and the computer must acquire a data byte. When that byte is accepted, a pulse from the computer can be used to also clear the flag using the Reset input. A timing diagram shows the operation of such a circuit. This circuit is used frequently to set an interrupt request with an external event and to clear it with a pulse from the interface.

A binary counting circuit can be built from a series of master–slave flip-flops in which the Q output of each is fed to the *Cl* input of the next as illustrated in Figure 6.14. Since each flip-flop changes state on only one edge of the input, it divides the number of pulses by two, and the Q outputs generate a binary count.

This particular counter arrangement is termed a *ripple* or *asynchronous* counter since each stage can only change state after the previous stage has

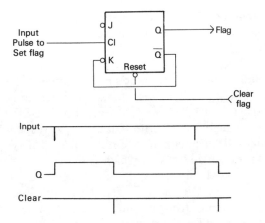

FIGURE 6.13. Using a flip-flop to set a flag. Some external occurrence can flag the computer interface by sending a pulse to the *Cl* input. No matter what happens to *Cl*, the flag will stay set until it is Reset by the computer.

changed. Slight delays occur at each stage, so the change from 01111_2 to 10000_2, for example, will occur sequentially from the LSB to the MSB; for a few nanoseconds, the output will read 01110_2, 01100_2, 01000_2, and 00000_2 before settling at 10000_2. References on digital logic provide details on *synchronous* counters in which all the stages change together; they should be used whenever it is necessary for all the bits to change simultaneously.

6.2.3. Decoders

A decoder generates a desired logic level at an output *only* when a predetermined combination of bits is present at its input. A decoder is required

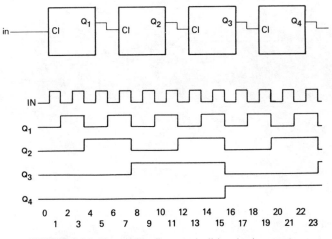

FIGURE 6.14. Use of flip-flops to build a ripple counter.

in a memory or an interface subsystem to determine when the address issued by the CPU matches the address of the memory or interface subsystem.

There are several ways in which an address may be detected by a decoding circuit. One is the EXCLUSIVE-OR circuit depicted in Figure 6.15 in which an address, possibly from the computer's bus, is compared with that selected by setting four switches in the subsystem. The selected address bit is **0** if the switch is closed, and **1** if it is open. Each EXCLUSIVE-OR gate compares the level of a single switch with that of the corresponding bus line (A, B, C, or D), and a **HI** results if they differ. Then, if a switch is closed to ground (logic **0**) for each address bit which is to be a **1**, the outputs of those XOR gates can be tested with a NAND gate. The output of the NAND gate is **LO** only if *all* of the XOR gates are **HI**. The result is **LO** *only* when the switches match the inputs from the bus. The four inputs in the figure, A through D, must all be **0** for Q to be **0**.

Integrated circuit decoders are more versatile. ICs, such as the 74138s depicted in Figure 6.16, can accept a 3-bit input and decode it. The output with a decimal number corresponding to the binary number at the input will be **LO** while all others are **HI**. If inputs A, B, and C are **0**, **0**, and **1**, then output 4 will be **LO** while all others are **HI**.

The example shows an 8-bit number, such as a port address, being decoded. To accomplish this, two 74138 decoders are used. Decoder X decodes 'he three LSBs using inputs A, B, and C. Decoder Y decodes bits $D3$ through $D5$. The chip-enable inputs are used to decode the last two bits and to cascade the two chips. X_O, is **LO** when inputs **A**, **B**, and **C** are 0, 0, and 1, respectively, *and* when the enable inputs are both true, that is, when the input to $G1$ is **1**, and the input to $G2$ is **0**. Y_O is **LO** when X_O is **LO** *and* chip Y inputs A,

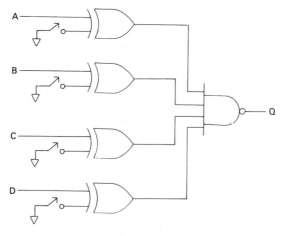

FIGURE 6.15. An address decoder. The output, Q, only is LO when the inputs, A–D, match the four switches.

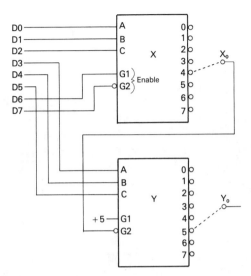

FIGURE 6.16. Use of a pair of decoder ICs to decode an 8-bit number. The output, Y_0, will go LO only when the value of the input is $6C_{16}$.

B, and C read 5 (1, 0, and 1, respectively). Therefore, the configuration in this example will read **LO** on output Y_O when the input data is 01 101 100$_2$ or $6C_{16}$.

6.2.4. Multiplexers

This useful circuit is used to carry out the digital equivalent of the analog multiplexer. According to the value of the control input, a chosen input signal is directed to the output. For example, a dual 4-line to 1-line multiplexer can be used to select from four input pairs; a 2-bit control input is required. The application in Figure 6.17 selects a single BCD digit from four by using a pair of dual 4-line to 1-line multiplexers.

The example is important. If a four-digit panel meter provides a fully parallel BCD output, it must have 16 connections, four bits for each of four BCD digits. That may be more bits than are available through the parallel digital interface. However, the digits may be read sequentially through a 4-bit input port to a computer by using two dual 4-line-to-1-line multiplexers as shown in the figure. Two bits of the computer's output port select the digit which is routed to four bits of the input port. In some cases, a single 8-bit parallel port can be configured with four bits of input and the remaining for output, so that port can acquire the entire 4-BCD digit datum.

We should remind ourselves that the multiplexer function (like the flip-flops and decoders, etc.) are available as integrated circuits, and that these functions are built up from simple logic gates. Part of a data sheet for a multiplexer is presented as Figure 6.18.

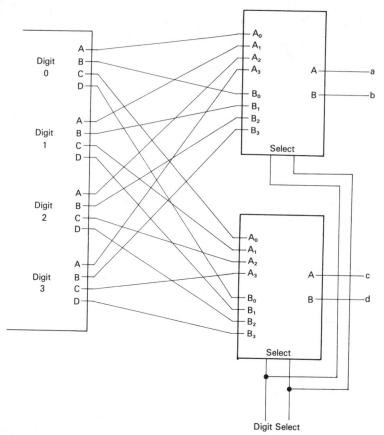

FIGURE 6.17. Application of digital multiplexers. Four BCD digits are routed, one at a time, to outputs, *a,b,c,d* using a pair of 4-in-2-out multiplexers. The Digit Select input determines which digit appears at the output.

6.2.5. Monostables

It is occasionally useful to take advantage of analog characteristics to implement digital functions, and the monostable provides a method of generating pulses whose widths are determined by the characteristics of a resistor and capacitor. As the name implies, the output has one stable state and one quasistable state. When triggered, the monostable enters the latter state and returns after its characteristic period. Some users call these devices "single-shots" and mark them with an "SS" on circuit diagrams.

As illustrated in Figure 6.19, monostables are triggered by an edge, like the clock input of an MS flip-flop; the values of R and C can be used to vary the pulse width from well under a microsecond to nearly a second. Like a flip-flop, a monostable has complementary outputs, Q and $\overline{Q}$. From there, several variations are available. Some are retriggerable; that is, if a trigger pulse arrives during the previous output pulse, the pulse period will begin again. If not retriggerable, input pulses are "locked out" during an output pulse. Some have Schmitt trigger inputs (see Chapter Five) to improve noise immunity; a nonTTL input also could be converted to a TTL pulse. For example, if the 60-Hz line voltage was transformed with a 10-V peak-to-peak transformer and rectified through a single diode, a monostable with a Schmitt trigger could produce pulses at 60 Hz.

Monostables are useful in interface circuits for stretching pulses which are too short or shortening pulses which are too long. They should not be used where precision is required, being subject to drift with temperature and age; often they are stable to no better than 5%. They also are susceptible to triggering by electrical noise in the environment, so all unused inputs should be wired to 5 V (this is good practice in all TTL design). The precision decreases with increasing pulse width; consequently, some versions are

TRUTH TABLE

SELECT INPUTS		DATA INPUTS				STROBE	OUTPUT
B	A	C0	C1	C2	C3	G	Y
X	X	X	X	X	X	H	L
L	L	L	X	X	X	L	L
L	L	H	X	X	X	L	H
L	H	X	L	X	X	L	L
L	H	X	H	X	X	L	H
H	L	X	X	L	X	L	L
H	L	X	X	H	X	L	H
H	H	X	X	X	L	L	L
H	H	X	X	X	H	L	H

Select inputs A and B are common to both sections.
H = high level, L = low level, X = irrelevant

PIN CONFIGURATION

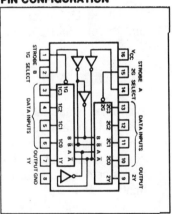

BLOCK DIAGRAM

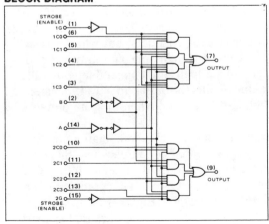

FIGURE 6.18. Part of a data sheet for the multiplexer IC in Figure 6.17. (Reprinted with the permission of Signetics Corporation.)

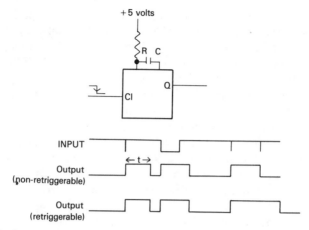

FIGURE 6.19. A monostable. Nonretriggerable monostables ignore trigger pulses which arrive during the semistable condition.

available which internally retrigger themselves for a preset number of pulses and thereby increase the pulse width digitally.

6.3. ISOLATION TECHNIQUES

Electrical isolation is frequently required in digital systems to isolate the system from noise. Isolation also is required of any connections made between an electronic component powered by moderately high voltage (the computer) and a human subject. Fortunately, this is a relatively simple process since digital information can be encoded as the presence or absence of light.

An LED and a phototransistor are the necessary components (Fig. 6.20). Light from the LED is detected by a phototransistor. The light falls on the base, and when the base is lit, the transistor conducts, but when it is not lit, the transistor is off. The current from the conducting phototransistor will turn on the following transistor, creating a logical **0**, compatible with TTL. The LED can be switched by the open-collector output of a TTL device; it is turned on by effectively connecting it to ground when the gate output is **LO**.

The LED and phototransistor pair is such a useful combination that it is available as an **optoisolator** in a single integrated circuit package. Since the phototransistor is not able to carry enough current to present a **0** to a TTL device, the second transistor shown in the figure is often integrated into that package (a Darlington configuration, see Chapter Eight). For a damaging electrical pulse to cross the gap between the LED and phototransistor, there would have to be electrical breakdown across that gap, and that will not occur if the pulse is less than several thousand volts.

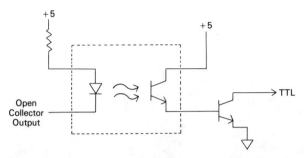

FIGURE 6.20. An optical isolator. The gap provides electrical isolation to about 2000 V. Light from the LED partially turns on the phototransistor; it provides enough current to turn on the second transistor which can present a LO to a TTL gate.

Parallel port units, particularly those for industrial environments, often employ an optoisolator on each bit of I/O. This protects the computer both from noise and possible catastrophic events in the external device.

BIBLIOGRAPHY

P. Horowitz and W. Hill, *The Art of Electronics*, Cambridge University Press, Cambridge, 1980.

H. V. Malmstadt, C. G. Enke, and S. R. Crouch, *Electronics and Instrumentation for Scientists*, Benjamin/Cummings, Menlo Park, CA, 1981.

CHAPTER SEVEN

Transducers: Temperature, Light, Electrochemical, and Electrical Power

The electronic hardware of the computer interface has been discussed. The task remains to pass information between the chemical or physical domains of the scientific experiment and the electronic and/or digital domains (Fig. 7.1). These conversions are accomplished by a wide variety of **transducers**. Although it is probably possible to obtain or devise a transducer for every observable phenomenon, to cover all of them would be impossible. In this and the following chapter, we will focus our attention on the most common transducers used for general instrumentation. However, since the range of experimental subjects for scientific inquiry is nearly infinite, we must keep in mind that many other transducers are required for highly specialized activities.

7.1. TERMINOLOGY OF TRANSDUCERS

According to a standard of the Instrument Society of America (Standard S37.1, 1969), all transducers "provide a usable output in response to a specified measurand, . . . a physical quantity, property, or condition which is measured." In most cases, the transducer is an *input* device to an instrument. Therefore, the measurand lies in the physical or chemical domains, and the output is electrical. Other devices that convert information from the computer's domain *to* the physical or chemical domain also can fit within an expanded definition of a transducer.

An input transducer often can be divided into two components representing its dual function. The physical or chemical property (or measurand)

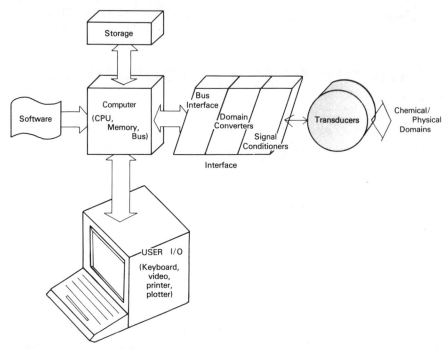

FIGURE 7.1. A laboratory computer system. Typical transducers are discussed in this and the following chapter.

acts on a **sensing element** which in turn acts on a second component, the **transducing element**, which generates the signal. These elements may not necessarily be physically separate components, but in concept the division is useful. The sensing element detects the phenomenon under study while the transducing element converts to an electronically useful value, voltage, or current. For example, in one method of measuring pressure (Fig. 7.2), a diaphragm may be used to *sense* the atmospheric pressure because it expands and contracts in response to changes in pressure thereby producing a displacement; the transducing element converts displacement to an electrical signal.

Transducers may be classified by their modes of operation as being either active or passive, represented schematically in Figure 7.3. An **active transducer** generates a signal as a self-contained energy source without an external electric power supply; for example, a dc generator, used to measure rotational velocity, develops a voltage by using the rotational energy as the energy source. An excitation voltage or current sources must be provided to a **passive transducer**, which acts on it to produce the output signal. The precision of the output is often, but not always, dependent on the stability of that power supply.

If the output is independent of supply voltage, we gain the advantage of

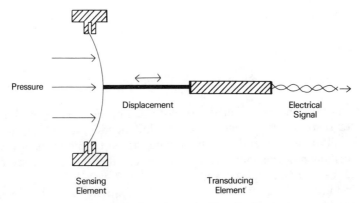

Pressure

Displacement

Electrical
Signal

Sensing
Element

Transducing
Element

FIGURE 7.2. Sensing and transducing elements. The diaphragm is an example of a sensing element that converts pressure to displacement. The transducing element converts displacement to an electrical signal.

having one less parameter to be controlled. On the other hand, the *necessity* to control this parameter also means *being able* to control it. For example, the magnitude of a signal from a photomultiplier tube light sensor is highly dependent on the supply voltage which must be carefully stabilized; however, the supply voltage can be used to control the range of this detector.

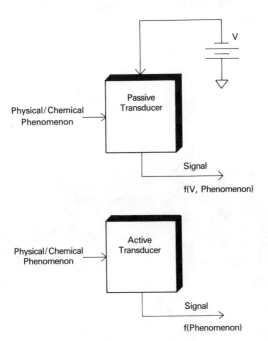

V

Physical/Chemical
Phenomenon

Passive
Transducer

Signal

f(V, Phenomenon)

Physical/Chemical
Phenomenon

Active
Transducer

Signal

f(Phenomenon)

FIGURE 7.3. Passive and active transducers. A passive transducer requires a power supply; the output depends on the phenomenon and usually on the magnitude of the power supply. The active transducer is independent of any external power supply.

The **transfer function** of a transducer is the relationship between measurand and the electrical output. Usually, it can be described by a theoretical curve, and ideally, the transfer function should be linear. Alternatively, the relationship should be linearizable via a known mathematical transformation such as a logarithm or an inversion. However, approximations are often required.

When measurements are made to plot a calibration curve (a plot of the transducer response as a function of the independent variable), all of the results ideally will lie exactly on the line described by the transfer function. The performance of the transducer is often described by the characteristics of that data set. Examples are shown in Figure 7.4.

The first graph differentiates low and high **sensitivity**; when the sensitivity is low, a small change in the response indicates a large change in the in-

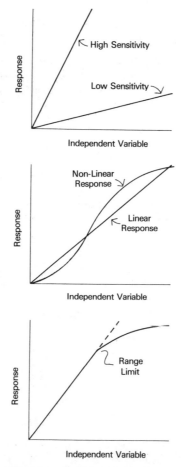

FIGURE 7.4. Limits to sensitivity, linearity, and linear range.

dependent variable. In practice, measurement uncertainty (noise) can obscure small changes in the independent variable; the quality of measurements is limited by scatter around the line making low-level measurand values difficult to determine.

In the second case, the response is nonlinear: A straight line cannot be fit correctly to the data by the method of least squares, since the data weaves back and forth around a straight line. The **linearity** of the transfer function is limited, so that (1) the sensitivity varies with the response, and (2) a computed transfer function, often an approximate, must be used.

In the last case, a straight line can be fit to the data at low values but not at high values. The response illustrates limited **range**. A limit can almost always be found at the transducer's high and low ends.

In addition, the speed or **response time** of the transducer may be in question. For some transducers, the response time is more important than the absolute response. For example, if vibration is measured, the transducer output must follow the measurand input (acceleration) very rapidly.

The discussion of transducers could be organized according to the technology of each type of transducer, and that would be suitable if an understanding of the technology itself were paramount. However, since the solution to an acquisition problem is the objective, the main headings will be the categories of information to be transferred. In places, cross-references are required since the transducing element of one type of transducer is often the transducer of another phenomenon.

7.2. TEMPERATURE

Temperature is a component of nearly every physical measurement. In calorimetric and heat transfer experiments as well as in live animal care, the experimental results are based on small changes in temperature. Where the temperature has significant effect on the experiment, as in chemical kinetics, it must be measured with precision. Often the measurement leads to some degree of control; the experiment is performed either at a series of temperatures or at some standard temperature. The ubiquitous laboratory water bath and temperature controller provide a means for maintaining *constant* temperature.

When a temperature-dependent series of experiments is performed, the typical approach is to set the temperature precisely and to use the precision control capability of the bath to maintain the preset temperature. An alternative approach allows the experiment to proceed at a roughly controlled temperature. Then the temperature is *dynamically* measured before, during, and after the experiment. Given the computational capabilities that the computer provides, extremely careful *control* of the temperature might then be replaced by careful measurement of temperature with subsequent inclusion of the actual temperature in following computations. This, of course, implies that the temperature dependence of the phenomenon is accurately known. (Many scientists also will need to overcome the compulsion to plot experimental results only as a function of round-numbered values: for example, 10.0°, 20.0°, . . . , and 80.0° rather than 8.83°, 19.13°, . . . , and 81.22°.)

In this section, four major and several minor temperature transducers will be considered.

7.2.1. Thermocouples

A thermocouple is produced by welding wires of dissimilar metals together. Whenever two dissimilar metals make contact, a voltage is produced as a result of the thermoelectric effect, making this an active transducer. The generation of this voltage takes places in most electrical circuits, but since a measurement cannot be made without a complete circuit (Fig. 7.5), there must be more than one such junction of dissimilar metals; if all are at the same temperature, the sum of all the potentials will be zero. However, a *difference* in temperatures at the two junctions will produce a net voltage. For example, if the two copper leads of a voltage measuring device are connected to a constantan wire at uniform temperature, equal opposing voltages will develop, but a net signal will result if the two junctions differ in temperature.

Three features of a thermocouple explain its usefulness: (1) it is a physically rugged transducer; (2) various thermocouples can be used over a wide temperature range; (3) in contrast with resistive temperature transducers (described later), it does not produce heat itself. The temperature–voltage

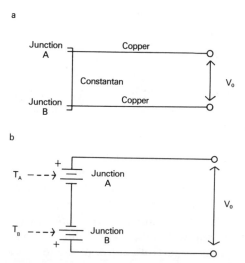

FIGURE 7.5. A thermocouple. (*a*) A voltage develops at each junction of dissimilar metals. (*b*) An equivalent circuit for a thermocouple. If $T_A = T_B$, then $V_0 = 0$.

characteristics of a thermocouple depend on the chosen pair of metals; all exhibit a nonlinear transfer function, but the regions of greatest curvature vary. The response curve of a useful thermocouple should have a large slope for high sensitivity in the applicable temperature region; some metal junctions are most sensitive below −200°C and others up near 2000°C. Thermocouple responses are provided in Figure 7.6 for two common thermocouples, iron–constantan (type *J*) and tungsten–tungsten-26% rhenium (type *G*). Tables of thermocouple response with subdegree resolution are available both from manufacturers and in National Bureau of Standards tables; with these tables, the temperature corresponding to an observed voltage can be interpolated with good precision.

If a voltage proportional to temperature is required, an analog linearizing circuit could be devised, and some are available. However, where the thermocouple is used with a computer, the standard table of voltages at a resolution as great as 30–40°C can be placed in software, and the processing capability of the computer can take responsibility for the interpolation. Nonlinear interpolation, using a simple fit of an equation to the data table, is a trivial burden for the computer.

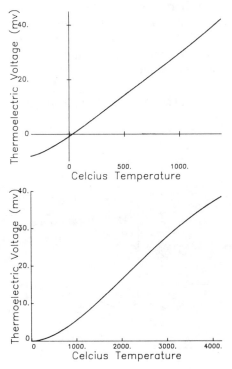

FIGURE 7.6. Thermocouple response as a function of temperature. Top, iron–constantan (type *J*). Bottom, tungsten–tungsten-26% rhenium (type *G*). The reference junction is at 25°C.

The key to an *accurate* measurement is control of the temperature at the reference junction. This point is called the **cold junction**, although it is not necessarily colder than the other. Since a thermocouple temperature measurement is inherently differential, the drift in the reference temperature must be small compared with the uncertainty permitted in the overall measurement. For example, where accuracy to within 5°C may be tolerable in a furnace at 1500°C, the minor variations that occur in a reference junction remaining at ambient room temperature are usually permissible; on the other hand, precision measurements to within a small fraction of a degree require equally precise control of the cold junction temperature.

Where the greater accuracy is required, either the reference junction temperature can be precisely maintained (in an ice bath or a thermostatic oven), or the variations in the cold junction voltage can be compensated with a separate temperature transducer.

With the computer as the readout device, two options for compensation exist. First, the cold junction temperature can be independently measured with a separate sensor which makes an absolute measurement, for example, a thermistor or semiconductor transducer. These sensors, described later, are more precise than the thermocouple, although they do not have the temperature range of a thermocouple. The look-up tables can be consulted to correct the reference junction to 0°C. Alternatively, compensation can be applied directly. In Figure 7.7, a semiconductor transducer (which passes a current proportional to temperature) compensates for variations in the reference junction voltage which is assumed linear over a small (~20°C) range.

The thermocouple signal is quite small, typically under 20 mV. Therefore,

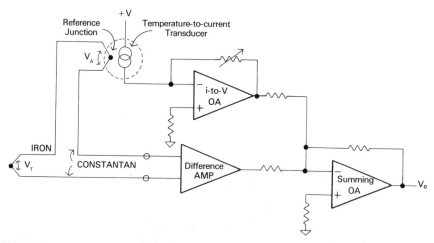

FIGURE 7.7. A thermocouple circuit with a compensated reference (cold) junction. A second transducer with an absolute output measures the reference junction temperature to compensate reference junction temperature changes.

care is required to protect the leads from picking up noise before the analog-to-digital conversion. Differential instrumentation amplifiers with gain can reject the common mode noise picked up in the leads and should be used unless the leads can be kept very short.

When thermocouple wire is purchased, it is usually two conductor wires of dissimilar metal. The wires are welded together at one end. The other end, near the amplifiers, is where the cold junction is produced. Thermocouples also are produced in dozens of probe configurations, from large probes for mounting into high-pressure pipes to miniature devices mounted in hypodermic needles for in vivo animal measurements.

Thermopiles are multiple thermocouples. In order to obtain more sensitivity, thermopiles are connected in series–like batteries; the sensitivity is proportional to the number of thermocouple junctions. The sensing junctions form a bundle which is in contact with the measurand. Alternate junctions are cold junctions which are bundled away from the measurand, and the same cold-junction corrections are required.

7.2.2. Platinum Resistance Temperature Detectors (RTDs)

The second group of temperature transducers utilize the phenomenon that resistance in any conductor is a function of temperature. A transducer based on this principle is a **resistance thermometer**. Although such transducers have been constructed with several materials, the only metal of importance is platinum.

Between standard reference points at $-182.97°C$ (the boiling point of oxygen) and $630.5°C$ (the freezing point of antimony), the resistance–temperature transfer function is defined by the Callendar–Van Dusen equation

$$\frac{R_T}{R_0} = 1 + a\left| T + \delta\left(\frac{T}{100} - 1\right)\left(\frac{T}{100}\right) - \beta\left(\frac{T}{100} - 1\right)\left(\frac{T}{100}\right)^2 \right|$$

R_T and R_0 are the resistances at the temperatures T and $0°C$, respectively, and a, δ, and β are constants which are characteristic of each unit. In practice, the data can be fit by a quadratic or cubic equation with very good results. Over a range that is approximately linear ($\delta \cong 0$, $\beta \cong 0$), a typical value for a is about $0.4/°C$.

Values of R_0 for popular RTDs are about $120\ \Omega$ although units are available over a wide range of resistances. Coils of fine platinum wire are often used, and in order to protect them and give mechanical and structural stability, they are usually sealed in glass or ceramic. For surface measurements, RTDs are die cut from platinum foil. For fast time-response, a length of conductor is produced by the vapor deposition of platinum on glass.

The great advantage of the RTD is exceptional stability, even at high temperatures; the stability follows from the properties of platinum, a very stable metal. Units that have been monitored and compared with standards at the Bureau of Standards have shown almost no change over 30 yr.

The RTD is a passive transducer. When a constant current source provides excitation, the voltage drop across the RTD is, according to Ohm's Law, proportional to the resistance. The precision of the measurement is directly dependent on the precision of the constant current supply. The sensitivity of the measurement is increased by raising the excitation current.

Retaining accuracy at high excitation current may become a problem in some environments due to **self-heating**, since at higher currents, the RTD itself becomes a heater. The RTD must dissipate the Joule heat that it produces according to $P = i^2 R$ where P is the heat in watts, i is the excitation current, and R is the RTD resistance. For a typical 100-Ω RTD to produce a 1-V signal, 1000 mW of heat are produced.

To characterize the problem, the ability of the medium to conduct the self-generated heat from a resistive transducer must be considered. If the thermometer is configured as a probe in a large volume of moving liquid, the heat is easily dissipated and the error is small, but when measuring the temperature of still air, which is a good insulator, the current must be kept as low as possible. RTD vendors typically provide specifications for self-heating in still air and in air moving at 1 m/s. Typical values are 0.2 and 0.07° increase in temperature per milliwatt generated, respectively.

A simple circuit is shown in Figure 7.8 for a 100-Ω RTD using an OA to provide the constant current source and voltage readout. In both amplifiers, current boosters are required to excite a low-resistance transducer. Amplifier A1 buffers the voltage reference; conversion of resistance to voltage takes place in A2. A third amplifier may be required to add an offset and match the range to that of the ADC circuit.

Self-heating can be controlled by applying current only while the measurement is being made. Before the temperature datum is to be acquired, the A1 output to the RTD is activated. When the datum has been acquired, the input to the RTD is turned off again. Also, use of bridge circuits (Chapter

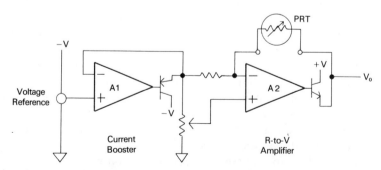

FIGURE 7.8. An amplifier circuit for a platinum resistance temperature detector. The resistance of the detector is only 100 Ω, so current boosters are required.

Five) increases sensitivity so that a lower excitation voltage might be possible.

7.2.3. Thermistors

Thermistors (*Therm*ally sensitive re*sistors*) also are passive resistive transducers, but they exhibit a much larger coefficient of resistance than do the platinum RTDs. Unlike Pt-RTDs, common thermistor sensors have a negative temperature coefficient (NTC). That is, when the temperature goes up, the resistance goes down. The useful range is from about $-80°C$ to about $150°C$.

NTC thermistors are made from semiconductor mixtures of iron, chromium, manganese, cobalt, and nickel oxides. As illustrated in Figure 7.9, they are typically sealed in glass probes.

As for other transducers mentioned thus far, the transfer function is nonlinear. The relationship of resistance, R_T at temperature T, to the resistance, R_0 at a reference temperature T_0, holds over any small temperature range:

$$R_T = R_0 \exp[\beta(1/T - 1/T_0)]$$

where β is a constant characteristic of the NTC thermistor material. Unfortunately, the constant β itself increases somewhat with temperature. The following empirical equation is more useful, fitting with high accuracy typically within $0.0015°C$ over the range $0-100°C$:

$$R_T = \exp[A_0 + A_1/T + A_2/T^2 + A_3/T^3]$$

A typical $\dot{R}-T$ curve is shown in Figure 7.10; the curve was expanded in the lower part of the figure in order to clearly show the response in the room temperature region.

A thermistor is made of a mixture of metallic oxides sintered into beads or disks and encapsulated in epoxy, glass, or mounted in probes. By control of the oxide mixture and thickness, resistances between $100\ \Omega$ and $450\ k\Omega$ are produced for a wide variety of applications. The reproducibility in manufacture is quite good so that interchangeable NTC thermistors cost only slightly more than the unselected variety. The time response of standard devices is on the order of a few seconds, but smaller, nonencapsulated "flakes" can respond in microseconds.

Linearized thermistors are networks consisting of a pair of NTC thermistors which, with the correct external resistors, exhibit linear response (Fig. 7.11). The linear response follows from the correct choice of thermistor

FIGURE 7.9. A typical thermistor package. The thermistor element is sealed in a glass probe.

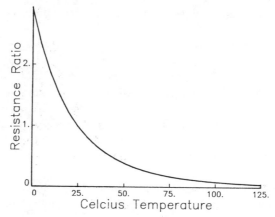

FIGURE 7.10. A resistance-temperature curve for a negative temperature coefficient (NTC) thermistor.

resistances which causes nonlinear terms to drop out of the series expansion of the transfer function. These devices can be quite useful in "nonintelligent" (nonmicroprocessor-based) instrumentation, but if a computer is already in use, the software can provide precision linearization at no additional cost.

Thermistors are passive transducers, applied in much the same way as

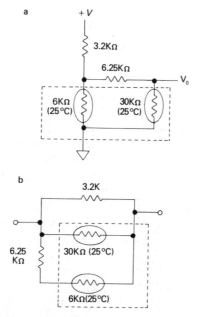

FIGURE 7.11. Linearized thermistors. (a) Configured for a linear output voltage. (b) Configured for linear resistance.

Pt–RTDs. A well-regulated excitation voltage, V_R, is required. However, since higher resistance units can be used, typically about 5 kΩ, the current is much smaller; to obtain a 1-V signal from a 5-kΩ thermistor, the current is only 200 μA, and only 200 μW of heat are produced. Therefore, the OA circuits require no booster transistors, and self-heating, described for RTDs, is less of a problem although still significant.

Two possible OA configurations are illustrated in Figure 7.12, showing the thermistor in two different positions. If the thermistor is placed in the feedback loop, the voltage output is linear with thermistor resistance; if the thermistor is the input resistor, the output is proportional to the reciprocal of the resistance. Visual examination of Figure 7.9 suggests that the thermistor transfer function resembles an inverse temperature relationship. Consequently, one can expect the output to be closer to being linear if the thermistor is the input resistor as shown in the lower part of Figure 7.12. As a result, the resolution of the ADC will be close to constant over the entire range. Even so, further linearization in software is usually required.

Bridge circuits, discussed in Chapter 5, are quite useful for the measurement of small *differences* between resistances. A bridge in which one arm is a thermistor compares the thermistor resistance with a fixed resistor. If matched thermistors are used in adjacent arms, as in the bridge circuit of Figure 7.13, the arrangement is the basis of a very sensitive differential thermometer.

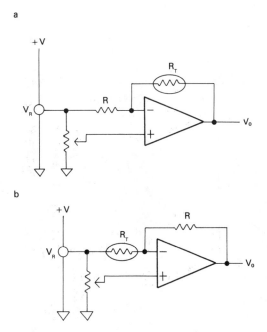

FIGURE 7.12. Operational amplifier circuits for thermistors. (*a*) The output is roughly linear with temperature^{-1}. (*b*) The output will be roughly linear with temperature.

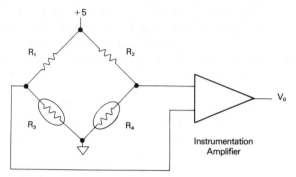

FIGURE 7.13. Bridge configuration for a *differential* measurement. V_0 is proportional to the difference between R_3 and R_4.

So far, only NTC thermistors have been described. However, certain barium and strontium titanate semiconductors exhibit a large positive temperature coefficient (PTC) over a narrow range; within that anomalous range, PTCs of 50%/°C to 200%/°C have been observed! The shape of a typical curve is shown in Figure 7.14. Since the temperature-resistance relationship is triple-valued over a wide temperature range, the value of a PTC is restricted to the narrow PTC range.

7.2.4. Semiconductor Transducers

Some of the most versatile temperature transducers are **semiconductor sensors**. Two useful types are the silicon semiconductor junction and an integrated circuit.

A simple silicon diode can provide the semiconductor junction; if the current through the diode is kept constant, the potential across the junction

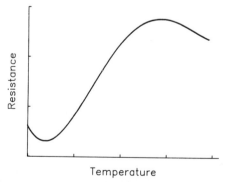

FIGURE 7.14. A resistance–temperature curve for a positive temperature coefficient thermistor. Thermistors are manufactured with the anomalous region at different temperatures.

varies at about 2.2 mV/°C. Therefore, any diode can be used with a constant current source as a temperature transducer.

The **IC transducer** is based on the elegant circuit shown in Figure 7.15. The basis of operation follows: Q3 and Q4 are identical transistors with the same voltage on both bases and at both emitters; therefore, they must pass identical currents. The base-to-emitter voltages of transistors Q1 and Q2 are proportional to the Kelvin temperature times the log of their currents, but Q2 actually consists of eight parallel transistors each carrying $\frac{1}{8}$ the current of Q1. Therefore, the difference in voltages at the emitters of Q1 and Q2, V_T, is

$$V_T = \frac{kT}{q}(\ln I_1 - \ln I_2/8)$$

where k and q are constants. Since $I_1 = I_2$, V_T is proportional to T. By careful adjustment of the resistance, the proportionality between the resultant current and temperature can be made 1 μA/°C.

Equally important, the current is independent of the excitation voltage. Since the transducer is a constant current device, additional resistance in series should have no effect. Long leads may be used without fear that lead resistance will cause error. The excitation voltage must be at least 4 V but should be kept as low as possible to minimize self-heating. Since the measurement is not bothered by series resistances, low-cost analog switches can be used to multiplex the transducers. A large number of transducers can be multiplexed when arranged in a two-dimensional array as demonstrated in

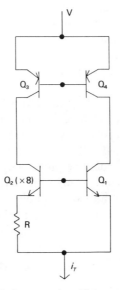

FIGURE 7.15. A simplified circuit diagram of an IC temperature transducer. The current, i_T, is proportional to Kelvin temperature.

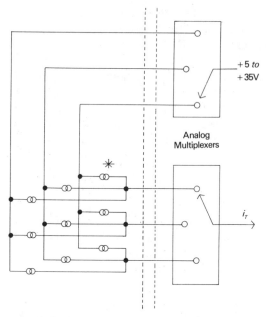

FIGURE 7.16. A matrix of IC transducers. Because the output is independent of the resistances of the switches, i_T depends solely on the temperature of the selected transducer (marked with an asterisk).

Figure 7.16; the excitation voltage is applied in one dimension, and the second dimension is used to select the transducer. The **ON** resistance contributed by the switch has no effect on the signal as long as the voltage drop across the transducer is at least 4 V.

By keeping the current low, self-heating of the temperature transducer is minimized. A maximum excitation voltage of 35 V is specified; however, the heat developed at 300 K would then be iV, $300 \times 10^{-6}A \times 35\text{ V} \cong 0.01$ W. Reducing the excitation voltage around 5 V would make the heat proportionately lower. The IC transducer's main limitations are the temperature range (-55–$+150°C$) and a relatively slow response time, about 1.2 s when coupled to an aluminum block.

The semiconductor transducer requires only a single OA, configured as a current-to-voltage converter, to produce a voltage suitable for an ADC. An offset current should be added to obtain a useful range. As illustrated in Figure 7.17, when -273.2 μA is added at the summing point, the signal is converted from Kelvin temperature to the Celcius scale.

7.2.5. Miscellaneous Thermometers

There are many other phenomena that are used to sense temperature for special applications. A few comments are added about several of them.

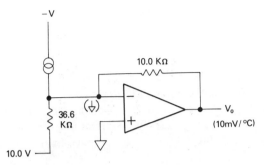

FIGURE 7.17. An IC temperature transducer configuration in which the output is proportional to Celcius temperature. The fixed offset current (10/36.6 K = 273 μA) converts Kelvin to Celcius temperature.

At cryogenic temperatures, **carbon resistors** are often used; the resistance increases dramatically as the temperature decreases below 4° K.

Germanium semiconductor resistors are very sensitive and provide reproducible response in the cryogenic region; they are particularly useful for temperature measurements between 2° K and 5° K which are otherwise difficult to make. Between 2° K and 40° K, the resistance of a typical specimen varies from 12 kΩ–4 Ω.

Moving to the other temperature extreme, **Radiation Pyrometers** make use of the characteristics of black-body radiation from solids. According to Planck's radiation law, both the wavelength of maximum emission and the overall intensity depend on the temperature. Several pyrometers have been developed which focus the radiation on thermopile detectors, and the output can be calibrated in degrees Kelvin up to about 4100° K. That thermopile output also can be made available for computer interfacing. The accuracy of optical pyrometers is limited by deviations to Planck's law due to internal reflection in the surface whose temperature is being measured. An empirical correction factor is required.

A pyrometer for plasma study has two fundamental improvements. First, a blackened, opaque screen is moved through the plasma modulating the radiating volume. Second, the measurement is made at two wavelengths. The ratio of intensities is compared with theoretical values. Where the electronic components of the measurement must be located far from the phenomenon being measured, the optical pyrometer can be coupled using fiber optics.

Expansion of various materials can be used to *indicate* temperature, and with some difficulty, to *transduce* temperature. The expansion of a metal, liquid, or gas converts temperature to position, a very suitable quantity to use for visual detection of temperature; however, the position must still be transduced to an electrical quantity. A few applications of mechanical thermometers interfaced to computers are still significant. For example, the **bimetallic strip** bends as a consequence of the difference in expansion coefficients of two metals bonded together. Such a coil could be used to position a potentiometer or other angular position detector for full analog detection of temperature.

Thermostats are *switches* which conduct on one side of a temperature set point but do not conduct on the other side. Some hysteresis is often desirable so that the switch does not oscillate at the set point. Most people are familiar with the bimetallic strip employed as a thermostat: The strip is positioned to activate a switch. Although adjustable bimetallic thermostats have limited precision, fixed bimetallic "snap switches" are available with a precision of <0.1°C.

Another type of thermostat uses a magnetically operated reed switch. A ferrite cylinder is formed around the switches but in contact with permanent magnets. The permeability of the ferrite changes dramatically at the "curie temperature" causing the switch to change state. The curie temperature is controlled by the composition of the ferrite.

7.2.6. Temperature Control

In most experiments involving temperature control, that control is provided by refrigeration and/or heating units, often with the capability of circulating fluids to the needed site. The exceptional stability of some of these units is derived in part from the inertia of a large thermal mass that makes temperature change a rather slow process. In most cases, the computer can enter the control loop of those devices by switching them on and off using techniques that will be described later in the chapter; solid-state relays controlled by logic·level voltages are easily introduced into the power circuit of a simple resistive heater.

The **Peltier-effect thermoelectric device** provides a solid-state alternative for maintaining the control of the temperature of small thermal loads, acting as a heat pump when electrical current is passed through it.

The Peltier-effect device can be described as an "inverse thermopile." The thermopile develops electrical current in response to a temperature differential between the hot and cold junctions, but the Peltier-effect device develops a temperature differential between the hot and cold junctions in response to electrical current flowing through it. Having no mechanical parts, they are quiet, small, easily mounted, and readily controlled.

Such a device consists of *p*- and *n*- semiconductors joined in a series by conductors as shown in Figure 7.18. When current flows, typically 0.1 to 5 V, heat is pumped from cold to hot junctions, resulting in a cooling of the cold junction if the heat is removed from the hot junction. Reversal of the current direction will reverse the heat flow direction and results in the simple conversion of a cooler to a heater.

Several variables control the temperature and heat capacity of a Peltier-effect device. As the figure illustrates, a large number of units of *p*- and *n*-material can be mounted physically in parallel, but electrically in series, to increase the number of junctions and the quantity of heat that can be ab-

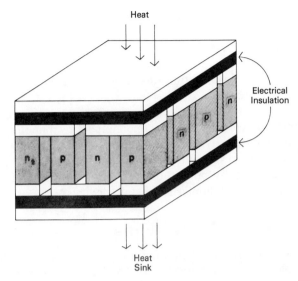

FIGURE 7.18. A single-stage Peltier-effect cooler. Current flow through the semiconductor material pumps heat from one side to the other.

sorbed. At steady state, the temperature differential is determined by the semiconductor material that is chosen and the magnitude of current passed through the junctions. The overall temperature differential can be increased by cascading additional devices in multiple stages where the hot side of one stage is placed in contact with the cold side of the next stage.

Peltier-effect devices, unfortunately, are not efficient as coolers. When operated in that mode, the hot junction is warmed both by heat pumped from the cold junction and by heat generated by Joule heating (heat derived by passing current through the resistance of the device); this heat must be removed from the hot junction. The coefficient of performance relates the heat pumped from the cold junction to that produced by Joule heating. Commercial units exhibit coefficients of about 0.1–0.2 for single stage units and 0.002 for units with 4–6 stages.

Peltier-effect devices are typically applied in applications such as the temperature control of imaging device-integrated circuits, microscope stages, and sample holders: wherever localized low-capacity cooling is required. Since they can be easily controlled electronically, they can give rapid response when used in a closed loop with a thermistor or other temperature sensor.

7.3. UV AND VISIBLE RADIATION DETECTION

The subject of radiation detection encompasses the entire range from gamma rays to radio waves. However, if we consider the scope of workers who do not specialize in a specific area of the spectrum, some limits can be drawn.

The detection of X rays, far infrared, and microwaves, for example, are specialized areas, but measurements in the ultraviolet and visible (UV–vis) portions of the spectrum are common to many laboratories and experiments. Consequently, the transducers that correspond to the latter form the major thrust of this section. Even within this region, the study will be restricted to *quantitative* detectors.

UV–vis detection systems fall into two broad categories: **photoemissive** detectors, which can eject an electron upon impact by a photon of sufficient energy, and **photoconductive** devices, chiefly semiconductors, which respond to the absorption of light by increasing the carrier population. The former are represented by phototubes and photomultiplier tubes, and the latter by photodiodes and similar solid-state devices.

7.3.1. Photoemissive Detectors

The emission of an electron from a photoemissive surface is characterized by Einstein's Nobel Prize-winning work. An electron can escape the surface if the energy of the incident photon exceeds the work function. An *ideal* surface would emit one electron for every photon of sufficient energy. By applying a negative potential between that surface (the photocathode) and a second electrode (the anode), resultant electrons can be collected as an electrical current.

Phototubes are the simplest devices based on the photoemissive principle (Fig. 7.19). Within an evacuated tube, a half-cylindrical photocathode surface faces the light input, and photoelectrons result from the interaction of photons of sufficient energy with the surface. Within that cylinder, an anode wire, at a very positive voltage relative to the photocathode, collects the emitted electrons. The photocathode is held about 90-V negative of the anode so that the emitted electrons are collected with high efficiency. Typical photocurrents are about 10^{-11} A.

No photocathode possesses that ideal characteristic of conversion of photons to electrons with 100% efficiency; the best surfaces are those of alkali metals and mixtures or compounds of those metals with antimony or silver. Their efficiencies approach 30% over a narrow wavelength range. The relative responsivity for several photocathode materials is plotted in Figure 7.20; the range of usefulness covers the ultraviolet, visible, and in some cases, the very near infrared. A common example is the "trialkali" $KNa_2Sb:Cs$, usually denoted as having an S-20 response; the ":Cs" in the formula indicates small quantities of cesium. Also shown are ERMA (Extended Red MultiAlkali) responses, achieved at the expense of UV response by making the photocathode surface thicker.

Photoanodic current can be measured by simply measuring the voltage across a dropping resistor (Fig. 7.21). However, the high sensitivity generated by a large resistor produces nonlinearity since as the voltage across the resistor increases, the voltage across the phototube must decrease. That

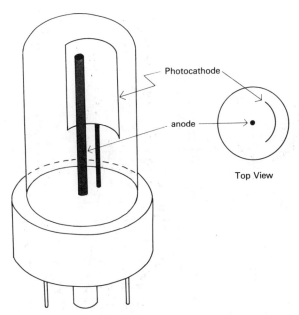

FIGURE 7.19. A phototube. Photons incident on the photocathode eject electrons which are collected at the anode, and the resultant current is measured.

problem is solved with an OA current-to-voltage converter, since the inverting input is maintained at virtual ground potential.

Photomultiplier tubes (PMTs) provide a large improvement in sensitivity. As in the phototube, photoelectrons are ejected from the photocathode of the PMT. However, 9–14 intermediate stages, called **dynodes**, are added between the photocathode and the anode, each held at a slightly more positive voltage than the former. Each surface is capable of secondary emission, the emission of two or more electrons as a consequence of a single electron's impact. A cascading effect is produced yielding a large current gain of up to 10^8 photoelectrons per photon.

For most applications, the voltages on the individual dynodes are controlled by tapping a simple voltage divider (Fig. 7.22) across a supply voltage between -600 and about $-1200--2000$ V.

The choice of resistor size in the divider chain is important. First, the current through the chain must be at least ten times the photocurrent or the dynode chain will act as another set of conductors in parallel with the resistors; then the relative voltages on the dynodes will change, resulting in nonlinear response. Second, the resistance, and therefore the voltage between the photocathode and the first dynode, should be several times that of the other stages as illustrated in Figure 7.22; this is because the collection efficiency of the first stage is more critical than that of the remaining stages.

When the incident light takes the form of discrete, separated pulses rather

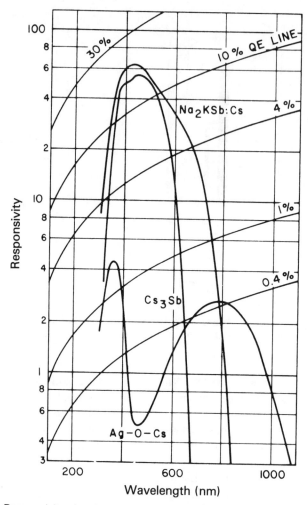

FIGURE 7.20. Responsivity of various photocathode materials as a function of wavelength. (Adapted from R.W. Engstrom, *Photomultiplier Handbook*, 1980, by permission of RCA Corporation, New Products Division, Lancaster, PA.)

than constant intensity, the *instantaneous* current may be several orders of magnitude greater than the *average* current. To accommodate the current peak, capacitors placed on the last few stages as charge storage devices, aid in maintaining constant dynode voltages.

The gain is strongly dependent on the supply voltage. A typical gain curve is shown in Figure 7.23; the gain increases exponentially with the supply voltage. The first consequence is that the supply voltage must be extremely stable, to at least 0.01% of full scale. Secondly, the gain can be easily controlled. Most PMT power supplies include inputs which accept a programming voltage in the 0–15-V range, and generate a proportionate high voltage.

The input voltage is within the capability of a computer's DAC.

By placing the computer within a complete feedback loop, as illustrated in Figure 7.24, the gain can be programmed to the needs of the measurement. If the photocurrent is too high, the DAC can be used to lower the gain. Typically, if high stability is required, the photocurrent should be limited to roughly 1 μA, or up to 100 μA for very short periods. By closing this feedback loop, gain control can be used to limit the current: the PMT is protected, and the signal is kept within the range of the ADC.

The designer must choose between many different PMT configurations. Most fall into two categories. The first and most common is the "side-on" tube (Fig. 7.25a). Tho photon passes through a transparent envelope and strikes an opaque photocathode; the dynodes are usually arranged in the "squirrel-cage" pattern shown. The second variety, "end-on" tubes, have a large semitransparent photocathode, up to several inches in diameter. Being very efficient in photon collection, they are more sensitive than the

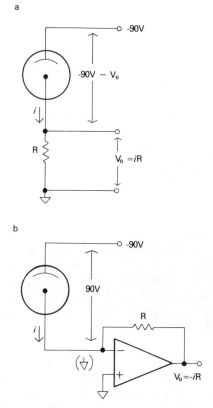

FIGURE 7.21. Conversion of photoanodic current to voltage. (*a*) Use of a dropping resistor causes the voltage across the tube to decrease by V_0. (*b*) When a current-to-voltage amplifier is used, the anode is always at ground potential.

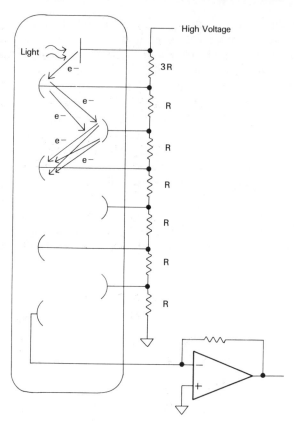

FIGURE 7.22. Gain stages of a photomultiplier tube. Light strikes the photocathode, and the resultant electrons cascade down the dynode chain. The voltage divider holds each dynode more positive than the last.

side-on variety. The tube shown in Figure 7.25*b* is designed for very high-speed operation as well.

The photocurrent, as we have seen, is then the product of several factors, the most important of which are the light intensity (incident photons per second), the quantum efficiency (photo-electrons per photon), and the PMT's gain (electrons per photoelectron). Other factors play only a small part; for example, the collection efficiencies of the dynodes are near unity and usually constant.

Response Time. The multiplication of a single photoelectron into a packet of electrons occupies a finite amount of time (the transit time). During that time, the resultant electrons travel varied distances and experience different electrical fields. Consequently, the group of electrons, which result from a single photon, can be expected to arrive at the anode spread over a finite time period (the response time).

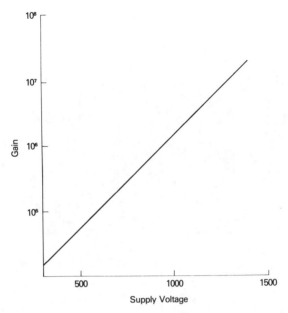

FIGURE 7.23. Gain of a typical PMT as a function of supply voltage. (Adapted from R.W. Engstrom, *Photomultiplier Handbook*, 1980, by permission of RCA Corporation, New Products Division, Lancaster, PA.)

The variation in the response time is determined both by the variance in the distances traveled by electrons and by the distribution in the initial velocity of photoelectrons ejected from the photocathode. The latter may be reduced by increasing the electric field between the photocathode and the first dynode. In an end-on tube, the distance spread for the various paths that a photoelectron may travel can contribute up to 10 ns to the response time, although the speed can be improved somewhat by using a curved faceplate and focusing electrodes so that all photoelectrons travel the same distance to the first dynode (see Fig. 7.25*b*).

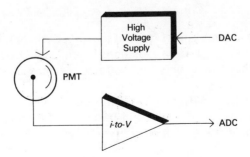

FIGURE 7.24. Closed-loop control of a PMT. A computer-controlled DAC sets the gain voltage on a PMT after the current is read by the computer using an ADC.

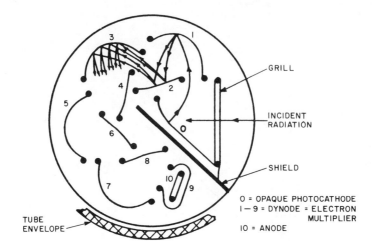

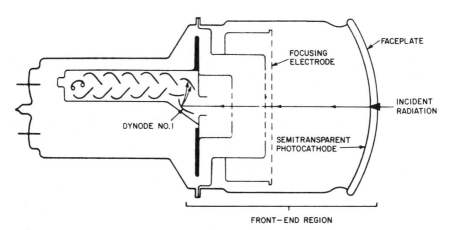

FIGURE 7.25. Two types of PMTs. Top, a side-on tube, top view. Bottom, an end-on tube. (Reproduced from R.W. Engstrom, *Photomultiplier Handbook*, 1980, by permission of RCA Corporation, New Products Division, Lancaster, PA.)

Dark current is current that is part of the anode current, but is generated even in the absence of incident radiation. It originates from three sources: ohmic leakage, thermionic emission, and regenerative effects.

1. Ohmic leakage is current flowing through impurities in the insulators within and without the tube and base; it becomes appreciable under unfavorable (dirty or high humidity) environmental conditions. Of the dark current sources, it predominates at low voltages and is proportional to the applied voltage.

2. Thermionic emission occurs primarily at the photocathode where a few electrons are randomly and spontaneously emitted without incident light. This emission, and the resultant current, can be reduced by cooling the tube.

3. Regenerative effects are produced only at high voltages and result from photons generated within the tube. For example, at very high current and gain, the dynodes will glow under electron bombardment; that light may be scattered back to the photocathode, regenerating photoelectrons. Similarly, stray electrons that strike the glass may produce fluorescent light.

Dark current increases dramatically after the PMT has been abused by excessive exposure to ambient light. When the gain voltage is on, overexposure results in large currents which can damage the dynodes. Even without applied voltage, a PMT can be affected by strong radiation, particularly in the blue and ultraviolet end of the spectrum. If the intensity is not large and the exposure is short, there may be recovery and the dark current will decrease over a period of hours to days.

Noise characteristics of PMTs must be understood in order to make measurements at the lowest possible light levels. It is possible to characterize the noise sources, and thereby understand how they can be removed or when the fundamental limits have been reached.

The first noise source is **shot noise**. Photocurrent is the result of photons that arrive at the photocathode as discrete events, and the probability of a photon arriving at any given time is described by the Poisson distribution. This probability sets the fundamental limit to the SNR of a light detector: the standard deviation, σ_P, in the number of photons, n_P, arriving in any given period is the square root of n_P. The maximum SNR is therefore proportional to the square root of the number of photons

$$\text{SNR} \propto n_P/\sigma_P = \sqrt{n_P}$$

It is important to keep in mind that this relationship is valid even though the discrete photons are measured as a current. (Interestingly, the first large-volume application of the PMT was as a noise generator for radar jamming.) Shot noise-limited SNR can only be improved by either increasing the rate at which photons arrive or the period over which they are measured; the latter occurs during low-pass filtering or integration.

Assuming a stable power supply and amplifier, the second limiting source of noise results from variations in the dark current. Thermionic emission can be reduced by cooling the tube, but dark current due to ohmic leakage remains, and is proportional to the gain. Detailed treatises on the subject of noise in PMT applications evaluate many other sources of variation in the photocurrent that are outside the scope of this discussion.

Photon Counting is a technique which can be useful at very low light levels. When the photon arrival rate is low, the signal consists of discrete pulses of current which can be observed on an oscilloscope; these pulses sum to a dc signal when the rate of photoemission becomes high enough. However, most of the dark signal, including that due to ohmic leakage and sources in the amplification stages, is essentially continuous current, not pulse-shaped. Photon counting is a method of discriminating between these sources.

If the dark current is subjected to pulse height analysis, the distribution of the heights of pulses that are observed is displayed in Figure 7.26. The plot shows a large number of dark current pulses of small magnitude in region A; these small pulses result from the dark current due to ohmic leakage and to thermionic emission that occurred late in the dynode chain. However, dark current pulses with greater height in region B are due to thermionic emission at the photocathode. They are the same height as pulses due to arriving photons, but their numbers are small and can be reduced further by cooling the tube. The task of photon counting is to collect only large pulses resulting from photons striking the photocathode and to discriminate against dark current.

A threshold can be fixed between pulse heights in region A and those in B, so that only those in region B will be counted, thereby distinguishing between these components of the signal. The current due to small magnitude pulses is removed by that discrimination level. If the current rises higher

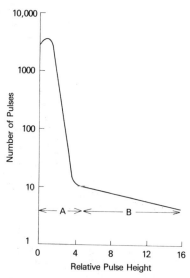

FIGURE 7.26. Pulse height analysis of dark current. In region A, the pulse height is small, so a photon counter discriminates against these pulses. In region B, the pulses are larger, but fewer. (Adapted from R. W. Engstrom, *Photomultiplier Handbook*, 1980, by permission of RCA Corporation, New Products Division, Lancaster, PA.)

than the threshold, a pulse is counted; if not, there is no pulse and no detected signal.

When the photoemission rate is low, less than about $10^7 - s^{-1}$, the pulses are sufficiently separated to be counted discretely. The PMT output is passed through an amplifier which shapes the pulses and a discriminator which sets a threshold for the pulses. They can then be counted by fast counters.

Besides discriminating against the dc dark current, this system also reduces a source of drift: The photon count is nearly independent of small changes in the gain; the gain changes only the height of the pulses, not the number of them. Only pulses due to actual photons and thermionic emission at the photocathode are counted, and since the latter are not subject to drift, they can be subtracted by making a separate dark measurement.

The same Poisson statistics apply to photon counting as were discussed earlier for the current measurement methods. However, because the system is resistant to drift in the gain voltage or dark signal, the counting can continue for long periods in order to improve the SNR; astronomers frequently count for days!

As the number of photoemissive events increases, the probability of coincident photons, or "pile up," becomes significant. A pulse due to two or more simultaneous photons would still generate only a single counted pulse. The upper limit to linearity depends partly on the speed of the amplifier and partly on the design of the PMT and its effect on the response time. To a certain degree it is possible to simplify that error and statistically evaluate the probability of coincident photons for a given photoelectron rate, but the upper limit usually lies somewhere around 10^6–10^8 photons per second, at least four orders of magnitude below the upper current capability of the PMT for dc operation.

Photon counting systems are usually purchased as units, but reasonably high performance has been demonstrated in systems constructed from readily available components, a video amplifier and a comparator. The front end of that system is shown in Figure 7.27. The 22-MΩ resistor converts current pulses to voltage pulses, and the following capacitor and resistor is a high-pass filter to remove any dc level. The video amplifier, which is quite fast, is a buffer for the circuitry which sets the threshold for the comparator that generates a digital pulse.

7.3.2. Photodiodes

These are solid-state semiconductor detectors. When light energy is supplied to the $p - n$ junction of a semiconductor, the carrier population is proportionately increased. This property gives the diode the potential of operating as a linear radiation detector. Packaged in a transparent case and optimized for speed, efficiency, and low noise, the photodiode has become very useful

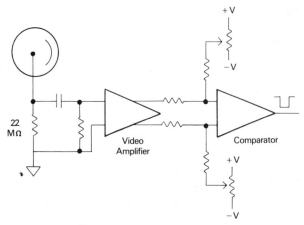

FIGURE 7.27. An input amplifier for a photon-counter. Any dc offset is removed by the input capacitor; the pulse is amplified by the video amplifier and converted to TTL pulses by a comparator acting as a discriminator. (Reprinted with permission from T. Nieman, *Anal. Chem.*, **53**, 350 (1981). Copyright (1981) American Chemical Society.)

as a detector in the visible and near infrared regions. Some have been prepared with enhanced response in the ultraviolet region as well.

Three properties of photodiodes define the applications in which they are useful:

1. Due to the samll size of the junction they have very fast response compared with the transit time for photoelectrons in a phototube.
2. They lack either the large active surface or the internal gain of the photomultiplier tube and consequently may be much less sensitive; they do, however, share the extremely wide dynamic range.
3. The spectral range reaches a maximum at a much longer wavelength than does the photomultiplier tube. These factors make them necessary for very high-speed spectroscopic applications such as in near infrared picosecond laser work.

The photodiode can be operated in two modes. The first is the **photovoltaic** mode in which it is an *active* transducer; current is generated at the junction and measured directly. However, faster response is obtained in the passive mode, by **reverse-biasing** the junction as shown in Figure 7.28. Essentially no current flows unless the carriers are generated by light at the junction. Response times as short as 10^{-11} s can be obtained.

7.3.3. Solid-State Imaging Detectors

The LSI techniques, which have succeeded in placing tens of thousands of transistors on a chip, also are suitable for semiconductor light detectors. By

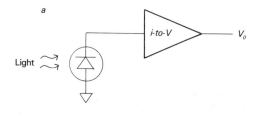

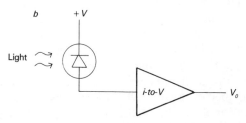

FIGURE 7.28. Photodiode operation. (*a*) Configured as an active transducer in a photovoltaic mode. (*b*) Configured as a passive transducer in a reverse-biased mode.

integrating semiconductor detectors in patterns and placing a window rather than black plastic or ceramic over the top, the integrated circuit can be used to detect an image, that is, intensity as a function of position.

As will be shown, the capability of storing the image during exposure also is inherent in these devices. The photosignal for each detector (*picture ele-ment* or **pixel**) is retained in the form of a charge until the output process takes place. Without the need for moving mirrors or other mechanical devices, the entire image can be acquired simultaneously.

Reasons for using array detectors fall into three main categories:

First, and most obvious, there is a solid-state advantage; measurements can be made at different positions without the uncertainty and unreliability of mechanically driven mirrors.

Second is the need to record an event which occurs too rapidly for a single detector to scan over the image mechanically. For example, if the cross section of a single laser pulse, the duration of which can be much less than a nanosecond is to be measured, all parts of that image must be acquired simultaneously.

The third reason is the **multiplex advantage** applied to a slower event. In most experiments, especially in spectroscopy, the quality of the measurement is improved as the measurement period increases, and maximizing the measurement period at a single position improves the measurement. An imaging detector requires the same time for acquiring the intensity at a single element as it does for the entire image. Therefore, the choice between measurement time and the quality of the measurement becomes unnecessary.

Intensifiers. Solid-state imaging devices are not capable of gain (although similar tube devices, such as vidicons, may have gain mechanisms included). A separate unit, called an **intensifier**, can be placed in front of the imaging detector to provide this gain. The best performance is obtained from a Micro Channel Plate (MCP).

MCP operation is similar to that of a PMT. A photoemissive surface faces the light converting photons to electrons. These electrons are collected by a bundle of microscopic hollow tubes illustrated in Figure 7.29 which act as electron multipliers; they have a resistive inner surface so that a large voltage gradient impressed over the length of the tube induces multiple collisions as does the PMTs discrete dynodes. Each time an electron strikes the inner surface of the tube, multiplication takes place, and a gain of up to 10^4 is possible. A phosphor surface can be placed at the end of the bundle to convert the electrons to visible light which can be detected by the imaging device. Normally, a fiber-optic bundle couples the rear surface of the intensifier to the surface of the imager so that the image is retained.

There are a number of devices which are or have the potential of being suitable for imaging applications. In this discussion, only the two most prominent imaging detectors will be considered: charge-coupled devices and photodiode arrays. Others being used for similar applications include charge-

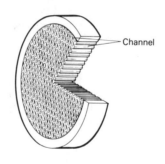

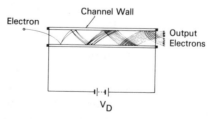

FIGURE 7.29. A microchannel plate. Electron multiplication occurs down the length of each channel in the plate; the optical image is retained. (Courtesy of Hamamatsu Corporation, Middlesex, NJ.)

injection devices (similar to charge-coupled devices), and vidicon tubes (currently used as TV camera detectors); hobbyists have even used large-scale memory-integrated circuits with the lid removed. The image dissector, a PMT with internal scanning capability but without internal storage, also is useful at low light level.

Charge-Coupled Device (CCD) technology has no exact analog in discrete devices such as transistors and diodes. As shown in Figure 7.30, transparent electrodes are integrated over bulk silicon semiconductor material. The potential on those electrodes creates a potential well which traps the electrons produced by photoevents. As photons continue to enter the region, the charge builds in that potential well until it "fills up," that is, is saturated with about 10^6 electrons.

The well is emptied when the readout sequence takes place. By activating sets of electrodes in sequence, as illustrated in Figure 7.31, the charge from each photodetector is shifted into a newly created adjacent potential well located in an area protected from the light. Another sequence moves the charge packets stepwise until they reach a charge amplifier that sequentially converts the charge, corresonding to each detector element, into a voltage pulse at an output pin. These charge packets can be transferred with remarkable accuracy, in excess of 99.99% for each transfer! While this readout sequence is taking place, a new photogenerated charge packet is building in the original potential well.

The CCD technology is being developed commercially for video applications, so the design of commercial devices is optimized for ambient light levels and visible light response. The sensitivity peaks in the near infrared. However, devices prepared primarily for military applications (read "applications without budgetary restraints") and astronomy have shown excellent response in the ultraviolet. A primary difference in the fabrication of these devices is that by constructing them on thin silicon substrates, they can be illuminated on the opposite side from the electrodes.

Commercial CCD arrays are produced as either linear or area arrays; a linear array is shown in Figure 7.32. Up to 4096 pixels are integrated in a

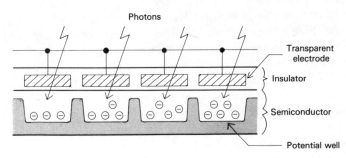

FIGURE 7.30. Light detection in a charge-coupled device. Potential on an electrode creates a potential well that traps photogenerated electrons.

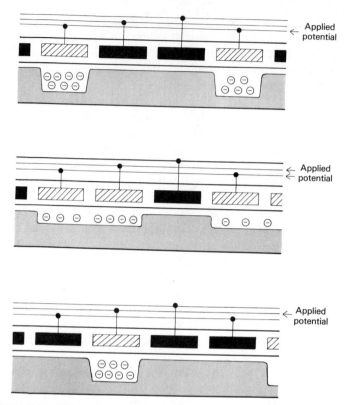

FIGURE 7.31. Simplified model of CCD signal transfer. The packet of electrons is moved linearly by changing the positions of the potential wells. Current CCDs require only a two-phase system.

linear array, usually spaced on 0.0005-in. centers. Two-dimensional arrays have comparable spacing, but up to 128,000 pixels are incorporated in a single device.

In the linear device shown, the charge accumulates at the individual pixels until a transfer sequence takes place; the charge packets are moved in parallel to one or two analog shift registers. From there the charge packets are moved step-wise to the output amplifiers which convert the packets into voltage pulses. A sample-and-hold amplifier integrated into the chip, generates a nearly continuous signal from the pulses. The upper limit to the clock rates is about 10 MHz. A 1000 pixel image can be read every 100 μs if a readout device that is equally as fast is available.

During the period between readout sequences (the exposure time), the signal is integrated in the potential wells. For the greatest sensitivity, a long integration on the chip is preferable to the co-addition of repeated shorter exposure times, since readout noise would only be injected once.

The exposure time is limited by the thermally generated dark signal which

FIGURE 7.32. Linear charge-coupled device detector arrays with 1728 pixels. (Photograph courtesy of Fairchild CCD Imaging, Palo Alto, CA.)

contributes to the charge in the potential wells. With the correct exposure time and even in the dark, the pixel will saturate; that limits the dynamic range of the measurement. The dark signal can be greatly reduced by cooling, typically by about half for each 10°C. A Peltier-effect cooler, adjusted for −10°C, can·perform this function well; a stream of dry air is necessary to prevent frost from forming on the window. At room temperature, the dark signal might saturate the CCD in only 500 ms, but at −10°C, the device is usable with a single exposure time of several seconds.

Because of the small area covered by each pixel and the front illumination employed by commercial devices, CCD arrays are not extremely sensitive, and the poor UV response (and glass window) limit the wavelength range. However, a CCD detector array can provide a SNR well over 2000:1 for multichannel measurements at high light levels. A very useful application is for detecting multiple wavelengths simultaneously in molecular absorbance spectroscopy or laser spectroscopy.

Photodiode Array (PDA) detectors are built around integrated circuits with up to 1024 photodiodes linearly integrated on a single chip; two-dimensional detectors are available as well. For each detector, a small capacitance is included (Fig. 7.33), and is charged to a set level. The current generated by the exposure of the photodiode discharges that capacitance. When that pixel is read out, the charge on the capacitor is restored, and the output voltage is proportional to the charge required. Therefore, similar to the CCD, the PDA is an integrating detector.

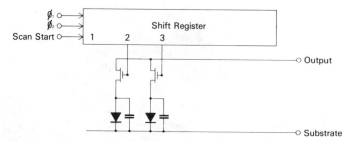

FIGURE 7.33. Schematic of a photodiode array. When light strikes the photodiodes, charge is removed from the capacitors. The output is a measure of the charge required to recharge the capacitor.

Unlike the CCD, the readout process effectively connects a single pixel directly to the output amplifier. Consequently, the exposure period lasts from one readout to the next for that particular pixel; although all elements have equal exposure periods, the periods are shifted in time.

PDAs can be obtained both as linear arrays with up to 1024 elements on 0.001-in. centers and as area arrays with up to 50 by 50 elements. Two outstanding features have generated great popularity. First, the UV response of the detector elements is good, and unlike the CCD, a quartz window is used in order not to limit that response. Second, a special version of the linear array has been produced with the spectroscopist in mind; the height of the pixels is increased to over a millimeter, vastly increasing the sensitivity.

There are several other ways in which the PDA is either similar to or different from the CCD array. A thermally generated dark signal also may limit the dynamic range when the exposure time is long and cooling is suggested. The output signal consists of signal pulses plus a pattern of pulses resulting from the internal switching circuitry; these latter pulses follow a fixed pattern which shows up as significant noise in the signal unless the data acquisition is synchronized to reject the pattern.

Fundamental limits to the SNR are better for the CCD technology, since the absence of capacitance makes a much lower theoretical limit to the SNR; we can expect CCD technology to dominate solid-state imaging in the near future. However, at this writing, commercially available PDAs are more often suited for scientific applications.

7.4. ELECTRODES

Electrodes are common transducers for biologists and chemists used both to monitor biological activity and chemical concentrations. One class of electrode is **potentiometric**. Chemical information is transduced to a potential between a pair of electrodes, and that potential can be measured. Nerve action, monitored by measurement of transient potentials, are measured in much the same way. On the other hand, **amperometric** techniques are growing, most notably with electrochemical detectors used in liquid chromatography. In these methods, the current is measured and develops in response to a potential being applied to the electrodes. The interfacing of these two classifications will be discussed in this section.

7.4.1. Potentiometric Measurements

Most of the electrodes for potentiometric chemical measurements use a membrane electrode in which two chemical cells are separated by an ion-selective membrane, and the resulting potential is related to differences across the membrane. The glass electrode used for pH measurements is the most ubiquitous. Other ion-selective electrodes, silver ion, copper ion, fluoride, nitrate, among others, are similar in their application.

The glass pH electrode (Fig. 7.34) has an internal silver electrode and a potassium chloride solution separated from the solution under study by a thin glass membrane. Since the glass membrane is responsive to hydrogen ions, a potential develops across the membrane related to the ratio of hydrogen ion concentration within and without the glass membrane. Other ion-selective electrodes vary by having different membranes. A doped lanthanum fluoride membrane is responsive to fluoride ion, and a cadmium sulfide membrane is sensitive to cadmium and sulfide.

Two electrodes are required to make a potential measurement. A reference electrode, a silver–silver chloride electrode or a mercury–mercury

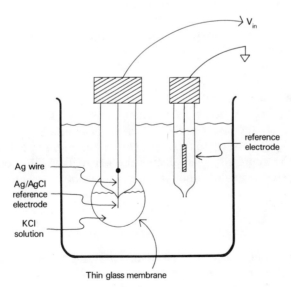

FIGURE 7.34. A potentiometric-electrode chemical-ion transducer for hydrogen ion. A potential develops across the glass membrane in response to the pH. This is one of many ion-selective electrodes.

chloride electrode whose electrochemical properties remain constant, is used for the second electrode and is connected to ground.

These electrodes are active transducers. A modified Nernst equation defines the potential developed between the ion-selective electrode and the reference electrode: assuming the solutions inside the reference and pH electrodes remain constant, the measured potential, E_0, between will be

$$E_0 = \text{constant} + \frac{RT}{F} \ln [H^+]$$

where $[H^+]$ is the concentration of hydrogen ion, T is temperature in degrees Kelvin, and R and F are constants whose quotient is about 2×10^{-4}. Since pH $= -\log[H^+]$, at room temperature the transfer function becomes

$$E_0 = \text{constant} - 0.059 \text{ pH}$$

In order to design the interface, one first notes that the value of the constant depends on the construction of the electrodes, and that the slope, nominally 0.059 at 25°C, varies with temperature. Second, one must note that the resistance of an electrode must be very high indeed since a glass membrane is part of the circuit; in fact the resistance of a pH electrode is on the order of 10 MΩ and the resistances of other ion-selective electrodes may be an order of magnitude higher.

A typical interface, shown in Figure 7.35, requires an input amplifier with very high-input impedance. The LF13741 OA (National Semiconductor) configured as a follower with a specified input resistance over 100 GΩ (10^{11} Ω) suffices. The second signal-conditioning stage amplifies the signal from about 0.84 V for a range of 14-pH units to the range of the ADC input. Finally, some offset is required so that the low end of the signal range will match the low end of the ADC input; the offset varies with the electrode. Consequently, a summing amplifier with gain is added as a second stage.

An alternate input amplifier, illustrated in Figure 7.35b, places the cell in the feedback loop. By doing so, the high-resistance glass electrode is held at ground potential; consequently, the noise susceptibility should be lower.

Circuits nearly identical to those shown are commonly used in commercial pH meters; the simplicity of replacing the pH meter with a one-dollar amplifier leading to the analog input of the existing laboratory computer is quite compelling.

7.4.2. Amperometric Measurements

Amperometric transducers are passive, providing the current that results from the application of a potential to the solution via another electrode. The electrochemical current that results is acquired through an indicator electrode which must be maintained at ground potential.

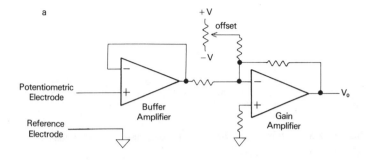

a

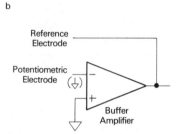

b

FIGURE 7.35. Input amplifier configurations for potentiometric electrodes. (*a*) The input stage is a simple follower amplifier; it is followed by a gain amplifier with offset capability. (*b*) An alternative input stage keeps the ion-selective electrode at ground potential.

The potential to be applied to the solution is directed, possibly from a DAC, to the potentiostat amplifier shown in Figure 7.36. That amplifier sets the potential on the working electrode. Current is produced in response primarily to electrochemical activity at the indicator electrode which is connected to an OA in a current-to-voltage configuration; this arrangement holds the indicator electrode at ground potential.

A third electrode, the reference, corrects for inherent errors due to resistance in the solution. Since current flows from working electrodes to indicator electrodes, there will be a voltage drop across the solution; by Ohm's Law, the voltage is the product of the current and the solution resistance. The voltage drop is the potential experienced by the indicator electrode, which is different from that impressed by the working electrode. Consequently, a reference electrode is introduced into the feedback loop of the potentiostat amplifier which makes continuous correction for the solution resistance.

For many applications, the current is very low so that careful attention must be paid to the problems of filtering and maintaining linearity at high gain.

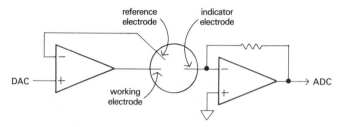

FIGURE 7.36. A three-electrode system for amperometric measurements.

7.5. ELECTRIC POWER CONTROL

In this section, we will consider how external electrical devices can be turned on and off digitally with a TTL pulse or logic level, and how high voltage and current can be controlled continuously. This can be a technically complicated topic, but there are some shortcuts that will be introduced. In so doing, the discussion will be kept well within reach.

7.5.1. Switching DC Power

To control dc power, a switch is required whose characteristics approach the ideal: zero resistance when **ON** and infinite resistance when **OFF**. Deviation from that ideal, particularly at high current, can cause catastrophic results because even a small resistance in a switch with high current flow will result in Joule (i^2R) heating in the switch. When we approach the choice of a circuit, which is used to switch a dc current through a load on or off, we must know the voltage and current being switched as well as the speed required.

For switching dc power, we will consider three methods: the bipolar (NPN or PNP) transistor, the electromechanical relay, and a type of FET capable of switching high power.

Bipolar Transistor Switches. In digital logic, the bipolar transistor is operated so that it is either ON (conducting) or OFF (not conducting). If the ON resistance is low enough and the OFF resistance is high enough, it makes a good switch.

The output stage of a TTL gate is itself a bipolar transistor (Figs. 6.4 and 7.37), and the gate can be capable of switching more current than just that amount necessary to determine the logic level itself. If both the voltage and current are small, a load can be switched directly. Using an open-collector TTL device (for example, 7438, or 75451 NAND gates), the output transistor can switch voltages up to around 35 V at currents as high as 50 mA, enough to control many small devices. Recall from Chapter Six that the gate output must *sink* the current.

A simple but very useful example is the control of LEDs as in Figure 7.37. The load (the LED and a current-limiting resistor) must be connected to the power source (a positive dc voltage) and are switched to ground by the transistor. When the logic state of the gate is HI, the transistor does not conduct and the LED is not powered. When the logic state is LO, the transistor does conduct; the current is determined by Ohm's law where the resistance is the sum of the LED, the transistor, and the resistor, the first two of which are comparatively small. Open-collector integrated circuit gates with relatively high current-sinking capability are sometimes called **interface gates**.

Because the transistor has some resistance, even when on, it generates heat when current passes through it. Consequently, in order to switch still higher currents, an additional transistor with greater capacity must be added. An NPN transistor classified as a general-purpose switch can be used as shown in Figure 7.38 where positive current flowing into the base of Q_2 causes it to conduct. When the output of the gate is a logical **0**, the gate transistor, Q_1, will conduct, and the current through the resistor will travel through Q_1 to ground; that effectively grounds the base of Q_2 so that Q_2 cannot conduct. When the gate output is a logical **1**, Q_1 does not conduct, and the current passes through the resistor to the base of Q_2 thereby causing Q_2 to conduct.

The transistor that switches the load must be capable of withstanding the supply voltage across its collector and emitter when it is not conducting, and it must be able to handle the current when it is conducting. In order to switch the load on, enough current must be given to the base to completely turn on the transistor. Otherwise, if the transistor is only partially turned on, it acts as a resistor and power is dissipated as Joule heat. Besides limiting

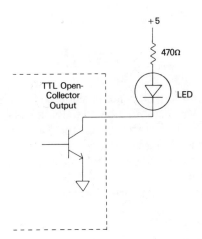

FIGURE 7.37. Switching a low-current device from an open-collector IC. The current is regulated by the series resistance that is dominated by the resistor.

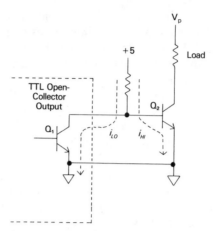

FIGURE 7.38. Control of an external transistor to switch a load. When the output is LO, the load is turned off.

the current through the load, the partially ON transistor can be damaged by the heat. The amount of current into the base is controlled by resistor R, which must be large enough to safely limit the current through the gate's output transistor, Q_1, but small enough to provide sufficient current to the base of Q_2.

The base current that is required can be determined from the gain which is specified as the **dc beta** or **dc forward current transfer ratio**; the symbols used for this parameter are h_{FE} or β_N

$$h_{FE} = \beta_N = I_C/I_B$$

Therefore, to turn the transistor completely **ON**, sufficient base current must be provided so that

$$h_{FE}\, I_B > I_C$$

If, under these limitations, it is not possible to switch the required amount of current, a second transistor can be added in a **Darlington** arrangement as shown in Figure 7.39. The current from the logic gate is amplified by the current gain of Q_2, providing enough current to drive Q_3. Darlington transistor pairs are often mounted in a single transistor package which can easily have an overall current gain, h_{FE}, of 10^3; if an open-collector TTL device can sink 50 mA, it can be used to drive a Darlington transistor which switches several amperes.

In the examples described so far, neither pole of the load is grounded, and sometimes a device must be grounded. Switching the power source instead of sinking current also is possible, though somewhat more compli-

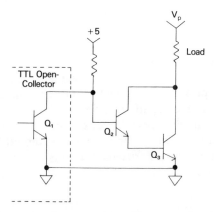

FIGURE 7.39. Increasing the current-handling capacity with transistors in a Darlington configuration.

cated. An example of a current source driver is shown in Figure 7.40. Q_1, a **PNP** transistor (which conducts current between collector and emitter in the opposite direction from an NPN transistor), is turned on when the gate output is LO, and the pull-up resistor turns it off when the gate output is HI. Q_1 can then provide substantial current to feed a Darlington pair.

Switching power with **negative polarity** requires a different approach since the configurations shown thus far are limited to the switching power of positive polarity with positive TTL logic. Additional transistors are required to shift levels from the positive logic of TTL to the negative power level. PNP

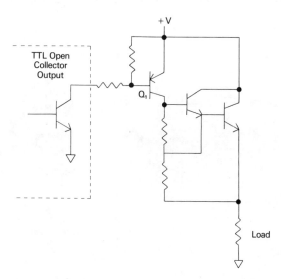

FIGURE 7.40. Switching the positive side of a load. To pull the base of the Darlington transistor pair high enough, a PNP transistor (Q_1) is required.

transistors serve to turn on and off a negative voltage level (Fig. 7.41). When the gate's output is a logic **1**, Q_1 conducts so that current can flow from resistor R through Q_2, turning it on, and that turns the load transistor, Q_3, off. When Q_1 and Q_2 do not conduct, the base of Q_3 becomes negative and the transistor will be turned on. The circuit is somewhat more complicated than circuits for positive power, but integrated circuits such as a DS3687 can be used as TTL-compatible negative power switches.

Some loads may require digital control over the *direction* of current flow; for example, the polarity of a dc motor determines the direction of rotation. Two approaches to switching the polarity can be considered; the equivalent circuits are drawn in Figure 7.42:

> One terminal of the load could be grounded and the other connected to either a positive or negative supply by a pair of current source switches (a "*T*" configuration)

> The connections of the load might be reversed by two pair of switches (an "H" configuration); switches 1 and 3 must be current sources as in Figure 7.40 while the remaining switches are current sinks.

Electromechanical Relay. One way of reducing the ON resistance and increasing the OFF resistance is to use an electromechanical relay. A small current through the relay's coil creates a magnetic field which pulls a mechanical switch and switches another load.

A relay has certain advantages. The resistive characteristics are nearly ideal; the contacts have near-zero resistance when closed, and the air gap between the contacts has nearly infinite resistance at low voltage when the

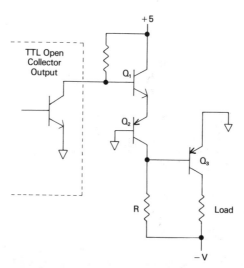

FIGURE 7.41. Switching negative power. Q_1 and Q_2 are required as level shifters.

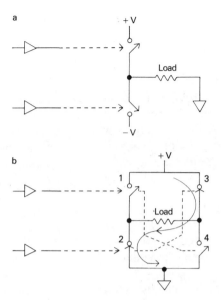

FIGURE 7.42. Bipolar switching. If the direction of current flow in the load is to be reversible, there are two approaches. (*a*) A "T" configuration; a bipolar power supply is required. (*b*) An "H" configuration; switches 1 and 3 use a circuit based on Figure 7.40 while 2 and 4 use the circuit of Figure 7.39.

relay is open. In addition, the coupling between the controlling signal and the power is magnetic; the load and power are electrically isolated from the driving circuit. Finally, very high current and voltage can be carried if the correct unit is chosen.

Although there are few applications that cannt be handled with solid-state electronics, the electromechanical relay can be quite convenient (Fig. 7.43).

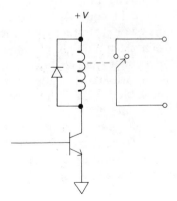

FIGURE 7.43. Control of a mechanical relay. The diode is required to shunt the inductive "kickback" that occurs when power is removed.

A relay can be driven with simple circuits for unipolar dc control. Of course the relay's coil voltage and current requirements must be met by the transistor. For low-power applications, small "reed" relays are often built in plastic integrated-circuit-like packages and can be controlled directly from open-collector TTL outputs.

However, several disadvantages also exist. The operating time (the time from when the current is turned on until the contacts close) is comparatively long, 0.1–100 ms. Also, the metal contacts bounce several times when they close creating multiple closures; for low-current applications, relays with mercury-"wetted" contacts may solve that problem. Finally, the contacts may spark on closure and generate electrical noise that can affect digital logic.

A precaution must be taken against the kickback generated whenever the current in a coil is switched off. A pulse of negative voltage will be induced by the change in the coil's magnetic field when the current is removed. Therefore a diode should be added across the coil as a shunt to short out that pulse.

If moderate-to-high voltages or currents are switched, a spark can be produced across the contacts that is damaging to the contacts and generates intense electromagnetic interference that can affect logic circuits. Consequently, the contacts should be protected by zener diodes or capacitors across the contacts.

Power FETs. A third approach to switching dc power is to use a VMOS (Vertical Metal-Oxide Semiconductor) power FET. Only recently has it become possible to manufacture FETs with high-power capacity, but VMOS power FETs are hailed as power transistors that require no compromise. While that statement may be somewhat exaggerated, it is true that they are nearly ideal devices for switching dc power: the **ON** resistance is very low; the **OFF** resistance is extremely high; the control input has a very high-input impedance, and the time required for the transistor to switch is several orders of magnitude shorter than for the equivalent bipolar transistor.

The FET is controlled by voltage rather than current. Current can flow from the source to the drain when the field produced by the voltage on the gate is sufficient to create a conducting channel. It is necessary to raise the voltage on the gate above a threshold to cause it to conduct. As was true for bipolar transistors, the VMOS FET, when used as a switch, it should be operated only fully **ON** or fully **OFF**.

To be turned **OFF**, the voltage on the gate (with respect to the source) must be between 0 and 1 V. To be turned **ON**, the gate should be raised to about 10 V. These requirements are easily met by an open-collector TTL gate with a resistor that pulls the output up to 10–15 V as shown in Figure 7.44.

A zener diode between gate and source is often included in the transistor package for protection. Because so little current is required to operate the

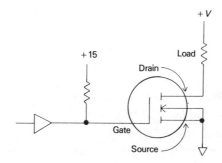

FIGURE 7.44. A VMOS FET for power applications.

FET, the high-voltage, low-current spikes from static electricity can be devastating. This zener diode will typically break down and conduct if the input voltage exceeds 15 V, and it also will conduct should the gate voltage go negative; in the normal input range between those limits it is uneffective.

Each VMOS power FET does have its limits; some of these can be understood by examining the electrical characteristics of a VFET listed in Table 7.1. The maximum voltage, V_{DS}, which may be switched by various devices is between 40 and 80 V; higher voltages may be controlled by using these transistors in series. Devices with drain currents of 1–12 As are common, but there are two routes to switching higher currents: (1) since negligible gate current is required, FETs can be added in parallel easily, and (2) some devices such as the IRF 150 from International Rectifier will switch 28 A at 100 V!

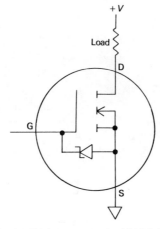

FIGURE 7.45. An integral Zener diode protects the VMOS FET from voltage spikes and static electricity.

TABLE 7.1. Electrical Characteristics of a VN66AF VMOS Power FET

Maximum Drain-Source Voltage (V_{DS})	60	volts
Maximum Continuous Drain Current (I_D)	2	amps
Maximum Forward Gate-Source (Zener) Voltage	15	volts
Maximum Reverse Gate-Source Voltage	-0.3	volts
OFF Resistance (V_{DS}=48 V; V_{GS}=0 V)	480	MΩ
ON Resistance (V_{GS}=10 V)	<2	Ω

7.5.2. Switching AC Power

The semiconductor device that is often used to switch ac power is the thyristor, a triac, or a silicon-controlled rectifier. Design with thyristors can be somewhat complicated and outside the scope of this discussion. Fortunately, a solid-state switch for ac power is so useful that a wide variety of easy-to-use **Solid-State Relay** (SSR) modules are commercially available. The SSR isolates the switch from the logic input often with a LED and phototransistor as shown in Figure 7.46; in such a case, they can be interfaced in exactly the same way as any other LED. The input terminals are connected to $+5$ V through a current-limiting resistor (about 390 Ω) at the positive terminal and to the open-collector output of the gate at the negative terminal. Some varieties use an oscillator and transformer for isolation, but the interfacing requirements are similar.

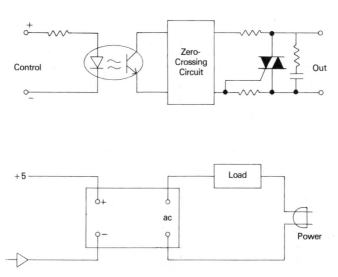

FIGURE 7.46. A solid-state relay for switching ac power. The relay is optically isolated, and it only switches when the ac power is at ground potential.

The remainder of the SSR module contains all of the circuitry to control the thyristor switches. Therefore, the designer need not be concerned with the remainder of the device's internal circuit except to determine that the SSR can handle the power requirements.

The SSR has many advantages over the electromechanical relay: (1) a solid-state device has a longer life expectancy than its mechanical counterpart; (2) the TTL logic compatibility is built into the device; (3) the switching time ranges from the low microsecond region to a worst-case of 8.3 ms, if switching is delayed until the ac voltage crosses zero; (4) contact bounce is eliminated; (5) electromagnetic interference (EMI) is reduced; and (6) they are physically robust.

Some SSRs are able to solve the problem of EMI generation by delaying the activation of the switch until the ac power crosses zero. For turning the power off, this is an inherent characteristic of a thyristor switch, since it only turns off at zero current. In order to turn *on* at zero crossing, additional logic is necessary to detect 0 V and is included within the package of a **zero-crossing solid-state relay**. When the input to the unit is activated, the logic waits until it detects a zero crossing before the thyristor is turned on. This will naturally require a delay of up to one-half cycle, $\frac{1}{120}$ s, but unless the delay cannot be tolerated, the zero-crossing versions should be used.

Solid-state relays are available for switching loads from only a few milliamperes to several tens of amperes at 120-V ac (Fig. 7.47). Similarly packaged versions for switching dc power also can be obtained.

A second option for switching AC power is the use of an electromechanical relay as was discussed previously. Because the exact time at which the contacts close cannot be precisely controlled, problems of arcing across the contacts and the attendant interference must be considered. A capacitor with a sufficient voltage rating should be placed across the contacts.

7.5.3. Controlling High-Voltage or Current

There are many cases in which it is necessary to use the computer to control the voltage at up to several kilovolts (for example, to control the gain of a photomultiplier tube) or to control current at up to 50 A (for example, to control a tube furnace for atomic absorption spectroscopy). We already noted in Chapter Five how the DAC can be used to control low voltage or current. However, the voltage and current limits are those imposed by the output OA, typically ± 15 V at a few milliamperes. The current can be boosted somewhat with the current-boosting methods which were described, but even so, OA circuits producing current substantially over about 1 A are difficult to design. Therefore, simpler alternatives will be presented.

Two methods of continuously controlling electrical power can be used. The power amplifier is an obvious answer, a device which linearly amplifies electrical voltage or current. An alternative is to rapidly modulate the power,

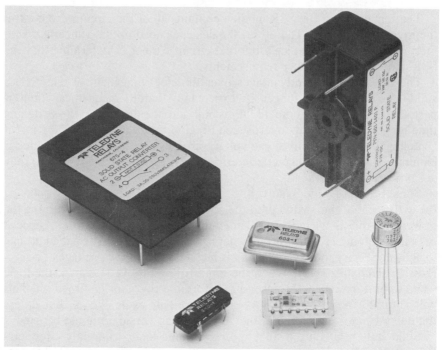

FIGURE 7.47. Various solid-state relay packages. (Photograph courtesy of Teledyne Relays, Hawthorne, CA.)

that is, turn it off and on so that the fraction of full-scale power obtained depends on the fraction of time that the power is on.

Power amplifiers with high capacity are difficult to design since the design can be complex, particularly as the voltage, current, or frequency goes up. Fortunately, at low frequency the design of general-purpose **variable, programmable, power supplies** solves this problem. That design can be represented by the block diagram of Figure 7.48. The reference and control section of the power supply uses only low voltage circuitry followed by a power amplifier. The consequence in most cases is that the internal voltage reference, possibly a stable reference voltage and a voltage divider, can be replaced by an input for external programming. Since the range for that input is often 0 to 10 or 0 to 15 V, it is suitable for a DAC. Programmability is usually found in the high-voltage power supplies used for photomultiplier tubes so that a 0–10-V input is typically translated in to a 0– − 1500-V output.

Low-voltage power supplies often have a programmable input; for example, the Kepco ATE 25-49M can linearly translate a 0–15-V input to either a 0–25-V output or a 0–40-A output. This type of power supply often is sold as an **operational power supply**, which has its similarity to an OA.

Where external programming of a power supply is not made available by

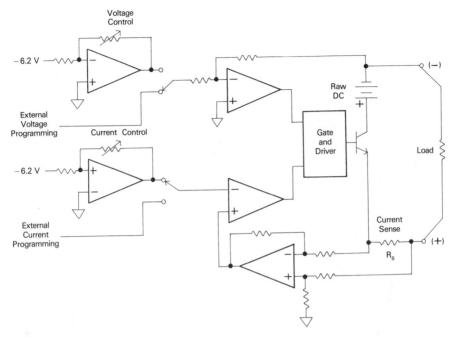

FIGURE 7.48. A programmable (operational) power supply. The control circuitry is at low voltage, and the operational amplifiers can be bypassed for external voltage or current programming.

external connection, an appropriate modification is often straightforward. The potentiometer and gain switches of many power supplies consist of a low-voltage voltage divider, the output of which is amplified to higher current or voltage. Frequently one can isolate that circuitry and replace it with input terminals. A single-pole double-throw switch is added to connect the terminals for an external reference.

Where high current is not required, the circuit of Figure 7.49 illustrates a method for obtaining high voltages which are linear with a low voltage input. DC-to-dc converters with various gains can be obtained that are very linear once an input threshold has been exceeded. The central component of a dc-to-dc converter is a transformer; transformers operate only on ac power so a chopper or modulator is necessary to convert the input to AC, and a rectifier and low-pass filter are required on the output to restore dc. If the input is high enough to activate the chopper circuitry, the output voltage wil remain directly proportional to the input voltage. Such a circuit is readily used for photomultiplier tube power supplies, which needs to provide a milliamp. One design factor requires particular attention; the input to the dc-to-dc converter must be able to give more power than will be drawn from the output since a small amount of power will be lost within the con-

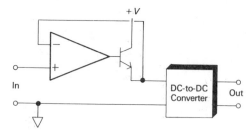

FIGURE 7.49. A dc-to-dc converter for high-voltage control from low voltage.

verter. That is,

$$I_{in}V_{in} > I_{out}V_{out}$$

where I_{in} and I_{out} are the input and output currents and V_{in} and V_{out} are the input and output voltages, respectively. If the converter is capable of producing 1 mA at 1500 V from a 15-V input, that input must be able to provide in excess of 100 mA.

Pulse-Width Modulation (PWM) offers a simple alternative to power amplifiers. This method regulates power by controlling the fraction of the time that the power is **ON** as compared to the time it is **OFF**, that is, by modulating the width of high-frequency pulses. Power is given to a load in maximum amplitude pulses of constant frequency but with varying **duty cycle**. The duty cycle is the proportion of the total time that the power is turned **ON**. Since the frequency remains constant, the overall energy provided to the load in a given time period is proportional to *pulse width*.

Of course, the device being controlled must not show a pulsed effect.

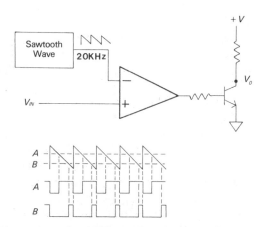

FIGURE 7.50. Analog pulse-width modulation. The level of V_{IN} determines the width of the pulses from the comparator and therefore the *average* voltage, V_0.

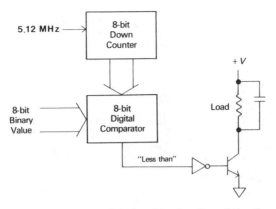

FIGURE 7.51. Digital pulse-width modulation. The fraction of the time ON is the input value divided by 256.

Therefore, the frequency must be high, and the pulses are filtered with filtering capacitors and/or the inertia of the object of the control. For example, the rotational inertia of a motor or thermal inertia of a heated water bath or incandescent lamp may sufficiently filter power pulses. Frequencies greater than 20 kHz are often used. However, switching large loads at high rates requires careful design to eliminate radio frequency and audio noise.

The PWM interface, generating pulses of variable width at a constant frequency, can be designed in several ways:

1. If the computer has no other tasks to perform while the motor is operated or if an interrupt system can be used, the computer can continuously time each **ON** and **OFF** period.

2. An analog PWM modulator can be based on a triangular wave generator; the width of the pulse, and therefore the average power, is directly dependent on the magnitude of the reference input, probably from a DAC. Such a circuit is shown in Figure 7.50.

3. The process can be off-loaded to a digitally programmable pulse-width converter such as the one in Figure 7.51; the length of each pulse is determined by a down-counter whose magnitude can be set from a parallel port.

CHAPTER EIGHT

Transducers: Strain, Pressure, and Translation

In this chapter we continue describing a variety of transducers that are significant in scientific investigation. These transducers are related by the frequent use of the measurement of position as part of the transduction process.

8.1. PROXIMITY DETECTORS

The need to test for the proximity of a component in an experiment is a common requirement. The problem may be as simple as determining the position (on or off) of a toggle switch on the front panel of an instrument, or it may involve synchronization of an event with the phase of a high-speed rotating device (i.e., When is the device in the required position?).

There are many transducers capable of proximity detection. Some are inherently digital and are capable of only two signal levels. Others are basically analog, and a threshold must be set to discriminate between two levels.

8.1.1. Mechanical Switch-Closures

Mechanical switch-closures are inherently binary: The contacts of the switch are either closed (making contact) or open (not making contact). Mechanical switches serve several needs. The first is as an interface to the operator; if a panel or keyboard switch is used by the operator to enter information, the computer must read the position of that switch. However, mechanical switches are also used as part of the interface to an experimental device.

Small precision switches, often called **microswitches**, are used to indicate with high reproducibility whether a component is or is not in position.

For example, a microswitch might detect that a bar has been pressed in an animal behavior experiment, as shown in Figure 8.1. When the animal presses the bar, the switch is closed, and that condition is made available to the computer for counting or for further action.

A microswitch placed at the end of the travel of a leadscrew carriage can be used to sense the limit of the screw; when the carriage reaches the end, the switch contacts are closed. This closure could be used as a safety device to prohibit further travel. It also can provide a means of calibrating the position of the carriage, its closure indicating that the carriage has reached a known position. In the latter application, these switches can be remarkably reproducible, often indicating position within a few thousandths of an inch.

Another microswitch application indicates the position of a closing door by including a microswitch in the path. Also, a floating bulb attached to a switch converts the switch to a level detector for a fluid.

A mechanical switch closure is easily converted to a TTL level. If one terminal, as in Figure 8.2, is connected to ground and the other to the TTL input, closure grounds the input thereby indicating a **0**, but when the switch is open, no current from the TTL input can flow to ground and the switch indicates a **1**. The resistor is not actually required in a true TTL system. However it should routinely be added to "pull-up" any noise spikes that might indicate a false **0** and to make the switch compatible with other LSI MOS chips.

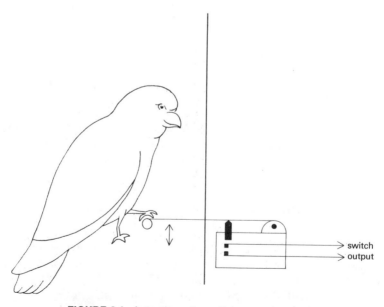

switch

output

FIGURE 8.1. Actuation of a switch as a transducer.

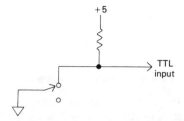

FIGURE 8.2. Conversion of switch condition to TTL.

A word of warning should be applied to applications in which the *change* of the state of the switch must be detected (for example, if we are to *count* the number of times that the door closes). Mechanical contacts will *bounce* when they change states making it appear that they have briefly reverted to the previous state; a half-dozen bounces are typical.

There are three methods of avoiding bounce error. First, if the switch position is read directly by a computer port, the acquisition program can include a time delay, executed after the closure is first noted; the delay masks the bounces (but also could mask the next real transition). Second, a "de-bounce" circuit (Fig. 8.3) can be added: an R–S flip-flop (described in Ch. Six) is connected to a two-position (double-throw) switch; after the switch is thrown and the flip-flop changes state, the output cannot change until the

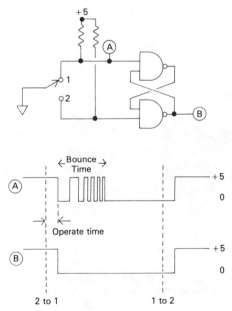

FIGURE 8.3. "Debouncing" a switch. An *R–S* flip-flop removes the errant pulses that result from contacts bouncing in a mechanical switch.

switch position is returned. Third, a monostable can be added, generating a single pulse whose length is longer than the bounce period, thereby masking the bounce.

8.1.2. Noncontact Proximity Detectors

In order to operate the mechanical microswitch, physical contact between the object and the switch must be made. This may be undesirable for any of several reasons: the switch must be outside the container; the environment might damage the exposed contacts, or the moving object might be affected by striking the switch. Several proximity detector methods, represented in Figure 8.4, do not require contact. These include the use of magnetic reed relays, mercury switches, optical interruption, and Hall-effect detectors.

Magnetically operated reed relays consist of contacts completely sealed in the case; a magnetic material is used for one contact so that within a magnetic field the contacts close. If a magnet is fastened to a moving object, the switch will close whenever the object is near.

A **mercury switch** is formed with a pair of immobile contacts sealed into a glass tube with a small amount of mercury. When the switch is rotated,

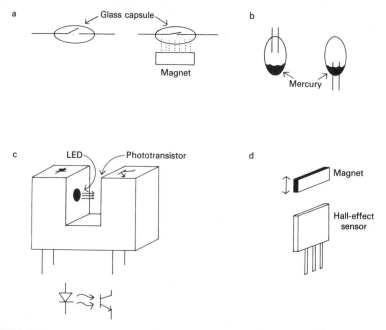

FIGURE 8.4. Noncontact proximity detectors. (a) Magnetically operated reed relay. (b) Mercury switch; the condition depends on the orientation of the switch. (c) Optointerrupter; light from the LED, detected by a phototransistor, can be blocked by a solid object. (d) Hall-effect transducer; presence of a magnetic field is detected by this solid-state detector.

the mercury covers the contacts and the circuit is closed. Thermostats often switch current through a mercury switch fastened to a bimetallic strip.

Transistors have already been introduced as switches, and as an input device, a **phototransistor** is quite useful. Of course, the phototransistor is readily used to indicate the presence or absence of light, but such a function has wider application than simply determining whether the room lights or spectroscopic sources are turned on. By coupling the phototransistor with a small light source, position is tested by determining whether the light beam is or is not blocked.

Assemblies are readily available that include both an LED as a light source and a phototransistor, integrated into a plastic housing for easy mounting. These devices, illustrated in Figure 8.4, come in a variety of geometries and are known as **optointerrupters**. They are functionally the same as the optoisolators in Chapter Six except for the addition of an accessible gap in which the light path can be interrupted.

A variant can be used to monitor the presence of an object capable of reflecting the light. The example in Figure 8.5 shows such a device used to determine the position of a sample holder. Reflective strips at the sample positions encode the sample number; when the strip is located in front of the detector, the light from the LED reaches the detector.

Light falling on the base of a transistor will cause the transistor to conduct. We will continue to avoid details of transistor's operation; it is only necessary to point out that a phototransistor can be turned on when sufficient

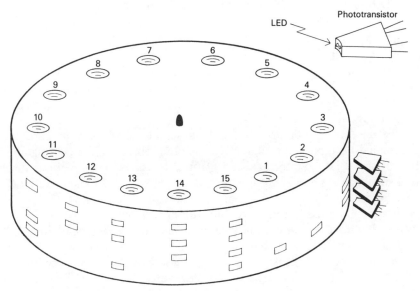

FIGURE 8.5. Use of a reflective optointerrupter to detect position. The phototransistor is illuminated when facing a reflective surface. The position number of the sample holder is decoded from the signals from these optointerrupters.

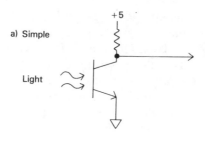

a) Simple

Light

+5

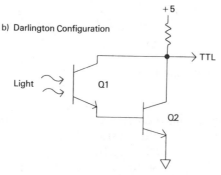

b) Darlington Configuration

+5

TTL

Light

Q1

Q2

FIGURE 8.6. Conversion of a light level to TTL. (*a*) The output is related to light intensity; to convert to TTL, a Schmitt trigger must be added. (*b*) With a Darlington transistor, it may be possible to drive TTL directly.

light falls on the base (Fig. 8.6). In Figure 8.6*a*, if the LED light source is either turned off or blocked, the transistor is off and generates a **1**, but if the transistor is illuminated, the phototransistor conducts and the output voltage drops.

The phototransistor usually has insufficient current-sinking capacity to ground a TTL input. Consequently, it is often operated in a **Darlington configuration** with the phototransistor (Q1 in Fig. 8.6*b*) directly feeding the base of a second transistor (Q2); this provides enough current gain so that Q2 is able to ground a TTL input. In many cases, the "Darlington pair" of transistors is integrated into the optointerrupter package.

Another noncontact proximity detector uses **Hall-effect** technology. When current passes through a flat conductor as illustrated in Figure 8.7, there will be no potential across the width of the conductor. However, the influence of a magnetic field deflects the electrons, distorting the current flow. As a result, a Hall voltage appears across the device; the voltage is dependent on the conductor material and shape and is proportional to the magnetic field and the current.

Although the Hall effect has been used for several decades for measuring magnetic fields, it has only recently been found useful for detecting proximity. A magnet is attached to the moving member, so that the sensor detects the field, and an associated circuit determines whether a threshold has been exceeded. Since one of the problems in manufacturing long-lasting keyboards has been the lifetime of mechanical contacts, many new keyboards use a Hall-effect sensor and magnet combination for each key, eliminating the weakest keyboard component.

In some of these switches, the condition to be monitored is not inherently a simple binary condition which is either on or off, but is a continuous phenomenon to be compared with a reference or *threshold*. Until a phototransistor or Hall-effect detector is saturated, the signal continues to increase as the light source or magnet moves closer; the output is in the analog domain. Single-bit ADC is the required solution.

Previously we met the single-bit ADC in the form of a **comparator** (Ch. Five): the output of a simple OA comparator is the positive or negative output voltage limit depending on which input is greater. However, since the input signal will be noisy to some degree, a Schmitt Trigger comparator should be used. An integrated circuit (7414) with 6 Schmitt Triggers on a chip may be useful if the signal varies between 0 and 5 V, as does the output of a phototransistor.

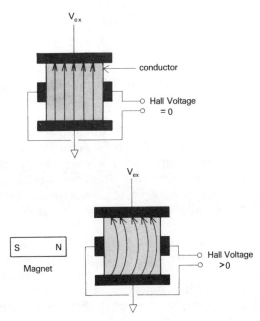

FIGURE 8.7. A Hall-effect sensor. The presence of a magnetic field distorts the current path thereby creating a potential laterally across the sensor. A sensitive amplifier is used to convert to TTL.

8.2. POSITION MEASUREMENT

The precise position of an object moving in a single dimension must be acquired in many experiments. Sometimes, data regarding the movement of the object under study may be required. In other cases, objects must be translated under computer control with precision; feedback of position is necessary. Complete control over the physical translation of an object includes both the **position** of that object and the **rate** at which the object is moved.

Most DC and AC motors are capable of precision movement even at very high speed when separate transducers feed back information about position. Stepping motors, considered in a later section, can inherently be controlled with precision. Sometimes a combination of DC and stepping motors is useful: the dc motor for speed, and the stepping motor for precision.

One type of position detection already has been introduced: the binary condition of being in or out of position. The applications that will be considered in this section will involve precision measurement of position in one dimension.

For positions that are measured continuously, position is usually controlled by a rotary device. This includes linear position; for example, a motor can drive a leadscrew, slidewire, or rack-and-pinion assembly; measurement of linear position may best be accomplished by measuring rotation of the motor. Two classes of devices for measuring rotation are useful: generally, they can be considered analog and digital.

8.2.1. Potentiometers

These are *passive analog* position transducers. Wired as a voltage divider (Fig. 8.8), the potentiometer voltage at the wiper is proportional to the position between the extremes.

Precision potentiometers are manufactured in several varieties, usually designed to be adjusted by hand. The total excursion may be less than one full turn or up to 30 turns. Single-turn potentiometers without stops at the extremes also can be found. They rotate continuously through the extremes so that they can be used, for example, in a wind vane; if the vane is attached to the wiper of the potentiometer, the direction can be obtained from the resistance.

Linear potentiometers also exist and eliminate the need for conversion of linear to rotary motion; the wiper of the potentiometer takes the form of a plunger. However, the length of motion is usually quite limited.

Implementation of potentiometers as position transducers entails both mechanical coupling by means of gears, pulleys, slidewires, etc., and excitation with well-regulated power supplies; measurement of the wiper voltage requires an ADC.

The immediacy of the output of potentiometers has earned them a suc-

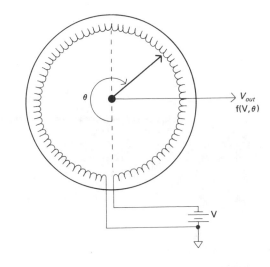

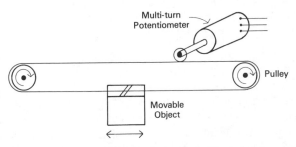

FIGURE 8.8. A potentiometer as an analog position transducer. Top, a single-turn poten-tiomenter wired as a voltage divider; V_{out} is proportional to the angle. Bottom, a multiturn potentiomenter used to transduce linear position.

cessful application in such servo devices as strip-chart recorders. They form a feedback element much like the feedback components included in OA circuits; the power to the motor results from the comparison of the input voltage controlling the position and the voltage produced by the potentiom-eter so that the motor is only powered when a discrepancy occurs. The resolution limit is about 0.05–1% of a rotation, depending on the precision of manufacture.

8.2.2. Linear Variable Differential Transformers (LVDTs)

The LVDT transducer measures linear position directly, without the need to convert to rotary position, and does so with great precision.

The basis of the LVDT is shown in schematic diagram in Figure 8.9. The

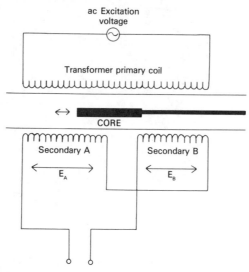

FIGURE 8.9. A linear variable differential transformer (LVDT). The core has high magnetic susceptibility. If it is placed symmetrically between the two secondary windings, they receive equal excitation, and the net voltage is zero. The difference between the two output voltages is a measure of the core position.

primary winding of the transformer is excited by a reference AC signal, possibly produced by an internal oscillator circuit. The core is movable, and if it is positioned symmetrically with respect to the pair of secondary windings, the output signals of the two secondaries will be equal, and the difference between their outputs will be zero. However, if the core is moved, the difference in the outputs will be proportional to the displacement. The output is reported linear to within $\pm 0.5\%$ over the working range which is typically $\pm 0.05 - \pm 3.0$ in.

When employed as a transducer for computer input, the LVDT output must be amplified, converted to dc, and acquired with an ADC.

8.2.3. Incremental Shaft Encoders

Incremental shaft encoders are inherently *digital* devices; pulses are generated when the rotation of a ruled or slotted disk is sensed by optical interrupters; these pulses are counted discretely. Then, with appropriate circuitry, an up–down counter can maintain a record of the absolute position.

A simple shaft encoder having only half-turn resolution, shown in Figure 8.10, produces a pair of pulse trains that are approximately 90° out of phase. If motion only takes place in a single direction, the pulses from one of those trains can be counted digitally as a measure of position.

However, if the direction of rotation is reversed, either because the device is bidirectional or because of mechanical vibration or backlash, the pulses

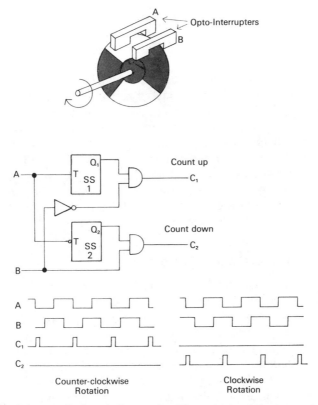

FIGURE 8.10. A low-resolution shaft encoder. The outputs of the two optointerrupters produce signals *A* and *B*. The circuit in the center uses output *B* to select the monostable pulses that correspond to the direction of movement.

produced in one direction are indistinguishable from those in the other. On the other hand, two outputs, which are out of phase with each other, give information about both the extent and the direction of the movement.

These two signals can be decoded into clockwise pulses and counter-clockwise pulses with the circuit of Figure 8.10. Output *A* is fed to a pair of monostables that generate separate pulses on the falling edges of *A*; output *B* selects the appropriate monostable by disabling one of the monostable outputs according to the direction of rotation.

The resultant pulses can either be counted by a hardware up–down counter (e.g., a series of 74193 TTL counter chips), or they can trigger interrupt circuitry in the computer, and the interrupt completion routines can increment or decrement a memory location.

The optical shaft encoder in Figure 8.10 requires only a pair of optical interrupters and a simple mask to generate the required waveforms; it is easily constructed. However, by photographic means, disks of very high resolution can be constructed; resolution up to 8000 pulses per revolution is attainable! The mask for a moderate-resolution shaft encoder is shown in Figure 8.11. With these masks, the encoder itself will seldom limit the resolution of the measurement; the limiting factor lies typically in the mechanical linkages.

A means of calibrating the encoder–counter combination is usually required; some position must be considered zero. Often, a limit switch can be set up as in Figure 8.12, and the motor can be operated until that limit is detected; at that point either the hardware or software counter is cleared.

Another automated procedure sometimes used with monochromators is to use a reference light source such as an alignment laser. The wavelenth is scanned under computer control until that peak intensity is found; the counter can be calibrated by the result. In some cases where recalibration is tedious, a battery-powered series of up–down counter-integrated circuits can maintain the count indefinitely.

8.2.4. Absolute Shaft Encoders

Absolute shaft encoders read out the position information for a single revolution by using concentric tracks for each binary bit of the data word. If, for example, the outer track is blazed with 256 apertures, the successive inner tracks would contain 128, 64, 32, 16, 8, 4, and 2 apertures. Eight detectors, one for each track, will generate a parallel output unique for every position. The "count" is not lost when power is lost, and the calibration need be performed only at installation.

At first one might expect the mask on an absolute optical encoder disk to generate a binary sequence, but such a pattern presents a problem on

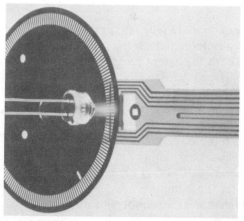

FIGURE 8.11. A moderate-resolution shaft encoder. Without the housing, the components are visible. An LED light source illuminates the mask behind which are a pair of phototransistors. (Photograph reprinted by permission of Sensor Technology, Chatsworth, CA.)

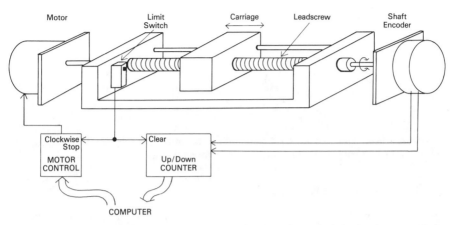

FIGURE 8.12. Calibration of linear position. At the end of travel of the leadscrew, a limit switch can be used to clear an up–down counter that counts pulses from the shaft encoder to measure the position of the carriage.

transitions such as that from seven (0111_2) to eight (1000_2). All 4 bits must change exactly simultaneously, and that requires unreasonable mechanical precision, particularly in very high-resolution units; (encoders with 13 or more output channels are available although expensive). Consequently, alternate coding schemes are used in which only 1-bit changes at a time, the most popular being the **Gray code** described in Chapter Two. Subsequent conversion of gray code to binary may be placed in TTL code-conversion circuits, but is better handled in software in most cases.

FIGURE 8.13. Absolute encoder masks. The mask on the left follows a binary sequence. Although not all of the markings are visible in this copy, it has 12-bit resolution. Using the mask on the right, the output follows the Gray code.

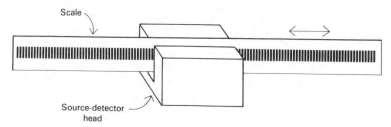

FIGURE 8.14. A linear encoder. If the length of travel is not too great or linear position is not easily converted to rotary position, this encoder is useful.

8.2.5. Linear Encoders

Linear encoders such as that shown in Figure 8.14 avoid the need for linear to rotary conversion; either absolute or incremental patterns can be used. Resolution up to 1000 transitions per inch are available with a length of up to a foot. An effective low-cost, low-resolution linear incremental encoder can be built from an optointerrupter and artists' decorative tape.

If we were to compare linear and shaft incremental encoders in a task where either could be used, the shaft encoder will usually be preferred since much greater resolution is achievable. For example, if an object, such as a disk drive head, a mirror mount, or a monochromator sine bar, is translated by a leadscrew, the position is already controlled by the number of rotations. With 32 threads per inch, a shaft encoder will require only 30 transitions per revolutions to provide the resolution of a more expensive 1000 transition per-inch linear encoder. Applications of the latter are limited to cases in which no conversion from linear position to rotation is available.

8.3. TACHOMETERS

Tachometers provide information about the *rate* of rotational displacement. The two primary techniques are based on the dc generator and the optical incremental shaft encoder.

8.3.1. DC Tachometers

DC tachometers are electrical generators from which a signal, not power, is drawn; therefore they fall into the active transducer category. Since no power is drawn, the windings can be very light in order to reduce friction.

Typically a dc tachometer is mounted on the same shaft as the motor that actuates the rotation; very precise speed control is possible when this type of tachometer is used in the feedback loop of an OA that drives the motor (Fig. 8.15). Otherwise the output can be fed to the input of an ADC.

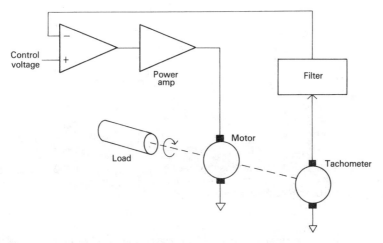

FIGURE 8.15. Closed-loop control of a dc motor. The tachometer feeds back rotational velocity information to stabilize the speed of a dc motor.

8.3.2. Optical Tachometers

Optical tachometers are based on the incremental encoder, although high resolution is not usually required. A useful variant of an optical encoder employs an optical interruptor in a reflective mode; in Figure 8.16, an LED and phototransistor are shown integrated into a housing that makes possible

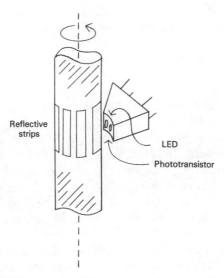

FIGURE 8.16. An optical tachometer. The pulses from the optointerruptor either can be converted to a dc level or can be measured as frequency.

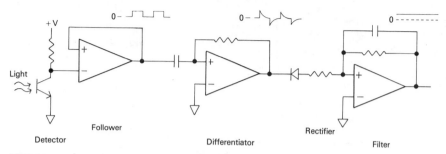

FIGURE 8.17. Conversion of optical tachometer pulses to a dc level. The pulses are differentiated to remove high and low dc levels, half- or full-wave rectified, and subjected to filtering with a long-time constant.

the detection of a reflective surface such as a piece of reflective tape on a rotating shaft.

In order to accurately and precisely convert the resultant pulses to a rate, the pulses can be counted for a preset period either with a hardware counter or by using interrupts and a software counter. This method is capable of very high precision if the count is made over a sufficient period, but counting does not provide an instantaneous measure. For faster response, the period between pulses must be measured.

If instantaneous *control* of rotational velocity is desired, the pulses can be converted to a dc voltage as part of a feedback loop similar to that shown for the motor tachometer. As shown in Figure 8.17, the square wave can be differentiated, half- or full-wave rectified, and low-pass filtered to produce a voltage proportional to the rate at which transitions are detected. Alternately, the pulses can trigger a monostable, and the resultant asymmetric square wave can be low-pass filtered to generate a dc level.

8.4. TRANSLATION CONTROL

Frequently, the positions of objects must be changed or adjusted in the course of an experiment. In some cases, such as when a valve is adjusted or the micrometer on a mirror mount is turned, the action is inherently rotary; in others, the object must be moved linearly, but this is accomplished with a screw type of drive (leadscrew) so that it may be considered rotary. Generally, positioning requires the use of motors as output transducers.

The final objective of the rotary action either may be to continuously drive an object such as a light chopper or a pressure pump or to move accurately into position an object such as a camera, mirror, or sample holder. The former problem can require accurate speed control while the latter problem usually requires accurate position sensing.

In this section, we will consider the output devices, various types of motors. There is a variety of motors that can be used for control of position or to drive other mechanisms. In this section, the control of speed and position of several useful types are outlined.

8.4.1. Synchronous Motors

The speed of these motors, usually operated from 120 V ac, is determined by the power line frequency. They are useful where a rather slow but reasonably constant rate of speed is required. The speed is fixed at manufacture and can only be changed by changing the line frequency.

Synchronous motors should be considered unidirectional, although change of direction is possible for some by using an appropriate phase-shifting circuit. They are easily switched with solid-state relays. Synchronous motors have been used to drive the chart on strip-chart recorders; often a change in chart speed required changing the motor.

When operating common synchronous motors, one should bear in mind that the line frequency is not a perfect constant. The electrical power companies are in the business of selling power not frequency, and so there may be a variation of up to about 1% in the period of a cycle. Measured over a 24-h period, the power company does make the frequency a constant by making periodic adjustments that accommodate the many clocks that use synchronous motors. Attempting to vary the speed of synchronous motors is possible, by generating the ac power from dc, but it is not a worthwhile endeavor when other types of motors are available.

8.4.2. DC Motors

Direct current motors furnish straightforward control of direction and speed since the direction is determined by the polarity of the device and speed is determined by the power provided. Direct current motors come in all sizes and varieties from miniature high-speed motors to larger motors of all speeds; gear systems are included in many motors produced for instrumentation, the latter being quite expensive unless purchased from surplus dealers.

The speed of a dc motor is related to the power that can be approximated as being proportional to the *time integral* of the voltage. Two approaches follow for **speed control**: variation of the dc voltage and pulse-width modulation. In either case, the speed is only controlled in a relative sense; it only can be increased or decreased. The applied voltage, current, and torque interact nonlinearly and these relationships depend on the type of motor. A separate tachometer is required to absolutely determine the speed.

The dc voltage can be continuously controlled by using a DAC. The provision of sufficient current requires an adequate power amplifier; linear amplifiers were described in Chapters Five and Seven. This approach also provides the start–stop mechanism directly, since the voltage can be lowered to zero. The motor can be stopped quickly with a circuit that can momentarily reverse the polarity of the power.

Servo motors are part of *positioning systems* designed on the principle of feedback. A drive signal is sent to the servo control system representing the desired position. The actual position is continuously measured by a transducer, such as a potentiometer configured as a voltage divider, and is fed back to the servo control system which develops the difference or error signal. That error signal is amplified to activate the motor. As the motor moves, the error signal moves toward zero, and when the difference between drive signal and feedback signal is zero, the motor stops.

The essential concept of a servo system is the feedback loop; the difference between the requested position and the actual position drives the motor. Usually, the loop is closed in hardware. The computer generates the drive signal with a DAC, and that voltage is compared with the output of the potentiometer that encodes the position. The loop also can be closed by the computer. The computer can read the position with a potentiometer and ADC or with a shaft encoder; the difference between that value and the desired position determines whether the computer should turn on the motor.

8.4.3. Stepping Motors

So long as high speed is not required, stepping motors are ideal translation devices to be placed under computer control. Stepping motors rotate *incrementally* to discrete positions that are fixed by the positions of coils within the motor. Each step of rotation is commanded separately, a move from one discrete position to the next, so complete control over position is obtained. Stepping motors are frequently the choice for robotics applications and other precision, interactive-control applications.

Since the steps are discrete, a stepping motor can be controlled with great precision; the error in moving to a position is noncumulative, limited to the uncertainty in the last single step. Maintaining the position information is simply a matter of keeping track of the number of steps that have been taken; no feedback mechanism for speed or position is required so long as the motor is able to respond to the step commands. Consequently, it is easily placed under digital control.

The operation of a stepping motor can be understood by referring to Figure 8.18, a schematic of a permanent magnet (PM) stepping motor with four phases per revolution. The stationary component or **stator** consists of four separate windings while a permanent magnet forms the rotating component or **rotor**. The rotor, moving in discrete steps, will seek the position that experiences the strongest pull. Since each stator winding is independent, that position is precisely determined.

In the simplest case, the windings are sequentially energized in a **wave excitation** sequence of states (Fig. 8.19); each winding develops a north-pole facing the permanent magnet, and the south pole of the rotor follows the

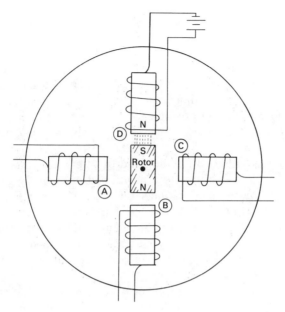

FIGURE 8.18. A stepping motor with four poles per revolution. The rotor is a permanent magnet which is attracted to the energized winding.

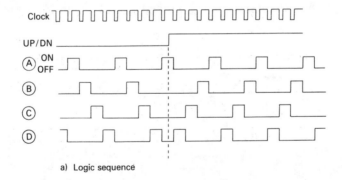

a) Logic sequence

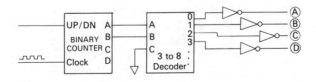

b) State generator

FIGURE 8.19. Wave excitation. (*a*) The logic sequence shows only one winding to be active at a time. (*b*) A state generator circuit for wave excitation.

sequence. The rate of the switching sequence determines the speed, and the direction of the switching sequence determines the direction of rotation. A simple circuit for generating the logic state sequence also is shown.

Energizing only a single phase at a time makes inefficient use of the windings when power and speed must be maximized. Improvement is gained by energizing two adjacent windings simultaneously in a **two-phase excitation** (Fig. 8.20). The equilibrium position for any state lies between the pair of windings that are energized (Fig. 8.21). This is the most common configuration.

The motor's resolution can be doubled by alternating the wave and two phase steps; the rotor is alternately positioned on a winding and between the windings. The **half-step excitation** sequence is shown in Figure 8.22. The caveat in this procedure is that not all steps will have equal power, and consequently where the load is large, they may not have equal size. Usually, this sequence will be applied only to smooth out the full steps, and the stopping position should always be a position where two windings are excited.

The process of increasing the power of the motor can be carried a step further. Under the sequences shown thus far, the south pole of the stator does not interact with any electromagnetic field; all of the windings could be energized simultaneously if the polarity of the electrical power also was controlled (Fig. 8.23). The **bipolar two-phase excitation** state sequence and the **bipolar half-step excitation** state sequence of Figure 8.24 provide that

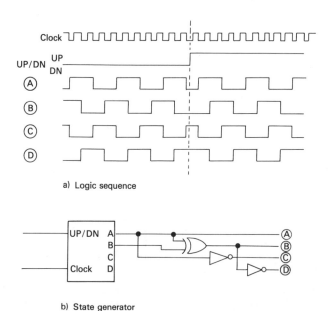

a) Logic sequence

b) State generator

FIGURE 8.20. Two-phase excitation. (*a*) The logic sequence shows two windings energized at any given time. (*b*) A state generator circuit for two-phase excitation.

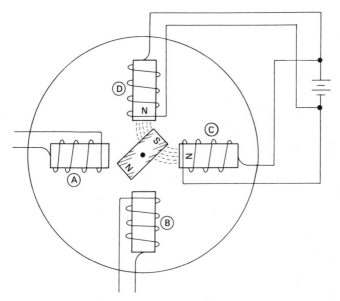

FIGURE 8.21. The position of the rotor during two-phase excitation.

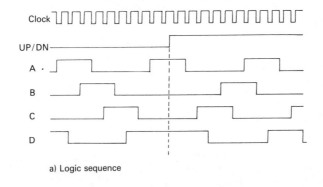

a) Logic sequence

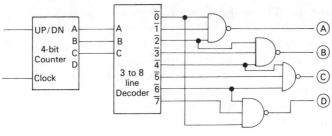

b) State generator

FIGURE 8.22. Half-step excitation. This sequence generates all the states of both wave and two-phase excitation. It produces higher performance during rotation.

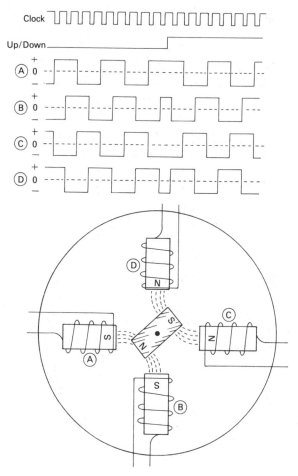

FIGURE 8.23. Bipolar two-phase excitation. The logic sequence is the same as for unipolar two-phase excitation except that both ends of the rotor are attracted, giving the motor greater power.

additional capability at the cost of significantly increased power-amplifier circuit complexity.

If there are only four stator windings per revolution, the motor will make large, full steps of 90°. The resolution is increased by employing repeating sequences of the four phases. In practice, the largest step size for commercial stepping motor is typically 15°, 24 sets of windings per revolution. Stepping motors are commonly manufactured having step angles as small as 0.72°, 500 sets of windings per revolution.

Since a stepping motor is driven in discrete steps at controlled frequencies, there can be frequencies at which the vibrations will destructively resonate, resulting in loss of torque. The speed-torque relationship for a typical

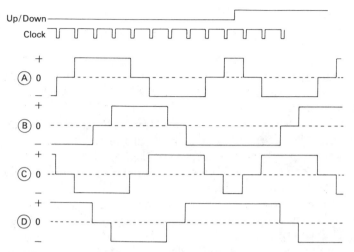

FIGURE 8.24. Bipolar half-step excitation. This is the bipolar analog of Figure 8.22.

stepping motor is illustrated in Figure 8.25; the curve represents the maximum torque developed while the motor is able to operate without losing a step. Two characteristics that limit the step rate are apparent: the general loss in torque with speed, and the location of nearly discontinuous regions of very low torque.

The drop in torque with increasing speed is a consequence of the reduced current that passes through the winding as the step rate increases; the sudden flow of current through this inductive load each time the winding is energized generates a back emf opposing the flow of current, so that the average current decreases with the step rate.

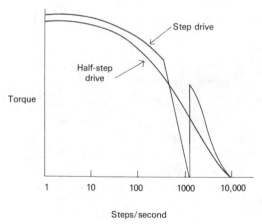

FIGURE 8.25. Speed-torque relationship for stepping motor. The curves for slew and for start–stop conditions will differ. The step-drive curve shows a discontinuity where the motor is in resonance.

The speed–torque curve is different for maximum step rates under slew (continuous high-speed) and start–stop conditions; of course, more torque is required during acceleration than while maintaining constant speed. In order to obtain the very highest step rates without error, the step rate must be ramped to maximum speed, and the stopping position must be anticipated so that the speed can be ramped to a stop. For a given torque requirement, the ratio of maximum slew speed to start–stop speed may exceed 10.

The discontinuity is due to a resonant frequency at which the motor will not respond. The resonance is the result of the oscillations that occur in each discrete step as the rotor overshoots the equilibrium position; the ringing (Fig. 8.26) can be reduced either by complex electronic means or more commonly by providing mechanical damping in the form of external friction pads or internal viscous fluid. The use of half-step excitation also can reduce the torque loss at the resonant speed because smaller steps result in less ringing. The natural resonant frequency decreases with increasing load.

The overall circuitry for unipolar drive is divided into the **state generator**, which develops the excitation sequence, and the **power amplifier** (Fig. 8.27), which converts from TTL level to the requirement of the motor.

The inductive nature of the load requires special attention; each cycle turns on and off a magnetic field, and when that field is turned off, a back emf is naturally generated. A resistor in series with the winding improves response since the motor is operated at a greater supply voltage, which opposes the back emf of the winding without increasing the steady-state current.

For example, a motor with a winding resistance of 8 Ω and a maximum current per winding of 1 A can be driven directly by an 8-V supply. Adding a 40-Ω resistance in series and raising the supply voltage to 48 V will maintain the 1-A steady-state current, but the 40 V "over-drive" opposes the back emf and the maximum step rates can be increased by as much as a factor of three; a ratio of series to winding resistance of eight typically gives a

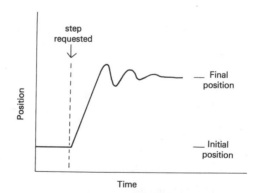

FIGURE 8.26. Ringing in the position after a step. When the step rate equals the ringing frequency, resonance quenches the torque.

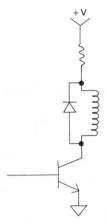

FIGURE 8.27. A simple power amplifier. For highest performance, the resistor, and consequently the voltage, should be as large as possible. Much of the electrical power will be dissipated as heat in the resistor, however.

factor of 12. The limitation occurs in the Joule (i^2R) heat which must be dissipated by the series resistor, in this case 40 W.

An alternative method of achieving a high overdrive without the power loss of the series resistor is to use a chopper. Similar in concept to pulse-width modulation, a unipolar chopper produces voltage pulses at a frequency much greater than the step rate such that the *average* current does not exceed the specified rating. Since this may imply a chopping rate over 10 kHz the design can be complex, suggesting the use of commercial units.

The circuitry that controls the power amplifier is the **state generator**. It can be designed from TTL logic in a variety of ways and examples of several are illustrated in Figures 8.19, 8.20, 8.22, and 8.28. However, the hardware of the state generator can be minimized by utilizing 4 bits of an output port and calculating the bit patterns for the desired excitation sequence. In fact,

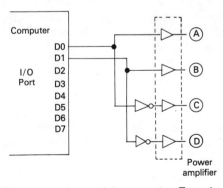

FIGURE 8.28. Using a computer as state generator. Two-phase control is illustrated.

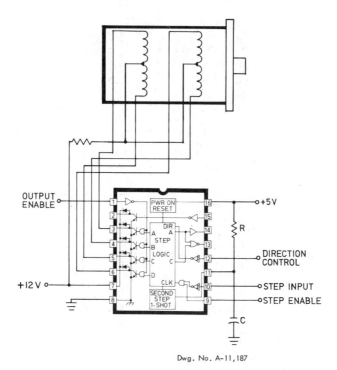

Dwg. No. A-11,187

FIGURE 8.29. An integrated circuit-stepping motor controller (Sprague UCN-4204A). (Courtesy of Sprague Electric Company, Mansfield, MA.)

only 2 bits are required for two-phase drives (Fig. 8.28) since pairs of windings have complementary states.

An integrated combination of the logic circuit and the power amplifier is available in an IC, (Fig. 8.29) that can handle 500 mA/winding. A single input pulse moves the stepping motor by a full step. A variation of the half-step mode is included to improve stability; the input pulse first generates a half-step of duration determined by the monostable after which the remainder of the full step is generated. If necessary, a separate higher capacity power amplifier can be used.

8.5. FORCE

When a force is applied to a solid body, the **stress** (force per unit area) produces a deformation of the body. The extent of the deformation (bending, compression, extensions, etc.) is measurable as displacement measured in terms of **strain**, the change in length divided by the overall length. According to Hooke's Law, the relationship between stress (and therefore force) and strain is linear within an elastic region.

The stress can be applied to the body in a variety of ways: tension (stretching), compression, bending, shearing, or torsion (twisting). As long as the strain remains within that elastic region, the change in length is reversible and directly related to the force. To exceed the limit is catastrophic: the elasticity is lost and the body will eventually break.

The change in dimension as a consequence of force is the basis for many devices that measure various forces: stress, pressure, weight, etc.

8.5.1. Strain Gages

To measure strain, we must measure very small changes in one dimension of a body. To achieve this, we start with our capability, using a bridge circuit to measure very small changes in electrical resistance. The resistance of a conductor is proportional to length, so the increase in length due to strain results in a proportionate increase in resistance. This change in resistance is the basis for a measurement of the forces producing the strain. A device for measuring stress in this way is therefore called a **strain gage**.

A strain gage is characterized by its **gage factor**, the ratio of the relative change in resistance, $\Delta R/R$, to the relative change in length due to strain, $\Delta L/L$, where R and L are the resistance and length of the strain gage. A typical strain gage material, constantan metal, has a gage factor of about 2.0: if the length can be stretched by 0.1%, the resistance will change by 0.2%.

Strain gages are manufactured in many different forms, and a useful example is the **foil strain gage**, illustrated in Figure 8.30. A metal foil is bonded

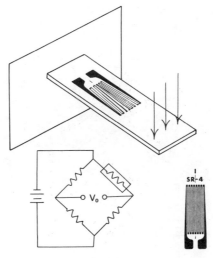

FIGURE 8.30. A strain gage. The gage shown is die-cut from foil and bonded to a bending beam. The bridge circuit can be used to measure small changes in resistance. (Courtesy of BLH Electronics, Canton, MA.)

to an insulating carrier layer and the pattern is photoetched; with suitable adhesives, it can be bonded to any medium subjected to stress. The pattern shown maximizes the overall conductor length in one dimension while minimizing the width. If this gage is cemented to a beam subjected to a bending stress, the resultant change in resistance can be used to measure the bending force.

Typical observed changes in resistance to be measured may be as little as one part in 350, but, as was shown earlier, a bridge is a useful tool for measuring small resistance changes (Chapter Five). However, we should keep in mind that the sensitivity of the bridge circuit would be doubled if we could place a pair of these variable resistances in opposite arms of the bridge. Two strain gages would then be used. Further sensitivity gains can be obtained by arranging the bridge so that the remaining arms are sensors that respond to the same phenomenon, but in the opposite direction. The bending beam shown in Figure 8.31 achieves this; the strain gages on top experience extension, while those on the bottom experience compression.

Strain is not the only phenomenon to produce changes in the resistance of a conductor; temperature effects also are very important. Whenever significant changes in temperature are to be expected, there must be compensation. In a bridge circuit, compensation is obtained if the bridge arm adjacent to the sensing arm responds identically to temperature but differently to the strain. An identical gage bonded to the same or an identical body can accomplish this. That requires either a dummy gage, responding only to temperature, or a gage configured for opposite response to that of the first gage, as shown in Figure 8.31. As long as these elements are at the same temperature, the temperature effect will cancel.

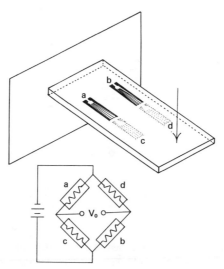

FIGURE 8.31. Strain gages mounted on a bending beam for full-bridge measurement. This arrangement is sensitive and is compensated for temperature variations.

A sampling of available strain gage configurations is shown in Figure 8.32. Gage *A* is used when strain is produced on a single axis only. The remaining configurations can be used to measure strain produced on several axes. The pattern of gage *B* is suitable for measurement at 90° angles or for measuring strain in one dimension with temperature compensation. Gages *C* and *E* use three elements to measure the direction of the strain. Torque in a rod is measured in a compensating configuration with *D*. Rosette *F* is designed as a 4-element compensated bridge for diaphragms.

Semiconductor strain gages utilize the resistive changes in semiconductor materials when force is applied. The strain gage factors are high, up to 30 times greater than those of metal gages, but the temperature sensitivity is much greater than that of the resistive strain gage. Consequently, its use is limited to applications in which temperature can be carefully controlled. The semiconductor surface may even be light-sensitive, so the gage might have to be shielded from light.

8.5.2. Force Measurement

Several examples of force measurements with strain gages are shown in Figure 8.33. In each a force in one or more dimensions produces strain. The magnitude of the strain is determined by the mechanical characteristics of

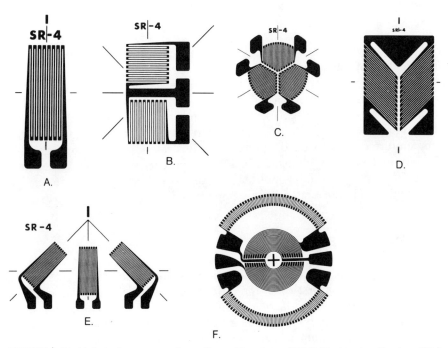

FIGURE 8.32. Foil strain gage configurations. (Courtesy of BLH Electronics, Canton, MA.)

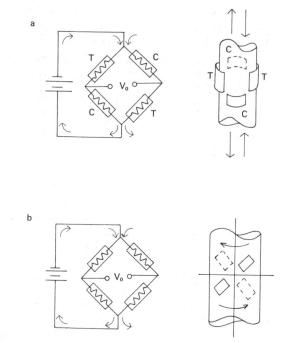

FIGURE 8.33. Strain gage applications. (*a*) Temperature-compensated measurement of compression or extension as in a load cell. (*b*) Temperature-compensated measurement of torque. (Courtesy of BLH Electronics, Canton, MA.)

the body to which the strain gage is attached, so that the force range is independent of the strain gage itself.

The first example is a load cell that produces compression in the pair of strain gages mounted vertically, while the horizontally mounted gages become compensating gages. The lower example is used to measure torque; the configuration is sensitive to the direction of the torque and compensates temperature effects. A bending force was illustrated by the strain gage configuration of Figure 8.30, using gages on either side of the bending element; since the responses will be opposite, the gages are placed in adjacent bridge arms and temperature effects are compensated.

8.6. PRESSURE

The techniques for the measurement of pressure fall into two categories, both indirect. In the first case, the difference in pressure across the transducer converts *force* into *displacement* which is proportional to pressure; that displacement can be measured by any of several methods. The second case, suitable for vacuum measurement only, involves the effect of molecular density on some transducer property, typically heat conductivity.

8.6.1. Pressure-to-Displacement Converters

Pressure-to-Displacement Converters sense a pressure *difference*; the displacement results from the difference in forces between the "outside" and "inside" as illustrated in Fig. 8.34. The outside may be the atmosphere for which the unit is commonly called a **gage pressure transducer**, a "zero" vacuum for which an **absolute pressure transducer** is used, or it may be some other reference pressure for which a **differential pressure transducer** is used.

Three types of converters, shown in Figure 8.35, are common. The **Bourdon Tube** can be used with the highest pressure, up to 10^4 lbs./in^{-2} (psi). The noncylindrical tube, configured either as a curve of less than 360° or as a helix, has a tendency to straighten as the pressure inside increases compared with the pressure on the outside; the displaced end of the tube can be attached to any transducer capable of responding to small displacement. The second type is the **diaphragm** which is a circular surface bonded at the circumference. The displacement of a diaphragm is proportional to the total force exerted on it, the product of pressure and area. Depending on the material of which they are constructed, diaphragms are typically incorporated in transducers with ranges from 0–15 psi up to 0–500 psi. The third group, **bellows**, are the most sensitive converter in this category, useful in ranges from 0–0.30 psi up to about 0 to 40 psi.

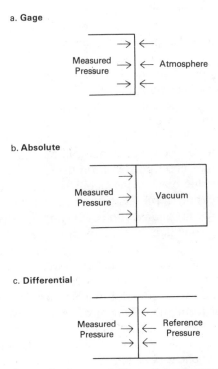

FIGURE 8.34. Gage, absolute, and differential pressure measurement.

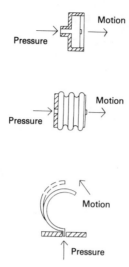

FIGURE 8.35. Sensing elements for pressure measurement. From top to bottom, a diaphragm, a bellows, and a Bourdon tube. Each requires a position or strain transducer.

8.6.2. Displacement to Electrical Transducing Elements

Displacement to electrical transducing elements must be capable of responding to very small displacements. Many methods have been developed, some of which were mentioned as general displacement transducers. The common types are based on strain or capacitance measurement or employ differential transformers. Many others are in use: potentiometers, optical encoders, and photoelectric sensors are coupled to Bourdon tubes, to name a few.

Resistance strain gage pressure transducers consist of strain gages, such as those described in the previous section, bonded to a diaphragm. A gage pattern suitable for a diaphragm was illustrated in Figure 8.32*f*. Using this arrangement, the entire bridge resistance can be bonded to the diaphragm. The two halves of the center portion form the opposite arms of the bridge, and the outside conductors, having about the same resistance but being subjected to compression, form the two remaining arms. Consequently, this passive transducer requires connection of only a regulated power supply and a differential input analog interface. Many pressure ranges can be obtained, the pressure limit being dependent on the construction of the diaphragm.

The **capacitance-type pressure transducer** utilizes the principle that the capacitance of a pair of parallel plates is inversely related to the distance between them. Therefore, if a metal diaphragm is stretched between two plates (Fig. 8.36), it forms one element of each of two capacitors. As the diaphragm deforms, the capacitance of one element decreases while the other increases. One simple method of measuring these changes in capacitance is to use the capacitors as part of an oscillator circuit; for example,

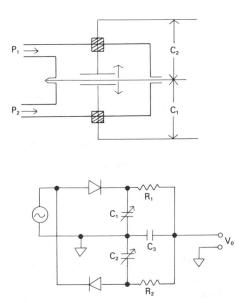

FIGURE 8.36. A capacitance pressure transducer. Small movements in the diaphragm create large changes in the two parallel-plate capacitors that are created.

the frequency produced in the voltage-to-frequency circuits described in Chapter Five is directly proportional to the capacitance in the feedback loop of the OA, and if a part of the transducer forms that capacitor, the measurement is reduced to a simple frequency measurement.

In the example illustrated in Figure 8.36, two capacitors are formed, one increasing and the other decreasing with the movement of the metallized diaphragm. In the accompanying circuit, the two variable capacitors are charged with the opposite polarity; any *difference* in the two capacitances will appear as a net voltage across capacitor C_3, and a proportional dc output.

LVDT pressure transducers combine the displacement produced by a Bourdon tube, bellows, or diaphragm, with the displacement measuring capability of an LVDT, (see Section 8.2.5.).

8.6.3. High Vacuum Transducers

The transducers for medium-to-high-pressure measurement are unsuitable for vacuum work since they are inherently differential measurement devices responding only to *differences* so that, for example, the difference between 10^{-1} and 10^{-5} torr, when compared by a gage transducer against atmospheric pressure, is negligible. Consequently, the high vacuum techniques are absolute transducers that respond to the number of molecules present.

Several measure losses of heat from a heated device; in a vacuum, heat lost by convection and conduction is dependent on the pressure of the atmosphere around it. As the vacuum increases, the loss of heat to the sur-

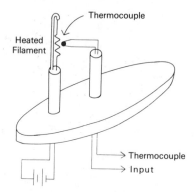

FIGURE 8.37. A thermocouple vacuum gage. Gas molecules carry off heat from the filament; therefore the temperature of the thermocouple can be used to monitor the vacuum.

roundings decreases, and the temperature must increase. These techniques are useful to about 10^{-3} torr.

The **Pirani gage** is dependent on the change in resistance of a heated filament; the temperature decreases at higher pressures as the environment removes heat. The **thermocouple gage** (Fig. 8.37) is similar, but uses a thermocouple or thermopile to directly measure the temperature of the filament.

Alternating current is applied to the **Hastings heated thermopile** so that it operates in a self-heating mode; the increase in temperature from this heat depends on the ability of the surroundings to remove that heat which is in turn dependent on the pressure. The increase in temperature produces a dc

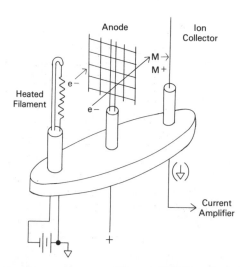

FIGURE 8.38. An ionization vacuum gage. The ion collector attracts ions, ionized by electrons from the heated filament and accelerated by the anode.

voltage component in addition to the applied ac, and that dc component is measured separately. **Thermistor vacuum sensors** use a similar principle, although they are simpler; the resistance can be measured in a bridge circuit with the thermistor operated in a self-heated mode. That is, the applied voltage is high enough that i^2R heating occurs in the thermistor, and the resultant increase in temperature depends on the amount of heat carried away by the environment.

To measure much lower pressures, **ionization gages** (Fig. 8.38) are used which are similar to triode vacuum tubes. A hot filament emits electrons that are collected at the positive anode. If residual gas atoms are present, the electron beam creates positive ions that are collected at a wire held near ground which is negative with respect to the anode. The resultant ion current is proportional to the density of gas molecules from 10^{-4} torr–10^{-10} torr.

8.7. FLOW TRANSDUCERS

The list of techniques for the measurement of flow rate is even longer than the previous lists. Representative varieties respond to mass, thermal mass, differential pressure, and tracers.

The force resultant from the moving mass of the flowing fluid can be converted to several other forms. In a **turbine flowmeter** (Fig. 8.39) a rotor is placed in the tube that will spin at a rate directly proportional to the flow rate. The rate of rotation can be sensed by a variety of means. If a permanent magnet is placed in the rotor, it can induce current in a coil around the transducer, or the periodic changes in magnetic field can be sensed by a Hall-effect transducer. Commercial units can be used with flow rates from 0.08 to 26,000 gal/min.

A vane placed in the flow (Fig. 8.40) will bend at an angle related to the flow rate; the degree of the bend can be determined by several methods. A potentiometer attached to the vane will measure the angle, or the strain in a cantilevered vane can be measured with a strain gage. By placing a radioactive source on the vane, the displacement is determined from the change in radioactivity discovered by a fixed detector.

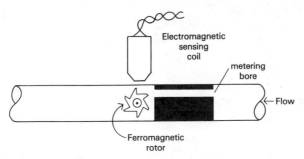

FIGURE 8.39. A turbine flowmeter. The rate of the rotor is a measure of flow rate.

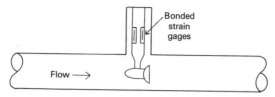

FIGURE 8.40. A vane flowmeter. Deflection of the vane is detected by bonded strain gages.

The flow rate measurement can be converted to a **differential pressure measurement** by constrictions such as is found in a Venturi tube (Fig. 8.41). The square root of the pressure difference is directly proportional to the flow rate. The differential pressure can then be measured with a strain gage-based pressure transducer.

Workers using low gas-flow rates, such as are encountered in gas chromatography, are familiar with the rotameter, a tapered tube in which the flow supports a sphere. Since the force produced by the flowing fluid decreases as the ball rises, the position is related to fluid velocity. The **LVDT rotameter** (Fig. 8.42) employs a differential transformer such as those described for translational measurement, but in which the core is suspended by the flow; small changes in position related to flow rate can be precisely measured.

The **hot wire anemometer** measures the cooling effect of a fluid on a heated

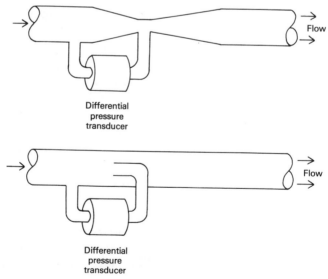

FIGURE 8.41. Differential pressure measurement. In each configuration, a pressure differential is created between the two inlets to the pressure transducer, and that differential is a measure of flow rate.

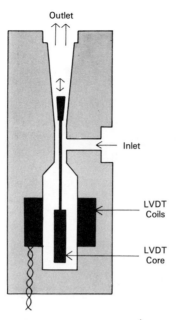

Outlet

Inlet

LVDT
Coils

LVDT
Core

FIGURE 8.42. An LVDT rotameter. Flowing gas lifts the core, creating a signal.

probe. A fine platinum resistance wire is supported across the tube and heated by passing a constant current through it, and the resistance changes with temperature. If a fluid flows over the wire, the temperature will drop and the resistance will change; the result (Ohm's Law) will be a measurable change in the voltage across the probe. Bonding to glass provides the mechanical support necessary for higher flow rate. If the fluid is electrically conductive, the wire must be insulated. Devices similar to the Hastings thermopile, discussed in the previous section, also can be suitable.

Time-of-flight methods are precise techniques used for fluids in tubing. A thermal-pulse time-of-flight liquid flow meter (Fig. 8.43) incorporates a pair of thermistors: one for heating and one for sensing. A 100-mW pulse delivered to the first is detected about 1.5-in. downstream by the second; the response is differentiated twice in order to discriminate against temperature drift.

In a gas chromatography carrier-gas line, a pulse of an alternative carrier gas can be injected into the line and detected by a thermal conductivity detector downstream. For example, a small pulse of helium injected at regular intervals into a nitrogen stream will have little or no effect on the separation, but the dramatic change in thermal conductivity can be detected easily downstream.

Finally, the **magnetic flowmeter** has applicability when a *conducting* fluid passes through the tube. The principle, illustrated in Figure 8.44, is based

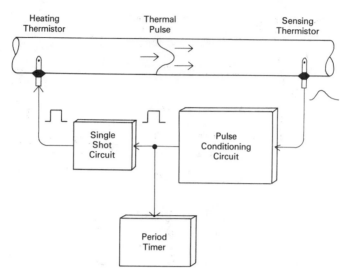

FIGURE 8.43. Time-of-flight measurement. A power pulse to the first thermistor creates a thermal pulse detected by the second thermistor. The time delay is measured.

on Faraday's Law of Electromagnetic Induction: a voltage will be induced in a conductor moving through a magnetic field. As the conductive fluid moves through that magnetic field, the electrodes sense the voltage induced in the moving conducting fluid. The signal is proportional to the flow rate and the strength of the magnetic field. The conductivity must exceed only about 10^{-5} Ω^{-1}/cm.

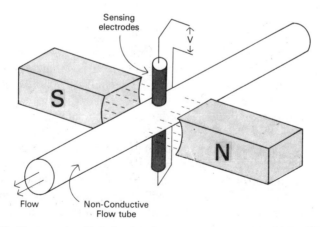

FIGURE 8.44. Electromagnetic induction flowmeter. If the fluid is conductive, the magnetic field induces a potential that can be detected by the sensing electrodes.

8.8. LABORATORY ROBOTICS

Almost everyone must be familiar with the Hollywood science fiction concepts of computers and robots; these images place heavy emphasis on anthropomorphic qualities. On the other hand in Chapter One, the evolution of the concept of a laboratory computer was discussed, and the contemporary robot is more an extension of that small computer than a relative of the Hollywood creation. The laboratory robot advances the computer's programmability to the area of physical manipulations. That is, a robot is a *programmable manipulator for flexible automation.*

The list of the advantages of using a robot to manipulate objects is similar to the list that was drawn up for on-line computer use:

1. Changes in the experiment are made by changing the program rather than by changing the hardware.
2. Procedures can be duplicated exactly, and the robot's reproducibility over long periods is better than that of a human experimenter.
3. The robot can operate for long periods of time without a break, unattended, and without introducing changes in the experiment.
4. The cost per hour, when amortized over several years, can be substantially less that the cost of employing a person for a single shift.
5. A robot can operate in environments that are hazardous to human technicians; for example, appropriately fabricated robots can operate in the presence of high radiation for extended periods.
6. Every step in the operation can be documented by the supervisory computer.
7. The sequence of operations can be closer to an ideal serial sequence than the batch mode usually followed by a human technician.

To explain the seventh item, consider four objects, A, B, C, and D which must undergo operations 1, 2, and 3; the operations might be the dilution of a chemical sample, followed by heating, and analysis. Typically, in order to avoid confusion, a technician will subject all four to operation 1, after which the four will undergo operation 2, and then operation 3; as a consequence, samples A and D will experience a different time period between heating and analysis, for example.

A robot-driven scheme could move the objects in a pipeline. At one point, while sample C undergoes operation 1, sample B and sample A undergo operations *2* and *3*, respectively. Upon completion, all move by one operation. Consequently, all experience the same timing.

A robot consists of three components. It must have a **hand** or **end-effector** (which may be a gripper or a specialized end-effector for performing other tasks), an **arm** (with a drive system to move the arm in the available directions), and a **controller** (to direct and coordinate movements).

In this section, we will review some of the coordinate systems used by robot arms, discuss the types of drive systems and hands, and explain how robotic systems can be installed. Most of the discussion will be related to two laboratory robots currently on the market, shown in Figures 8.45 and 8.46.

8.8.1. Arm Coordinate Systems

A combination of positioning components are required for a robot to be able to reach objects located in random positions around it, and many different combinations have been developed. In this section, two will be described.

The **cylindrical-coordinate robot** of Figure 8.45 has the capacity to reach objects within a hollow cylinder (not including the space occupied by the robot itself.) The hand can be moved in and out; the arm can be moved up and down; the unit can be rotated on its base.

The **anthropomorphic robot** of Figure 8.46 has a spherical reach volume. However, movements are made using an articulated arm with some similarities to a human arm.

A brief comparison would suggest that the anthropomorphic robot is more versatile, since it can reach over objects in order to reach its target. However, control of that robot is more difficult than control of the cylindrical-coordinate robot, because movement in several joints must be coordinated

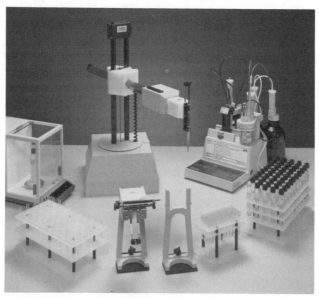

FIGURE 8.45. A cylindrical-coordinate laboratory robot. The end-effector is a dispensing pipette. Clockwise from the robot are a dispenser, tube racks, hand-holders, another tube rack, and a balance. (Photograph courtesy of Zymark Corporation, Hopkinton, MA.)

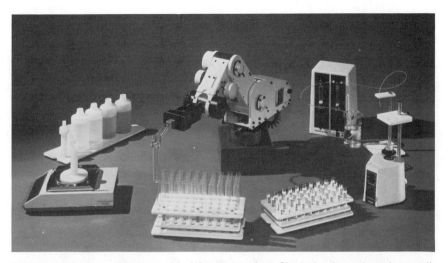

FIGURE 8.46. An anthropomorphic laboratory robot. Clockwise from the robot, a dispenser, vial-capping station, tube racks, and a balance are shown. (Photograph courtesy of Perkin-Elmer Corporation, Norwalk, CT.)

in order to achieve smooth and efficient movement; continuous computation of position coordinates is required.

Books on robotics (or one's own imagination) can show that many other configurations are used. Some robots operate on tracking systems to produce lateral motion. In others, the robot arm is the radius of a sphere within which the robot can reach. Additional combinations are used as well.

8.8.2. Drive Systems and Hand Design

The definition of programmable manipulators was applied to robots earlier, and robot development has always included a trade-off between adaptation for a specific set of manipulations and being as general as possible. The latter leads to the risk of being a "jack of all trades and master of none," but by moving too far away from generality, programmability is lost. In this section, comparisons will be made of the approaches to several components, drive systems, the hand's wrist, and the hand's fingers. The problem of verification of the operations also needs attention.

Drive Systems. Robots use a variety of drive systems, dependent on the cost, the speed that is required, and the payload that must be moved. Large payloads and high speeds, both of which are required in many manufacturing environments, are achieved by using pneumatic and hydraulic power systems. The table-top systems appropriate for laboratory operations are driven either by servo motors or by stepping motors.

Servo motors can be made to operate at high speeds for large displacements and moderate loads. Stepping motors are suitable only for lighter service, since speeds greater than a few revolutions per second are difficult to achieve, and moderate torque requires large power supplies and motors. However, stepping motors are easier to control digitally, since each movement is a precise step, and errors do not accumulate. If the load is kept well under the motor's capacity, a stepping motor can be operated in an open-loop configuration, that is, without feedback from a position transducer.

Hands. The robot hand has two components: the wrist and the fingers, and the more degrees of freedom that can be provided, the greater the flexibility in programming.

A fully functional wrist has three degrees of motion, but many robots possess fewer. The motions can be best explained by reference to a human forearm, wrist, and hand, held outstretched with palm down. **Roll** occurs when the palm is rotated to a palm up position. **Yaw** describes the movement of the hand from side to side without moving the forearm. **Pitch** is the angle of the palm with the forearm as the rigid palm and fingers are rotated up and down. Of these, for example, the robot of Figure 8.45 possesses only roll motion, whereas the robot of Figure 8.46 is capable of roll and pitch, and therefore is able to grip objects from more angles.

Changeable hands provided an alternative to full flexibility in a hand. For example, one common problem in laboratory robots is to use a pipette to dispense liquids; this tool cannot easily be manipulated by a robot hand. Consequently, in some robot systems, the gripper hand can be exchanged automatically for a hand with a built-in pipette.

Fingers. A great deal of effort has gone into creating a finger design which will handle many types of objects, including balls, eggs, and tubes; multijointed and multifingered hands can be used. Commercial laboratory robots concentrate on manipulating containers and specialized tools, so the end-effectors can be relatively simple. Containers are handled with the simple grippers seen in the figures, but greater flexibility can be obtained by allowing the robot to change hands under program control so that single-purpose end-effectors without fingers can be mounted. For example, the gripper hand might be replaced with an automated syringe for one particular operation.

Grippers may be operated either in a continuous mode, so that they can be closed, fully opened, or partially opened to any degree, or they can be controlled in a ''bang-bang'' mode, open or closed. The former requires a motor for actuation while the latter utilizes simple solenoids under single-bit control.

Verification. Manipulation errors may occur, due to the robot's imprecision or due to operator error, such as a missing component to be manipulated. Before advancing to the next step, the robot must have a way of detecting and handling errors. To simplify the discussion, we will consider the handling of a tube by a simple gripper; the problem is to determine when the tube is in the grip of the hand.

The most flexible, but most expensive, approach uses *sight*. An imaging system, mounted on the arm or separate from the robot itself, can be used to find objects. However, not only is the imaging system expensive, the real-time computational overhead in converting the pixel arrays to useful information will probably be overwhelming for most small computers.

One of the simpler methods involves *monitoring the gripper* by measuring the current to the motor that operates the gripper. When the grip on the object tightens, the motor begins to stall, and it draws more current. The rising current, measured with an ADC, provides grip information.

Many types of *touch sensors* are being developed. For example, arrays of microswitches or pressure sensors can decode the presence of an object. Silicone rubber pads, with embedded conductive matrices, can sense shapes and position.

In the specific example of the tube, either *a pushbutton* switch mounted on the table or *an optical interrupter* (Chapter Eight) with a large gap can be interfaced. After picking up a tube, the robot can verify the tube's presence by either using the tube in its grip to depress the switch or passing the tube through the interruptor's gap.

8.8.3. Installation

Before we move too quickly and buy a laboratory robot (currently costing $20,000 to $50,000), we should note that installation typically requires man-months to man-years of effort. Some of the time is needed to modify the experiment for manipulation by robot. Then the sequence of operations must be fixed. All of the ancillary equipment must be firmly fixed within the robot's reach and in a position that is efficient in terms of robot movement. Finally, the robot controller is programmed.

Ancillary equipment is fixed firmly to the table top within reach of the arm. A laboratory robot designed for chemical manipulations could require the following or other devices:

1. A bar-code reader for keeping track of individual objects and for identifying resultant data.
2. A balance so the robot can place an object on the balance pan after which the balance will inform the computer of the mass of the object.
3. A dilutor which, for when the system verifies that a tube is located under the tip, can be commanded to deliver a precision-measured volume of reagent.
4. A capping station where the tube can be placed within its grip; after the gripper clasps the tube's cap, the station rotates the tube until the cap is removed or replaced.

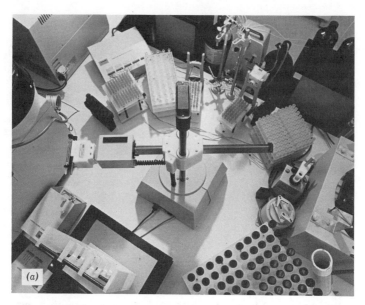

FIGURE 8.47. A system layout for a toxicology research. Note that the layout diagram is rotated 180° from the photograph. (Photograph and diagram courtesy of J. Rollheiser and K. Stelting, Midwest Research Institute, Kansas City, KS.)

5. Shakers in which two-phase mixtures are placed to aid dissolution and liquid–liquid extraction.
6. Centrifuges for removing suspensions from an analyte solution.
7. Automated instrumentation for making measurements. Besides balances, laboratory robots are typically mated with chromatographs, colorimeters, and nuclear magnetic resonance instruments.

The diagram in Figure 8.47 illustrates the layout for a robotics system used for chemical analysis. Note the positions of sample tube racks, the shaker, and the liquid chromatograph. Each must be firmly fixed to the table so that the robot arm will find it during its operations.

Once positioned, the robot must be trained. There are two ways of training a robot. One is to write into the program the exact Cartesian coordinates to which the object must be moved, and to specify the feedback information that is required before the next step can be undertaken. One of the risks that is involved results from the robot taking the shortest path, whether or not something is in the way; additionally, finding the precise coordinates is tedious.

The programming process is usually aided by use of a **training pendant** that allows an operator to remotely control each movement of the robot. Using buttons or joysticks on a hand-held console, the object can be moved safely either in straight lines or by circuitous routes that avoid collisions.

ROBOTIC SYSTEM

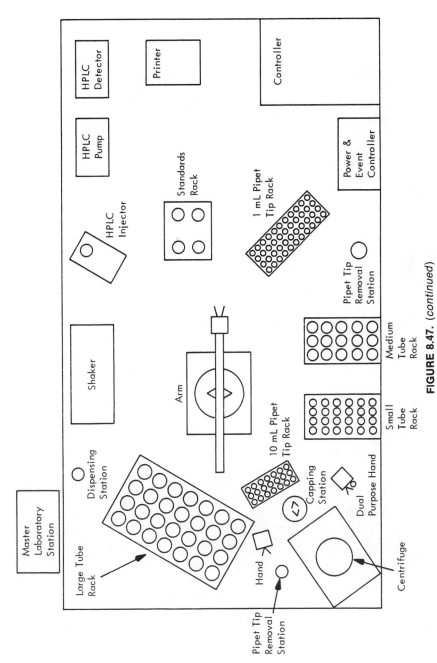

FIGURE 8.47. *(continued)*

Each component of an overall movement is "remembered" by the computer for playback later. Specific operations (such as to open the gripper) can be effected by commands in the robot-controller's language.

Laboratory robotics is yet in its infancy; as such, robots are bound to become easier to use, less expensive, and more flexible. Continuous advances in making difficult manipulations and sensing the experiment's status will help the robot to take over more and more routine work.

CHAPTER NINE

Data Communications

In an ideal world of computers and instruments, any instrument or I/O device would connect to any computer using standard cables and without modification. The software component of the interface for communication—the device drivers—would be part of each OS. Such compatibility requires absolute and well-defined standards that can be followed by all computer system developers. This ideal world has not yet arrived, but in this chapter we will examine the ways that computer-related devices can transfer information using *standard* methods.

The test of a communications standard is whether devices, purchased from different manufacturers, can be connected together for successful communications *without* a custom cable, custom software, or any changes to either device. In our personal communications, we are used to being able to buy telephones from any manufacturer and expecting them to plug right into the socket in our homes. The number of cases where that is possible for computer-related communication is small, but some moderately successful attempts have been made to follow a standard.

Parallel and Serial Transfer. The communications environment determines the type of standard to be used. Information is transmitted by standard interfaces 1 byte at a time, usually 8 bits at a time. In one method, the transfer is performed in **parallel**; the cable must include eight conductors to carry the byte *plus* at least one conductor for electrical ground *plus* conductors for the two devices to announce the sending and receipt of information (handshaking). The alternative to parallel is **serial** communication; the 8 bits of each byte are converted by the sender to a sequence of bits; upon receipt

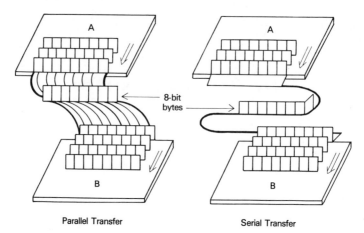

Parallel Transfer   Serial Transfer

FIGURE 9.1. Parallel and serial data communications. Parallel transfer sends the bits in each byte together, using a line in the cable for each bit. Serial transfer sends and receives the bits in each byte sequentially.

by the receiver they are converted back to a complete byte. These two approaches are illustrated in Figure 9.1.

Serial communication carries the advantage that a byte can be sent over as little as a single pair of conductors. Serial communications save cable in local systems and can also take advantage of a ready-made worldwide conductor network, the public telephone system. However, the timing and handshaking for a serial transmission can add a complication. On the other hand, parallel communication also has advantages: in the absence of parallel-to-serial-to-parallel conversions, the hardware is less complex and the transfer time is much shorter.

In the following sections, three types of communications are covered. The first, **serial**, is used in small computers for communication with other computers, terminals, and many printers and plotters. An example of a long-distance network protocol used between computers and laboratories follows. Then we look at two types of parallel communications. The **IEEE-488 General-Purpose Instrument Bus** (GPIB) is a standard for building local networks of instruments and computers and has applicability *within* the laboratory. The **Centronics parallel interface** is used to connect some local peripherals to the computer: printers, plotters, and similar instrumentation.

9.1. THE SERIAL INTERFACE AND THE RS-232-C STANDARD

The serial transmission of information *via* a standard protocol requires several levels of uniformity if the goal of a simple connection between devices is to be achieved. Such a standard must speak to all of the variables in serial transmission.

First, the parallel byte must be converted to a serial representation with the following variables:

- The order in which the bits comprising a byte are to be sent, LSB or MSB first,
- The method for signaling the beginning of a new byte,
- The timing definition; after detecting the arrival of the first bit, the receiver must know when each successive bit arrives, and
- A method for error detection in case of electrical noise in the connection.

Fortunately, these functions are nearly always carried out within a single LSI circuit embodying the standard.

Second, electrical and mechanical compatibility requires definition of the following:

- The electrical specification defining a logical '1' and a logical '0'
- The type of connector and assignment of the pins

Finally, it is good to include techniques for the two devices to inform each other of their "ready" or "not-ready" status. If the receiving device cannot keep up with the sender, it should have a way to tell the sender to wait. The term "handshaking" is used to describe this exchange of status.

9.1.1. Synchronous and Asynchronous Transmission

If data bytes are to be sent serially over a single conductor, the method of determining the beginning of the byte must be encoded within the stream of bits. There are two methods of defining the bytes: data are transmitted serially either asynchronously or synchronously (Fig. 9.2).

When the transmission is **asynchronous**, each new byte can begin at any random instant without reference or synchronization to any external event. Consequently, the receiver must detect and time each byte independently.

Under the heading of **synchronous** transmission, there are many different arrangements (protocols), but in each, the data are transmitted in concatenated packets of bytes with no space between the bytes. Several nondata bytes (a header), that describe the information packet which follows, immediately precede the data. One or more bytes may follow to signal the end of the packet and to provide information for error checking (a checksum).

Of the two techniques, synchronous transmission is more efficient for the transmission of large amounts of data since the synchronization needs to take place only once per packet. However, asynchronous transmission is used commonly in a local *interactive* environment where each character must be sent as it is generated. The remainder of this section considers only asynchronous transmission.

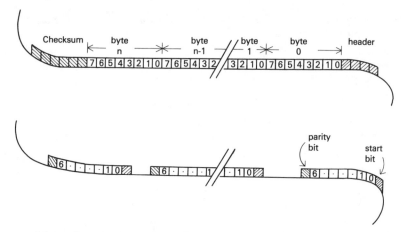

FIGURE 9.2. Synchronous and asynchronous communication. During synchronous communication, a packet of bytes are concatenated. Asynchronous communication allows each byte to be sent without regard to what went before or what follows.

9.1.2. The Parallel-to-Serial-to-Parallel Conversions

The first step in understanding serial communication is to learn how the full byte is converted to a series of bits that can be unambiguously interpreted by the receiver.

According to accepted standards, the *idle state* of the interface is the same as that of a data bit at logic '1,' often termed "mark" in data-communications terminology. When a byte is transmitted, as was demonstrated in Figure 9.1, it is transmitted *serially* by sending a byte with the *LSB first*. However, if the LSB happened to be a '1,' it could not be distinguished from the idle state, so there must be a method for the receiver to detect the first data bit. Therefore, the first data bit to be transmitted must be preceded by one **start bit** (Fig. 9.3) which is a single bit at the logic '0' or "space" level. Consequently, the receiver, previously in an idle state, will prepare to accept a full data byte by checking for a complete start bit.

The data bits follow the start bit. Most serial interfaces can be configured to send 5, 6, 7, or 8 data bits following the start bit. Eight-bit transfers are preferred in small computer systems. However many systems, particularly large-shared computers, use 7-bit configurations since 7 bits are sufficient

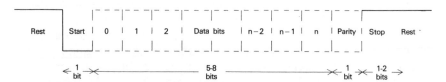

FIGURE 9.3. The bit sequence for asynchronous communication.

for transmitting alphanumeric characters using the ASCII code. Unfortunately, this makes transmission of binary-coded data difficult.

At the end of the data bits, a **parity bit** is optionally appended. Parity is used for error detection, but is used in several different ways. The sum of all of the '1's in the data bits generates the parity status. That sum will be either odd or even. If the interface is configured for "odd parity," the sum of all the data bits plus the parity bit must always be odd. Therefore, if the sum of the bits in the data byte turns out to be even, the parity bit must be set to a '1' to make the entire sum odd. Then if the receiver collects an even number of 1s, it knows that there was an odd number of errors, probably one error. Conversely, if the sum is odd and the interface is configured for "even parity", a '1' is sent.

It is worth noting that microcomputer OSs seldom check the parity bit since the error rate in devices connected directly is low. However, large computers often do since users typically connect via the phone lines, and some estimate an error rate of about one in 10^5 bytes on a typical phone line compared with less than one in 10^7 for local transmissions making direct connections.

After the start bit, the data bits, and the optional parity bit have been transmitted, the transmitter idles for some minimum period of time before another byte can be sent, again at "mark" condition. This period is defined as a number of **stop bits**. Usually 1, $1\frac{1}{2}$ or 2 bits can be specified. At the end of the stop bits, the sender can transmit another byte, if one is available.

The stop bits can buffer some differences in settings. For example if the transmitter (a terminal) sends 8 bits plus parity with 2 stop bits, and the receiver (a computer) expects 7 bits with parity, the eighth bit will fall in the receiver's parity period, and the parity will fall in the receiver's stop period. The receiver could detect a parity error and an "over-run" error, but most small computers ignore those types of errors. If the terminal is sending only alphanumeric data, so that only 7 bits have meaning, there is no problem.

More frequently, a problem occurs when the transmitter sends 7 bits and the receiver expects 8 so that the sender's parity bit is interpreted as bit 8 by the receiver. In such a case, about half the characters may appear to be incorrectly transmitted (if the receiver is checking parity or if the eighth bit has validity); if the receiver is a small computer, the error is probably ignored.

The *rate* at which data are sent is specified in "bits per second" (bps); although technically not the identical, that rate is for most practical purposes the same as the **baud rate**. Stable and accurate oscillators are required at both ends of the communication which ensures that both ends expect the time for a single bit to be the same. Most devices, which can generate serial interface signals, can operate at one or more of the following baud rates: 110, 300, 600, 1200, 2400, 4800, 9600, or 19,200 baud. From the baud rate, the maximum rate of transmission of bytes can be estimated: at 9600 baud, assuming a start bit plus 8 data bits, a parity bit, and 1 stop bit, the transmission rate would be 9600/11 = 873 characters per second.

It is usually necessary that the parity and word length be set the same at both ends, but it is imperative that the baud rate be the same on both ends. Some computers and some plotters have the capacity of testing the first byte and self-setting the baud rate: a known character code is sent to the computer which analyzes it to determine the rate. In the simplest form, the user might be required to send carriage returns until the computer responds; the computer keeps changing the baud rate until it identifies the receipt of a correct carriage return, and therefore can tell that the baud rate is correct. Some other devices are capable of timing the discrete bits of the first byte received, and they "auto-configure" from that information.

The functions described so far are normally handled by a single integrated circuit, a **Universal Asynchronous Receiver-Transmitter** or **UART**. (If the chip also is capable of synchronous operation, it may be called a *USART*, the 'SA' signifying *Synchronous–Asynchronous*.) At the minimum, a UART, such as the one represented in Figure 9.4, must be able to accept a byte from the CPU and send it with the appropriate start, stop, and parity bits, and it must receive a byte serially, converting it to parallel for acquisition by the CPU. The parity and word length options in the simple UART are set by grounding the appropriate pins; the baud rate is determined by an external frequency source (clock).

Most new UARTs, designed as companions to microprocessors, are multifunctional, having the ability to be configured for word length, parity, and

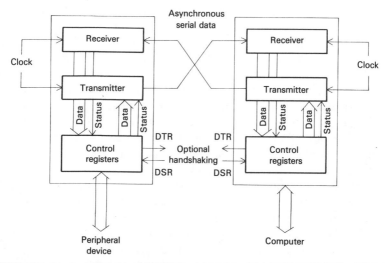

FIGURE 9.4. A simple universal asynchronous receiver-transmitter. Using the external clock for timing, the transmitter converts a byte to a sequence of bits and sends them with proper timing; the receiver reassembles the byte to be transferred to the computer.

even baud rate under software control. One such example is the Motorola 2661. It is connected directly to the computer bus and to a crystal which generates the clock frequencies. A series of internal registers are used to select the baud rate and all other operational parameters.

9.1.3. The Physical RS-232-C Link

The electrical and mechanical definition of serial data transfer come from the RS-232-C (Recommended Standard number 232, revision C) standard, promulgated by the Electronic Industries Association (EIA) with some generally accepted extensions from the International Organization for Standardization document ISO 2110. These define some of the physical variables in serial transmission: pin and connector assignments and voltage levels. They are virtually identical to the standard of the Consultative Committee in International Telegraphy and Telephony (CCITT) of the United Nations, V.28.

The entire definition of the RS-232-C standard is seldom implemented. Consequently, designers infrequently say that the interface follows the standard, but it is described as "RS-232-C-*compatible*." Compatibility means that the interface does *not conflict* with the standard, not that it fully follows the standard. (Even the "compatible" specification is misused almost as often as not.)

In describing the RS-232-C, one must first define and name the two ends of the transmission. In most cases, data flows in both directions, and the connection must be able to distinguish the two ends. The terminology is left from the days when connections were made only between terminals and computers. **Data Communications Equipment, DCE**, and **Data Terminal Equipment, DTE** are defined. For all practical purposes, we can simply state that at one end (the DCE), the device "looks like a computer," and at the other end (the DTE), the device "looks like a terminal."

While most RS-232-C-compatible equipment is clearly "computer-like" (mainframe computers and modems) or "terminal-like" (terminals, printers, plotters, digitizers, etc.), the DCE/DTE definition does not relate directly to the serial interface on a microcomputer or to a scientific instrument with a serial interface. These devices might be connected to external terminals in some application and therefore be "computer-like" or be connected to computers and be "terminal-like"; manufacturers have treated them both ways.

The ISO standard specifies a 25-pin connector (usually denoted DB-25 or D-type 25) whose physical specifications are shown in Figure 9.5. The connector at the computer end (DCE) should be (and usually is) a female connector. The connector at the terminal end (DTE) should be a male connector; here the convention is less often followed, probably because male connectors tend to be less robust. Consequently, the port cannot be identified as DCE or DTE simply by the sex of the connector. In fact, DB-25 connectors are

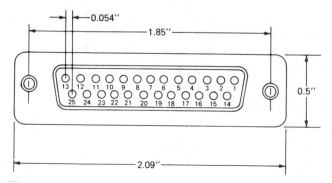

FIGURE 9.5. DB-25 connector specified for RS-232-C communications.

frequently used for other types of computer ports; for example, they are becoming quite common for connection between parallel devices. The manual for the device should always be consulted.

Twenty-one of the DB-25's twenty-five pins are defined. These are listed in Table 9.1. In the following sections, they will be considered by Signal Type.

Lines listed in **Category I** are the minimum signals required for all bidirectional applications. First we must have a ground line. The **Signal Ground**, pin 7, is *required for all implementations*. Next, the data are transmitted over the data lines: the **Transmitted Data** line, pin 2, and the **Received Data** line, pin 3. This nomenclature is ambiguous since obviously all data are transmitted at one end and received at the other. The RS-232-C defines these

Table 9.1. RS-232-C Connector Definitions

Pin	Definition	Signal Category
1	Protective Ground	III
2	Transmit Data	I
3	Receive Data	I
4	Request to Send	II
5	Clear to Send	II
6	Data Set Ready	II
7	Signal Ground	I
8	Carrier Detect	III
20	Data Terminal Ready	II
22	Ring Indicator	III
24	External Clock	III

Categories:		
I	Fundamental Signal required for all send-receive applications.	
II	Control Signals often used between small computers and peripherals.	
III	Signals rarely used in small computer applications.	

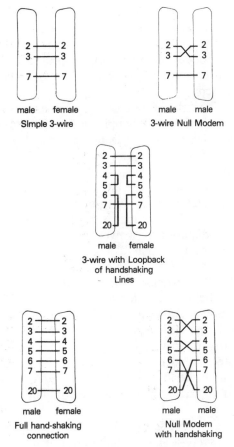

FIGURE 9.6. RS-232-C cable configurations. The choice depends on (1) whether handshaking is required by either or both ends, and (2) whether the ends are configured as DCE or DTE devices. Even so, the list is not all-inclusive.

lines *with respect to the terminal* equipment; that is, data are *transmitted* to the computer from the terminal via pin 2 and is *received* by the terminal from the computer via pin 3. In many applications, a simple "3-wire" cable, connecting only pins 2, 3, and 7, will suffice, handling the simple minimum of data transmission and ground connection.

Note that where a small computer is configured to "look like a computer" (a DCE) but is to be used as a terminal to a second computer, *both* ends are configured as DCEs. Consequently, the cable must cross lines 2 and 3. Historically, a cable that connected DCE to DCE was used to test serial ports on a single computer by joining them; as a result, it received the name **null modem**, a name still commonly used.

Thus far, no method of handshaking has been explained. If, for example, the DTE device is a plotter, it could easily happen that the plotter would fall behind the computer and drop some of the data. Some method of allowing the receiver to signal its status is required, and two ways are possible; one is to require the receiver to send a byte back to the sender telling it to stop sending until further notice (the XON–XOFF protocol) and the second is to allocate additional lines for the purpose. The following paragraphs describe the latter protocol.

The group of signals marked Category II in Table 9.1 is composed of four **control signals** which are frequently implemented. The first pair is the **Data Set Ready (DSR)** line, pin 6, and the **Data Terminal Ready (DTR)** line, pin 20, and the second pair is the **Request to Send (RTS)** line, pin 4, and the **Clear to Send (CTS)** line, pin 5.

DTR and DSR signal equipment readiness. It is rather common in small computer OSs for the status of pin 20 to be queried before a byte is sent. A logic '0' indicates that the terminal is ready to accept characters. This handshaking may be required for newer printers and terminals that frequently have character buffers capable of accepting data at a rate much faster than the data can be processed. If the computer gets sufficiently ahead of the peripheral so that the peripheral's internal buffer is filled, further transmissions are likely to be lost; then the terminal sets the DTR line to a logic '1,' signaling the computer equipment to stop transmission. When the buffer is only about half-full, the DTR is returned to a logic '0.' This feature must be enabled by the driver software in the OS of the computer, but unfortunately many computers do not use it. On the other hand, when the computer requires the DTR line to be '0' but the terminal equipment fails to send it, a jumper from pin 6 to pin 20 will satisfy the computer at the possible risk of lost characters.

The second pair of control signals, RTS, pin 4, and CTS, pin 5, are less commonly used, but serve a similar function. The RTS line should be set to '0' by the terminal equipment which then waits for the CTS line to be set to '0' to indicate that it is ready. However, modern terminal equipment less often monitors the CTS line.

Category III lines are infrequently used except for the **Protective Ground**, pin 1, which can be used to connect the equipment frames of the two devices. A pair of signals occasionally used with modem equipment are the **Ring Indicator (RI)**, pin 22, and the **Carrier Detect (CD)**, pin 8. The RI signal, originating from auto-answer phone modems (which are DCEs) indicates that the phone is ringing. The CD signal is moderately useful, indicating that a connection to a remote computer has been established and the modem carrier has been detected. The remainder of the signals are rarely used; for further descriptions, the references should be consulted.

The **electrical characteristics** of the RS-232-C include the voltage ranges for the two logic conditions and the signal driving characteristics. These conditions are illustrated in Figure 9.7. The voltage swing between logic levels '0' and '1' is much larger than for TTL. The increase in the voltage range generates a substantial and necessary improvement in noise immunity.

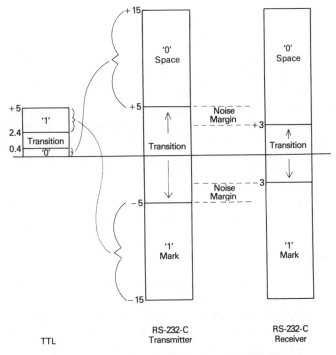

FIGURE 9.7. Conversion of TTL levels to RS-232-C levels.

Also note that the minimum voltage magnitudes for the transmitter are ±5 V whereas the minima for the receiver are ±3 V. These levels allow a 2-V loss between the driver and receiver.

The RS-232-C is specified to operate at distances up to 50 ft and at rates up to 20 kbaud. In fact, higher speeds and greater distances can be achieved in many circumstances, but differences in electrical ground levels between buildings, for example, may limit operation.

9.1.4. The RS-449/RS-422A Standard

In order to circumvent the ambiguity concerning the number of necessary lines and the speed–distance limitations of the RS-232-C, the RS-449 was promulgated in 1973 with two implementations, the RS-423-A and the RS-422-A; the latter is the most important. Several improvements were made while keeping a substantial level of signal definition compatibility with the RS-232-C.

One of the limitations of the RS-232-C is that only a single line handles the ground return for all of the signals, and all of the signals are detected with respect to that ground. A voltage drop may develop over the length of the cable, due to the magnitude of the current carried by that line; consequently, the received voltage levels may be shifted, and errors will result.

As shown in Figure 9.8, the RS-422 allocates a twisted *pair* of conductors for each signal and requires that the receiver respond to the *difference* between them. As a consequence, the differences in ground levels between the two ends of the connection can be accommodated. Furthermore, the use of twisted pairs of wires reduces cross-talk between signal lines. The result is a transmission rate up to 10 Mbaud at 10-m distances and 100 kbaud at 1000 m; the trade-off between speed and distance is shown in Figure 9.9.

The electrical levels also are changed; the transmission range, $+2$ to $+6$ V for logic '0' and -2 to -6 for '1,' is easier to provide from small computer power supplies. With the differential inputs for the receivers, the transition range between '0' and '1' can be smaller.

DE-9 and DC-37 connectors, similar in shape to the DB-25, are specified, the choice depending on the completeness of the implementation. For further detail, references on serial communications should be consulted.

9.1.5. Terminal Communications

In the process of communication between computer and terminal, there are two limitations in the system that may require accommodation. The first is the possibility that reliable transmission cannot take place in both directions simultaneously. The second is the possibility that the terminal device will not be able to accept data as fast as they are sent.

Half- and Full-Duplex Operation. From Chapter Two, we recall that a terminal is composed of two essentially independent devices, a keyboard and a display (or printer). When a user at the terminal connected to a small computer presses a key, it normally appears instantly on the screen; how-

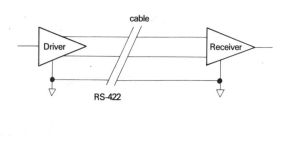

FIGURE 9.8. Comparison of RS-232-C with RS-422 driver–receivers. Each signal is transmitted over a pair of wires.

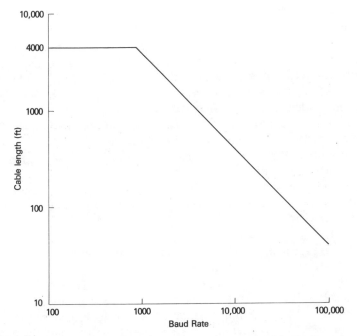

FIGURE 9.9. RS-422 speed–distance trade-off. Compared with RS-232-C maxima of 50 ft, 20 kbaud, the RS-422 driver–receiver can transmit much faster over greater distances.

ever, the character must be sent from the keyboard to the computer where it is processed and sent back (echoed) to the terminal. Since the user of that system has the full capacity to transmit in both directions simultaneously, this operation is called **full-duplex**.

However, access to large computers frequently utilizes telephone lines

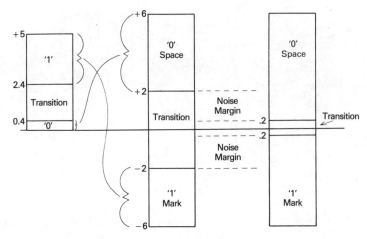

FIGURE 9.10. Conversion of TTL levels to RS-422 levels.

through which information cannot reliably be transmitted simultaneously in both directions. Consequently, the main computer requires that the terminal be responsible for both sending the character to the computer and echoing it to the display as depicted in Figure 9.11. This arrangement is termed **half-duplex** operation. Frequently, when half-duplex operation is employed, the computer also sends a control character each time that it becomes ready for input. Then the terminal can transmit a full line uninterrupted.

One can readily determine whether the computer is operating in full- or half-duplex mode. If the computer is in full-duplex but the terminal is set up for half-duplex (or local echo), messages will be transmitted from the computer correctly, but characters entered at the keyboard will be displayed on the screen twice. In the opposite configurations, messages from the computer will appear normally, but characters entered at the keyboard will not appear at all.

XON/XOFF Handshaking. When the computer has the capacity to send information faster than a terminal device can accept it, a method of **handshaking** is required, and the use of the RS-232-C lines DTR (Data Terminal Ready, pin 20) or RTS (Ready To Send, pin 4) has been discussed. However, in many systems, only the fundamental three lines, Transmitted Data, Received Data, and Signal Ground, can be active. This is usually the case, for example, when a telephone connection is made, and we are simply limited to a single signal-carrying conduit.

One of two control character protocols can be used instead of the DTR line to signal a computer to stop sending data. In some older equipment, an *ETX/ACK* control character pair is used, but in most cases an **XON/XOFF protocol** is now common. If a terminal, plotter, or printer has a character

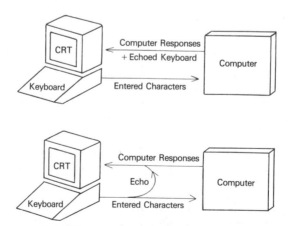

FIGURE 9.11. Full- and half-duplex communication. Top, characters can be both sent and received simultaneously in full-duplex transmission, and the computer echos each character. In half-duplex, the terminal must echo the character that it sends.

buffer, it can store characters when it falls behind in handling the characters that it is sent. However, if that buffer approaches being full, and the device must signal the computer to stop sending more data, it may do so by automatically transmitting an *XOFF* character, a control character with the ASCII code 13_{HEX} (control-S). Even in such operations as printing a program or text on the screen, the display can be temporarily stopped by manually keying the *XOFF* which is sent from the keyboard as a control-S. When the terminal device catches up again, it can command transmission to proceed by sending an *XON*, a control character with the ASCII code 11_{HEX}. From the keyboard, *XON* is sent by control-Q.

One must be warned that while the *XON/XOFF* protocol is honored by most small and many large computers in communication with a terminal or another computer, it is not nearly as often honored when the computer is communicating with printers and plotters. It would be well if both console and printer interfaces honored both the *XON/XOFF and* the DTR handshaking, and authors of peripheral drivers should be encouraged to include them.

9.1.6. Modems

Communication between widely separated computers is increasingly a necessity, but stringing wires between every pair of computers that might have need to communicate is not a practical path nor can the standards discussed so far send data over unlimited distances. The public telephone network provides an inexpensive network for communication which is already in place. However, the telephone network is essentially analog, and the transmission of digital information requires that the signals become electrically compatible with the telephone lines.

A **modem** is a device that converts the serially encoded digital data into a form compatible with phone lines designed for voice transmission. To do so, an audible signal must be modulated according to the bit values. The term modem is a contraction of the words **modulation/demodulation**.

Binary logic is usually expressed as voltage levels, separate levels for logical '1' and '0.' However, the phone system is set up to handle voice by detecting the sound vibrations which are converted to electrical *frequencies*. The telephone response is limited to an approximate range of 300–3300 Hz. A modem is required to convert between the logic level and the frequency representations of data.

There are a number of different ways that the logic states of the data bits can be encoded: the modem may modulate (vary) the frequency, the amplitude, or the phase of a sine wave. The common technique at low baud rates (0–600 baud) that will be described here is a frequency modulation method called **Frequency-Shift Keying (FSK)**. The FSK encoding technique must use different frequencies to represent binary '1's and '0's; in addition, since the same electrical carrier is used in both directions, a separate pair

Table 9.2. Bell-103 Transmission Frequencies

Transmit space	(logic 0)	1070 Hz	
Transmit mark	(logic 1)	1270 Hz	From the <u>originate</u> modem.
Receive space	(logic 0)	2025 Hz	
Receive mark	(logic 1)	2225 Hz	From the <u>answer</u> modem.

of frequencies is required for each direction. In "Bell 103-type" modems, the two ends are called the **originate** modem and the **answer** modem. During communication with large computers, the large computer will be the answer modem. When,two small computers communicate *via* modem, either the users at each end can decide and switch their modems accordingly, or if a new multifunctional modem is used, the modem may have the capacity to automatically answer the phone and communicate in answer mode or automatically be in originate mode when originating a call.

The frequency assignments for Bell-103 compatible modems are listed in Table 9.2. Waveforms generated by both an originate modem and an answer modem are illustrated in Figure 9.12.

Obviously the Bell-103 modem cannot handle baud rates above 600 since at 1200 baud, a complete cycle of 1070 Hz is not possible in the time for a single bit. Other schemes, most involving changing the *phase* of the signal rather than the frequency, are suitable for 1200 and 2400 baud. Although there are several types of 1200-baud modems, Bell 212A-type are most common; standardization of 2400-baud modems will be achieved soon. Telephone modems for transmission at even higher rates usually requires spe-

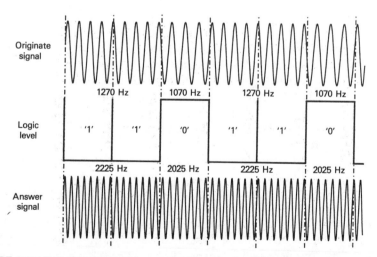

FIGURE 9.12. Bell 103—300-baud encoding frequencies. The two logic levels are differentiated by a frequency shift. Originate and answer frequencies differ by nearly 1 kHz. All of these frequencies can be transmitted over standard telephone lines.

cially conditioned lines leased from the phone company, and these are outside the scope of this introduction.

When a direct wire is available between two sites up to several miles apart, special dedicated units to drive and receive the signals are used. These are denoted **short-haul modems**. Since short-haul modems are used in pairs, usually within a plant or university, the only requirement is that they match on both ends. Some modulate the signal while most simply provide higher level drivers and differential receivers (and as such are not actually modems). Short-haul modems using fiber optics are becoming common because of potentially lower cost, higher speed, better noise immunity (light is unaffected by magnetic fields or electrically noisy environments), and better security (light pulses in a fiber cannot be noninvasively detected whereas electrical pulses can). Infrared laser beams are also in use as information-carrying media, for example, between office buildings.

9.2. AN EXAMPLE OF A PROTOCOL FOR A LONG-HAUL NETWORK, X.25

Although the telephone system appears to reach every corner of the earth, there are many times in which simple asynchronous modem access is insufficient. Among the reasons, we might note that

- There is no built-in error checking and handshaking except for the byte parity.
- The public phone lines do not have the capability of handling higher frequencies necessary for high baud rates.
- The telephone charges are quite high, particularly when one considers that when the computer is used interactively, there is no transfer taking place most of the time.

The X.25 protocol of the CCITT arranges the data into packets, each with an address, the information, and error-checking data. When one terminal or computer (A) is communicating with another (B), it follows a four-step process (Fig. 9.13):

1. Computer A signals to B, sending an **ENQ** character, and requests permission to send data.
2. B acknowledges by sending back an **ACK** character that it grants permission.
3. Computer A then sends a block of data followed by the error checkword.
4. If the data were received without error, B acknowledges receipt of the data block.

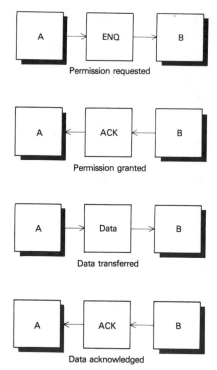

FIGURE 9.13. Handshaking sequence for X.25 communication. Data are sent only after the destination acknowledges readiness.

The protocol may be enacted either by the computer itself—rarely in small computers—or by a dedicated microcomputer called a **Packet Assembler–Disassembler (PAD)**. The small computer communicates with the PAD using an RS-232-C interface, and the PAD assembles the packet and sends it to its destination; when data are returned, the PAD disassembles the packet and sends them back through the RS-232-C port. To the small computer, there is no apparent difference from being connected directly to the remote computer by an RS-232-C line.

We have not yet considered how this protocol becomes the basis of a network. Each PAD may have a number of small computers or terminals connected to it, typically up to 16. Each PAD computer site has an address. The PAD and any larger computer supporting the protocol then become **nodes** in a network as illustrated in Figure 9.14, having the job not only of packet assembly and disassembly but also of routing the packets. When an X.25 packet is received the node must determine, from the address information, whether the packet has a local address or an address at another node. If the address is local, the node must disassemble the packet and deliver the information; if the address is not local, the node must determine which other node should receive the packet next.

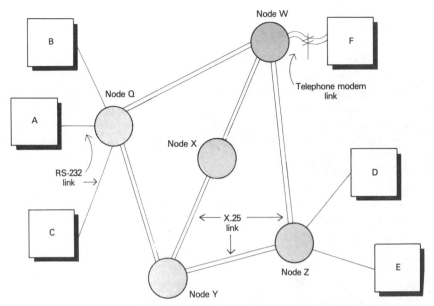

FIGURE 9.14. A typical topology for an X.25 communications network. The nodes may be large minicomputers, mainframe computers, or PADs. After connection to a node, any user can communicate with any other user or computer.

With this arrangement, many users can share a high-quality line between distant sites even though they do not all communicate with the same destination. In a network with the topology shown in Figure 9.14, there is also some redundancy. If B wishes to communicate with E and node Y is saturated or has failed, the information can be routed through W to get to node Z.

When a user needs to hook up to a computer in another part of the country, he can dial, with a local phone number, into a shared network which connects many computers. The code of the desired distant computer is then entered and a connection is made which appears to be a direct connection.

The X.25 also is used on university campuses or at research facilities to connect large minicomputers and mainframes. Through PADs, individual small computers can gain access to any computer in the network independent of the phone system.

9.3. THE IEEE-488 GENERAL-PURPOSE INTERFACE BUS

The standard defined by the IEEE Standard 488-1975 was developed with the purpose of allowing computers and *instruments* to communicate with each other rapidly over a *single* parallel data path. Initially that path, called the **GPIB**, was developed for test instruments, multimeters, logic analyzers, waveform generators, etc. However, as the GPIB has gained more acceptance, compatible interfaces have been designed into plotters and tape drives. Recently the GPIB has become standard in mid-level scientific instrumentation, specialized instrumentation, and many other devices.

In contrast with the RS-232-C, this standard is defined precisely. It includes the definition of each of the 24 lines (Table 9.3), the connectors, the electrical characteristics, the control commands that can be sent, and the protocols for transfer of data and commands.

The physical bus contains 16 signal lines. Eight lines (DI01-DI08) carry the data for *parallel* transfer of data bytes. Another eight lines handle the handshaking procedures. The unique connector, shown in Figure 9.15, has both plug and socket molded into a single stackable unit; as many as 14 devices can share the bus via stacked or daisy-chained connectors. Even the threads on the screws are stipulated. A single cable can be up to 2 m in length.

In order to be standard and flexible, the operation is more complicated than that of the other interfaces described in this chapter. This discussion provides only an introduction, and is greatly simplified.

The system logic requires that each device (up to 15) connected to the bus be defined as a **talker**, a **listener**, a **talker** *and* **listener**, or a **controller**.

Table 9.3. Pin Definitions for the IEEE-488 Instrument Bus

	Pin	Signal	Definition
Data Byte			
	1	DI01	Data bit
	2	DI02	
	3	DI03	
	4	DI04	
	13	DI05	
	14	DI06	
	15	DI07	
	16	DI08	
Handshake Control			
	6	DAV	DAta Valid
	7	NRFD	Not Ready For Data
	8	NDAC	Not Data ACcepted
General Interface Management			
	5	EOI	End Or Identify
	9	IFC	InterFace Clear
	10	SRQ	Service ReQuest
	11	ATN	ATtentioN
	12	Shield	
	17	REN	Remote ENable
	18-24	GND	Ground

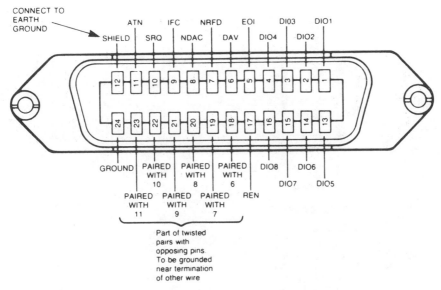

FIGURE 9.15. Connector used with devices following the IEEE-488 standard.

1. A listener can accept data. A digitally controlled power supply is a listener, accepting data and producing a corresponding voltage.
2. The talker is capable of providing data; an example is a digital voltmeter.
3. A talker and listener can do both. If that voltmeter is capable of being configured by commands received through the interface (change of range scale, etc.), it would be both talker and listener. Most laboratory instruments are talkers and listeners in an IEEE 488 environment, capable of being controlled and providing data.
4. The **controller** is capable of talking and listening, but also is capable of *sending commands* to any device, ordering it to accept data, to trigger an operation, or to transfer its data. Only one controller is allowed within the system, and it is nearly always a small computer. Even one can be unnecessary if, for example, a continuously operating tape recorder is used to log data from an instrument.

IEEE-488 Specifics

The bus organization is shown in Figure 9.16 showing the eight data lines, the three Data Handshake Control lines used to coordinate the transfer of data bytes, and the General Interface Management lines used to organize communications. The Data Handshake Control lines and the General Interface Management lines have acronyms listed in Table 9.3.

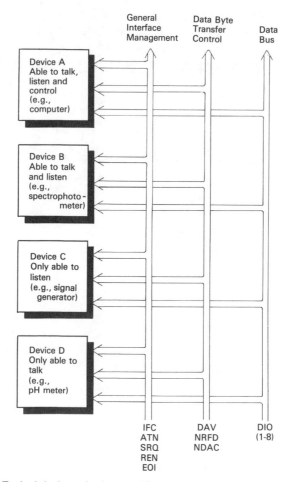

FIGURE 9.16. Typical devices sharing an IEEE-488 bus. The signals are defined in Table 9.3.

The sequence followed by the protocol to transfer a byte from talker to any listener is outlined in Figure 9.17.

1. The talker must find the bus free, as indicated by the status of the three handshake lines.

2. The talker, now ready to transmit, places the datum on the Data Byte lines and waits for the listener to become ready; this is indicated by the rising level of the Not Ready For Data (NRFD) line which, as a TTL line, will be **LO** as long as any listener on the bus holds it **LO**.

3. When the NRFD line indicates that *all* listeners are ready, the talker asserts the Data AVailable (DAV) line.

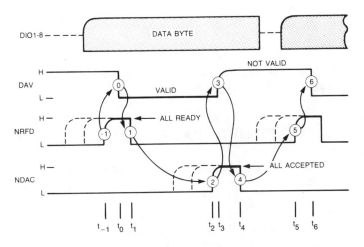

Data byte transfer

FIGURE 9.17. Handshaking sequence for an IEEE-488 data-byte transfer. When *all* devices signal ready on the NRFD line, a datum can be sent together with a data valid (DAV) indication. All devices respond on the NDAC line.

4. When the listener(s) receive the DAV signal, they respond by dropping the NRFD line *and* end the assertion of the Not Data ACcepted (NDAC) line. The NDAC line is also **LO** as long as any single device holds it **LO**.

5. When *all* listeners indicate that the data are accepted, the DAV signal is terminated, the data are removed from the Data Byte (DIO) lines, and the listeners return to the **LO** level of the NDAC line.

The sequence as listed, since the NRFD and NDAC lines can be held **LO** by *any* listener, will always wait for the slowest listener to respond, and therefore the GPIB can be a reliable databus system. It also means that the transfer is only as fast as the slowest listener on the system.

The addition of a *controller* to the system allows *directed* communication. The ATteNtion (ATN) line, asserted by the controller with an address on the Data Byte line, can be used to order the appropriately configured talker to "talk"; if the voltmeter is that talker, it would respond with the sequence just described. A number of such "messages" are defined by the standard to trigger, configure, or clear specific devices. Through the Service ReQuest (SRQ) line, the talkers and listeners also can interrupt the controller for high-priority activity.

The IEEE-488 is a true and well-defined standard, and as such, it is popular among manufacturers of small test instruments. This standardization is rapidly extending into scientific instrumentation with the advantage that up to 15 diverse instruments can be interfaced to a single computer port and can communicate in an orderly fashion.

The disadvantage of the IEEE-488 is that it is relatively complicated, *both* in hardware and in software. However, manufacturers that provide interface boards for small computers typically also have the necessary drivers.

9.4. THE CENTRONICS PARALLEL INTERFACE

One de facto standard has become so entrenched among manufacturers of microcomputers and their peripherals that a brief description is in order. Centronics, a printer manufacturer, defined a connector, electrical characteristics, and protocol for unidirectional data transfer from the computer to the printer. It has received a high degree of acceptance by manufacturers of equipment which accepts data from the computer: printers, plotters, and even some terminals.

The so-called "Centronics interface" provides a simple method of transferring data, in parallel, 1 byte at a time, with handshaking. As was discussed in Chapter Four, the parallel interface must be able to recognize whether or not the receiving device is ready to accept the byte, place the byte on an output port, and provide a signal that a new datum has been made available. Two methods of handshaking could be used, and both methods are included in the Centronics interface. Either the receiving device could send a pulse back each time that it has received a byte or it could maintain a bit that stated whether or not it was busy.

The connector commonly used in Centronics interfaces has the same style as the one shown in Figure 9.15, but has 36 conductors. The definitions of the lines are listed in Table 9.4. As was true for the RS-232-C, not all defined lines are uniformly used. Lines 1–11 are implemented uniformly in microcomputers; the others are used much less often and with little uniformity.

The normal data-transfer sequence is shown in Figure 9.18. First, the computer checks the BUSY line; if the BUSY line is TRUE, the computer must wait. When the BUSY line becomes FALSE (NOT BUSY), the computer can respond by placing the datum on the eight data lines (2–9). After a least a few microseconds to allow the lines to settle, the computer sends a STROBE pulse to the peripheral and indicates that a byte is available. The peripheral will probably respond by making the BUSY line TRUE. Before the computer sends another byte, it must wait for the BUSY line to become false again.

The peripheral also makes an alternate response. This second response is an acknowledge (ACK) pulse to tell the computer that the byte has been accepted, and that the peripheral is ready for another byte. In most microcomputer implementations, the ACK signal is ignored, and the computer simply examines the state of the BUSY line. However, with the ACK sequence, an interrupt-driven output driver can be implemented: when the

Table 9.4. Pin Definitions for a Centronics Interface.

Pin	Signal	Signal Direction
1	Strobe*	From Computer
2	Data 0	From Computer
3	Data 1	From Computer
4	Data 2	From Computer
5	Data 3	From Computer
6	Data 4	From Computer
7	Data 5	From Computer
8	Data 6	From Computer
9	Data 7	From Computer
10	Acknowledge*	To Computer
11	BUSY	To Computer
12	Paper Empty	To Computer
13	Auto Feed*	From Computer
16, 19-30, 33	Ground	--
31	INITialize*	From Computer
32	ERROR*	To Computer

* Indicates that the active logic level is **0** or **LO**

computer generates characters faster than the peripheral can accept them, the OS stores them in a buffer; each time the ACK is sent, an interrupt is triggered which fetches the next character to be sent. However, this sequence is not usually found in small computer OSs.

Using this straightforward handshake procedure based on the BUSY line, data can be transferred simply and efficiently. If the printer has a data buffer, this interface allows it to accept characters extremely quickly until the buffer is full and the BUSY line is raised. More complicated protocols such as XON–OFF or DTR-ready are not required.

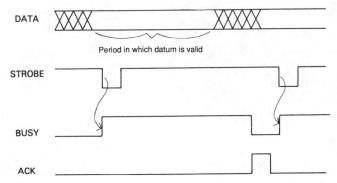

FIGURE 9.18. Handshaking sequence for a Centronics printer data-byte transfer. The computer must determine whether the printer is BUSY. If not, the computer outputs the datum, and signals with the STROBE that there is a new datum. The computer raises the BUSY line until it is ready for a new datum. ACKnowledge handshaking also is possible.

The Centronics interface specifies several other lines to be used by peripherals to communicate the source of error in the peripheral or to make some changes in the operation of the printer. For example, one line (pin 12) indicates "out of paper" status, pin 32 indicates that the peripheral is in a fault state; some OSs monitor these lines separately. Pin 31 is commonly used to initialize the peripheral to the state it had when the power was turned on; pin 14 directs a printer to generate an additional linefeed after each carriage return that it receives. However, these assignments are much less standard and are less often a part of a microcomputer's interface.

BIBLIOGRAPHY

E. A. Nichols, J. C. Nichols, K. R. Musson, *Data Communications for Microcomputers*, McGraw-Hill, New York, 1982.

John E. McNamara, *Technical Aspects of Data Communication*, Digital Equipment Corporation, Bedford, MA, 1977.

CHAPTER TEN

Graphics

The old line that "a picture is worth a thousand words" is no less true when the information to be pictured is scientific data. The cost of graphics is now so low, that even the smallest new computers are designed with integral graphics capability. This means that increasingly we can switch from dealing with printed lists of data to more intuitive interaction with images for greater effectiveness in data representation.

In this chapter, we introduce the "Why?" and the "How?" of computer graphics. A discussion of graphics software follows and a description of hardware concludes the chapter.

10.1. IMPORTANCE AND FUNCTION OF GRAPHICS IN LABORATORY SCIENCE

To a greater extent, contemporary scientific investigation requires that useful information be extracted from large and sometimes awkward quantities of data. The computer is an important tool for amplifying the ability of a researcher to interpret that data, but the computer operates best with digitally encoded numbers while a human works best with images. That is, we perceive and analyze much more *efficiently* with images than with lists of digits. Fortunately, the same immediacy that makes small computers essential for real-time experimentation can facilitate efficient, graphics-aided, data manipulation. As a result, graphics have become an essential part of efficient small computer use in scientific applications.

An instructive example of the difference in interpreting data is seen in the use of a digital watch. If we wish to attend a luncheon appointment at noon, we get from a watch or clock the current time, encoded either as an analog value—the position of sweep hands—or as a string of digits. When an analog watch is used, a quick glance that finds the minute hand extending left lets us know that "a little while" must elapse before the lunch period arrives. Often a person will have a sense of the remaining time without being able to quote the time after looking at the watch. However, if a digital watch is used, the observer must *read* the discrete digits (say, 11:45), subtract from 12:00, and convert those digits into the message "wait a little while." Through the use of "graphics," several interpretation steps are bypassed to get to the desired meaning.

A principal function of graphics is as an aid in data interpretation. One might suggest that in scientific investigation, an ideal automated experimental system would include a computer system having the intelligence to perform all of the data interpretation without the need for direct examination of the data by the investigator. However, in the absence of such a perfectly defined system, the investigator must actively be involved. By representing data as images, formatted according to the investigator's needs, the computer amplifies the pattern recognition capabilities of the investigator. Thereby both computer and investigator can make optimal use of their respective capabilities.

Consider a determination of the concentration of a substance by fluorescence spectroscopy. In the development of the method, the intensity of the fluorescent light is plotted against the concentration and, if all is ideal, a straight-line relationship should result. If a computer were attached to the fluorescent instrument, it could store a value for each concentration, perform a least squares fit of a straight line to the data, and produce numbers representing the slope and something about the quality of the fit (such as the correlation coefficient).

However, slope and fit information alone would not tell us whether any lack of quality were due to random scatter of the data about a line or whether it were due to an "S-shaped" curve rather than a perfectly straight line. This difference is illustrated in Figure 10.1. Two sets of data have the same slope and correlation coefficient, but clearly one set can be described as "S-shaped," while the scatter in the other appears random.

If the curve and the raw data can be flashed on a graphics screen, the type of limitation in the linearity and/or stability of the method can be assessed immediately. The computer extends the investigator's skill by providing data *rapidly* and in a *usable format* (ie., as an image) and by performing rapid computations to provide numeric data such as the slope or intercept. Thereby, the investigator's skills can be used fully and efficaciously.

It could be argued that the heuristics used to determine the type of error in the case above could subsequently be incorporated into the computer program. For example, we could compare the fit of a straight line with a fit

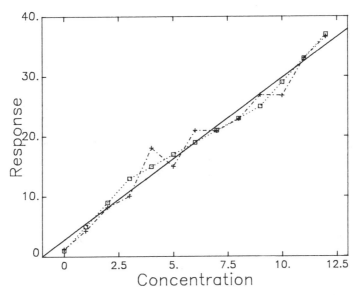

FIGURE 10.1. Importance of graphical aid in the interpretation of data. By regression analysis, the solid line is the best fit to both sets of data, the squares and the diamonds. The correlation coefficients are also equal. However, the error in only one is apparently random.

of a higher order polynomial; if the higher order made significant improvement, meaningful curvature would exist. However, as the process of interpretation becomes more complex, the ability of the investigator to make informed inferences becomes more important; that skill is enhanced by the computer's ability to provide information in a useful form.

When we discuss the display of information using a computer, several classes of application can be identified. We might display information for purposes of either manipulation or presentation; that information might be raw data, concepts, or representation of multidimensional objects and models.

The first category is **real-time display of experimental data.** When data are collected from an experiment, particularly if a control variable (time, position, wavelength, etc.) is manipulated during the experiment, the data should be displayed on a screen *during* the collection process if at all possible. This allows the experimenter to easily detect trends or errors and to intervene during the experiment. Line graphs, scatter plots, or sometimes bar graphs are suitable display forms.

Second, **interactive interpretation of information** requires a computer. The information is produced by modeling or by projection of three-dimensional objects: molecules, anatomical figures, line graphs, data surfaces, relief maps, or buildings. In some cases, we might wish to enlarge selected regions of a graph to study minor features in the data. In other cases, the computer might be used to display a three-dimensional image and the computational capabilities harnessed to rotate that image about any axis. When fitting equations to data, performing digital filtering, computing transforms, etc., it is important to see the results as they are generated.

Third, the computer is an excellent tool for producing **presentation graphics,** particularly now that very high-quality alphanumeric characters can be drawn with plotters and overhead transparencies, and publication-quality line drawings can be produced *directly* from a data file.

Fourth, **concept and model presentation** is facilitated by computer-driven graphics. Diagrams and charts can illuminate conceptual relationships between components of a project, instrument, or organization. The block diagrams and charts for publications and oral presentations are easily handled with suitable hardware and software.

If we accept that graphical output either is or will be an integral part of most laboratory software, we should find out what is necessary to develop the software and to display the results in an effective manner.

In the following two sections, information will be provided about the tools available for writing flexible graphics software and the equipment offered for the display.

10.2. GRAPHICS SOFTWARE

When preparing to write programs for applications of the types discussed in the previous section, a programmer should be able to focus on the application itself and not on the details of putting the dots or lines on a screen or paper. This degree of abstraction requires graphics software that allows the programmer to specify the image in the terms of the problem, not in terms of the device-dependent details of the graphics output unit. Graphics software is developing rapidly along these lines toward standardization, sophistication, and enhanced utility.

This concept is analogous to the comparisons made in Chapter Three concerning the use of higher level languages rather than assembly language or the use of available interfaces rather than the building of an interface from the bus level as described in Chapter Four. In each case, some speed in execution is relinquished in order to gain versatility and to expedite development.

The objective of graphics software is to allow the programmer to

1. Specify image components in convenient, easy-to-manipulate terms.
2. Achieve device independence, that is, to generate the same image on totally different output devices (screens, plotters, printers) without changes in the description of the image.
3. To accept graphics-mode input from a wide variety of input devices.

In this discussion, we will look at the layers of software required to proceed from an application program to a graphics device. We will work from the lowest level of software—that required to effect single step moves on the display device—to the high-level commands used by the author of the applications program.

10.2.1. Graphics Software Packages

Software to create graphical images on standard graphics devices has come from three basic sources.

1. The makers of plotting equipment, who are some of the *major vendors* have produced packages written in FORTRAN: Tektronix developed PLOT–10, and Hewlett–Packard generated PLOT-21. No attempt was made, however, by those vendors to support other output devices aside from their own single families.

2. A wide variety of *special-purpose software* is now on the market that includes graphics drivers for some of the more popular graphics devices; they include several data-plotting packages and some spreadsheets. These do not allow the user to write dedicated graphics programs.

3. An attempt has been made to make *standards* for graphics software which can be independent of the I/O devices. One of the more viable approaches is the **Graphical Kernal System (GKS)** which has been ratified by the ISO with the concurrence of the American National Standards Institute (ANSI). The GKS system was designed to present a software interface to the programmer which is independent of the hardware. In this section, we will look at a subset of the GKS command set. First, however, the commands to the device itself must be examined.

10.2.2. Device Primitive Commands

We can begin to understand the problem of writing device-independent software by considering the devices on which the images are created, the plotters and screens. (The hardware will be described in some detail in Section 10.3.) These devices have greatly varying attributes; the number of resolution elements on the horizontal axis varies from about 200 for some raster displays to at least 1000 elements per inch for moderate-cost plotters. Some devices will draw a straight line in response to a simple command while others require that the software in the host computer form an image of a straight line with each dot. These differences must be made transparent to the programmer through a graphics system such as the GKS.

For the following discussion, we want to understand how software might communicate with a typical output device to draw the shape of a diamond.

The X–Y coordinates of the diamond (in micrometers) are 0.0, 0.0; 5.0, 1.0; 0.0, 2.0; −5.0, 1.0. The desired image will appear, at maximum size on the output device, as four lines connecting the midpoints to the edges of the display.

A typical plotter might specify position in the range of 0 to 4000 in the X dimension and 0 to 3200 in the Y dimension. Then the diamond could be drawn by moving the pen to the X–Y coordinates 2000,0; then commands would be made sequentially to move the pen with the pen down to the following positions: 4000,1600 on right; 2000,3200 on top; 0,1600 on the left; and 2000,0 on the bottom. These coordinates are called **device coordinates.**

Using Figure 10.2 for reference, one can see that, for a typical plotter with line-drawing capability, the character string necessary to draw the figure could be

$$\text{M2000,0; D4000,1600; D2000,3200; D0,1600; D2000,0;}$$

In this example, our plotter accepts commands in a format typical of current graphics devices: an "M" to move the pen with pen up or a "D" to move the pen with pen down. The absolute coordinates follow, with commas and semicolons used as delimiters in the character string that comprises a command. These are device primitive commands.

In order to execute those primitives, the plotter itself must have even more primitive internal software. The plotting device must accept the coordinates of the line's limits and move the pen in a straight line. Typically,

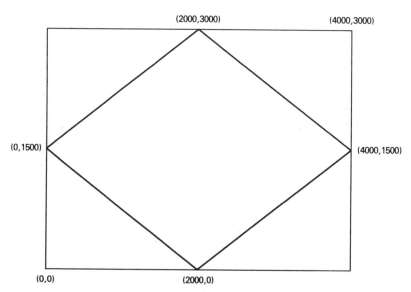

FIGURE 10.2. A diamond drawn on a typical plotter. The absolute plotter coordinates are show in parentheses.

stepping motors are used, and the number of steps in the vertical direction for each step in the horizontal direction must be internally computed, not a trivial task.

If, for example, the terminus of the line is positioned seven steps in the Y direction and 10 steps in the X direction from the current position, then the driver must determine for each step in the X direction when a step in the Y direction should be made. Similar algorithms are required to modify the matrix of dots in most CRT displays. On most output devices, the line will take a "stair-step" shape, but if the steps are sufficiently small—as they are in most plotters—the steps are invisible.

When the output device is the graphics screen of a small computer, such as an IBM PC rather than a plotter, the computer's graphics software must be one step more primitive: the computer must compute the position of each dot that comprises the line rather than simply transmit the coordinates of the termini of the line.

Only the move and draw primitives have been mentioned so far. The devices may have additional capabilities to fill an enclosed area with a color or pattern and draw characters in a range of sizes and orientations.

Other common device primitives include circle or arc drawing and color changing. Many others are found in advanced plotters; some CRT devices support primitives that approach in scope a high-level graphics language.

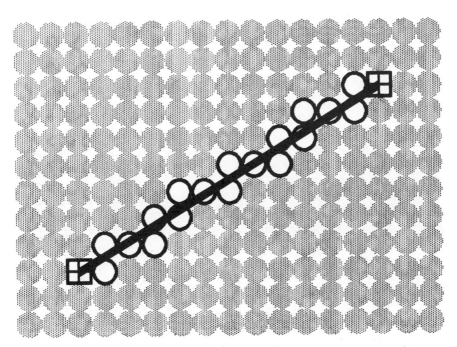

FIGURE 10.3. A straight line drawn from a matrix of dots. To connect two points, represented as squares, the line takes a stair-stepped appearance.

10.2.3. GKS Primitives

Clearly, for the programmer to have to write the code that will generate the device primitives (which are different for every device) is an inefficient process. The GKS system solves the problem by supplying primitives that create a standard interface to the device primitive commands. These commands must be customized by the developers of the GKS software package for the particular devices that the developers support, but the code that the programmer writes does not depend on the output device.

The line drawing primitive in the GKS is the function

$$POLYLINE (N, XPTS, YPTS)$$

The coordinates of N X–Y points are placed in two arrays, $XPTS$ and $YPTS$. The polyline command then connects the points. The data are specified in *world coordinates,* that is, in the Cartesian coordinate units of the original data.

To draw the diamond, as in Figure 10.4, the arrays are filled with the data (in world coordinates) and the polyline function is invoked

$$XPTS(1) = 0.0$$
$$YPTS(1) = 0.0$$
$$XPTS(2) = 5.0$$
$$YPTS(2) = 1.0$$
$$XPTS(3) = 0.0$$
$$YPTS(3) = 2.0$$
$$XPTS(4) = -5.0$$
$$YPTS(4) = 1.0$$
$$XPTS(5) = XPTS(1)$$
$$YPTS(5) = YPTS(1)$$
$$POLYLINE (5, XPTS, YPTS)$$

To draw that figure, the GKS primitive command must make the conversion from world coordinates to device coordinates. Additional, higher level commands in the GKS package produce dashed–dotted lines instead of solid lines.

FILL AREA is a second primitive which is invoked almost identically. Using the same syntax, a diamond can be drawn in which the area is filled with a pattern (which if not solid could be defined elsewhere)

$$FILL\ AREA\ (N, XPTS, YPTS)$$

When the *Fill Area* command supports a relatively intelligent device, it may be able to send to the device a simple command to fill that area. Otherwise, it either may have to generate the commands for a series of lines on a plotter to fill an area or might need to set each pixel in a raster screen image to the requested color. In any case, the programmer needs only to deal with the *Fill Area* command; the GKS package handles conversion to the particular device.

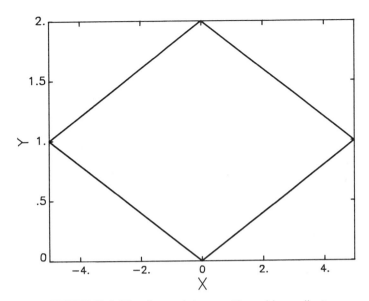

FIGURE 10.4. The diamond drawn with world coordinates.

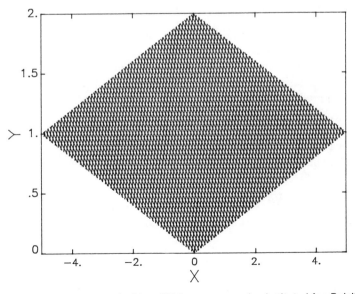

FIGURE 10.5. The diamond with a *Fill Area* command substituted for *Polyline*.

357

A third GKS primitive command places characters on the image using the syntax

$$TEXT\ (X,\ Y,\ STRING)$$

where the text is positioned at the $X-Y$ coordinates and *string* is a string of alphanumeric characters. Depending on the device, the *text* command might either use the character set internal to the graphics device so that only character codes would be transmitted or use a character set defined in the GKS software, sending commands to draw the lines needed to produce individual characters.

Additional commands are available in the GKS package to set the font, size, and orientation of the text.

10.2.4. Coordinate Systems

The "diamond" example left an important relationship undefined, that relationship which maps the world coordinates to the device coordinates. The diamond must be positioned on the page and scaled appropriately.

To establish the scale of the image, a **window** is defined using the command

$$SET\ WINDOW\ (N,\ XWMIN,\ XWMAX,\ YWMIN,\ YWMAX)$$

so that *XWMIN-YWMIN* defines the lower left-hand corner and *XWMAX-YWMAX* defines the upper right-hand corner in world coordinates. (*N* can be ignored for this discussion.) To draw the image in Figure 10.2 so that the points of the diamond lie on the edge of the window, the command would read

$$SET\ WINDOW\ (N,\ -5.0,\ 0.0,\ 5.0,\ 2.0)$$

Next, the position of the image window on the page can be defined. To do so, another set of coordinates are defined, the *normalized device coordinates*. Whatever the actual device coordinates of the device, they are mapped to a full scale of 0.0–1.0 in each direction. Then the size and position of the window is defined by the function

$$SET\ VIEWPORT\ (N,\ XVMIN,\ XVMAX,\ YVMIN,\ YVMAX)$$

Each parameter, except *N*, falls in the range 0.0–1.0.

If the entire surface is to be used to draw the diamond, the viewport is delimited by the command

$$SET\ VIEWPORT\ (N,\ 0.0,\ 1.0,\ 0.0,\ 1.0)$$

If the diamond is to be drawn in the upper left-hand quadrant, the viewport is set by

$$SET\ VIEWPORT\ (N,\ 0.0,\ 0.5,\ 0.5,\ 1.0)$$

These functions allow a user to write a routine which will draw the diamond at any place and any size so long as the object's coordinates fall within the window. If the device primitive coordinate lies outside the limits of the device (off-screen or off the plotter surface), the path of a line may be unpredictable. Consequently, if a coordinate lies outside the window, the line should stop at the boundary. Drawing that line so that it stops correctly at the boundary is **clipping**.

An example of the diamond drawn in different windows and viewports is found in Figure 10.6. Note the clipping when the window is too small for the coordinates.

In the GKS, clipping may be overridden with a function, *Set Clipping Indicator (NOCLIP)*.

10.2.5. Other GKS Functions

The discussion of GKS has only introduced the format of the fundamental set of commands. Many others are defined which can be learned from the references, but it is useful to describe two other groups of commands.

Several commands of the type *Set XXX* have been detailed; these set attributes such as windows, viewports, and the clipping indicator. Others are used to specify the text font, size, writing angle, and character slant. Attributes for *Polyline* set the color and line thickness. For the *Fill Area* command, the pattern used to fill the area can be defined.

Another group of commands accepts and formats the data for the graphics input devices described in Section 10.3.3. The position of the drawing or pointing device is returned in world coordinates or the condition of the buttons on the mouse, digitizer, or light pen can be monitored.

10.2.6. Three-Dimensional Plotting

This chapter has emphasized the requirements for plotting line graphs, usually presented in two dimensions. Three-dimensional graphics are frequently necessary for the representation of the models of physical objects, and when an additional parameter is needed to depict data, a third dimension in the line graph is useful: a surface will be displayed. The problems of showing a three-dimensional image of a rat brain or a molecule are the same as those that exist in the presentation of data as a surface.

The drawing of a three-dimensional image begins with arrays containing the X–Y–Z coordinates of each datum. In order to develop the image from that data, it must be converted to two-dimensional X–Y coordinates which encode the Z dimension. That encoding involves both the viewing angle and a method of depth perception.

When an observer views a surface, the image seen depends on the apparent position from which the surface is observed. If the data take on the appearance of a "mountain range" such as that in Figure 10.7, the visible features depend on the viewing angle. When viewing a mountain range from

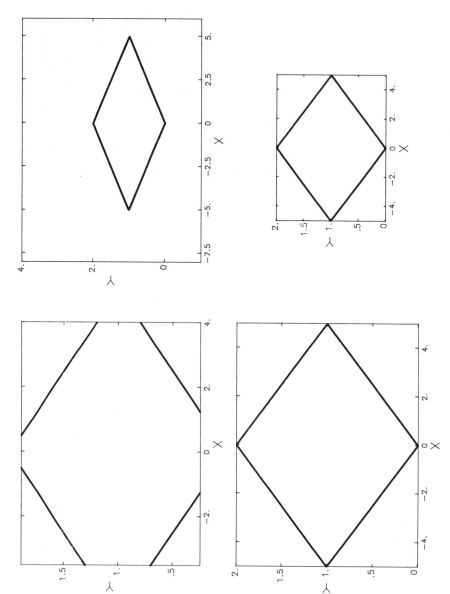

FIGURE 10.6. The diamond drawn with the same commands as in Figure 10.4, but with various windows and viewports.

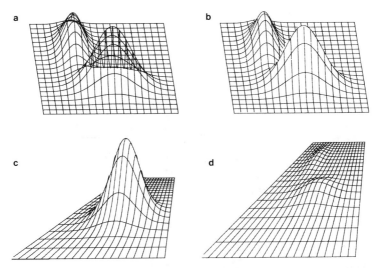

FIGURE 10.7. Three-dimensional graphics. (*a*) A surface with no perspective and with hidden lines included. (*b*) As in (*a*), but with hidden-line removal. (*c*) and (*d*), the same surface with perspective and different viewing angles.

the ground, an observer will find some features hidden behind others. When viewed from above, those features are visible, but the magnitude of the mountain peak is lost. In order to maximize the available information from a 3-D plot, the **viewing angle** with respect to the X, Y, and Z axes must be chosen with care and perspicacity.

Data surfaces are often drawn in "pseudo three-dimensional" form; this is the case in the first example, surface *a* in Figure 10.7. Along the axis, which is directed back into the paper, each plane is shifted both up and horizontally. This method only simulates the variation in viewing angle, and the plot does not suggest a real surface. However, the computational requirements described in the next section are reduced.

One major difficulty in three-dimensional graphics is the "hidden line problem," obvious in Figure 10.7*a*. If the data are represented as an opaque surface, the viewer should not be able to see elements hidden behind part of the surface that appears in the foreground; removal of the hidden lines, accomplished in surface *b*, clarifies the image. Several algorithms have been developed to remove the parts of the image that would not be visible if the object or surface being displayed were a real, opaque object. Graphics texts should be consulted for details.

Particularly when an object is displayed—but also when a surface is plotted—the **depth perception** is important. If the observer is close to the image, the realism is improved by scaling the part of the image at the back smaller than segments at the front. If the scaling factors remain constant, the image appears to be far away. The lower surfaces in Figure 10.7 depict the same data as those drawn above, but have been transformed to give depth perception. In surface *c,* the viewing angle is small enough that the farther peak is obscured by the closer peak. Surface *d* takes another extreme; the viewing position is so high that the peaks appear insignificant. The optimum viewing position can be difficult to find without the ability to rapidly rotate and translate the image.

Some of the high-resolution equipment used for engineering and molecular modeling gives depth cues by reducing the intensity for segments of the image which are to appear farther from the viewer. Sophisticated raster graphics devices have the capability to render surfaces. Adding shading, which gives the appearance of being lighted from a single point; further enhances depth perception. Rendering requires analysis of the angle that the surface makes with the axes of the image.

10.2.7. Calculation of Transformations

If we want to use interactive graphics for analysis of data—particularly three-dimensional data, but also two-dimensional data—we would like to rotate and translate the image as though it were held in our hands; the viewing angle changes with the movement, and the X–Y coordinates on the display need to be updated with every move. The transformations should take place in perceptively real time; that is, rotations on the graphic device should approximate the way real objects are seen to rotate, without jerks and lags. Currently this capability—for all but simple images—is not realizable on most small laboratory computers because of the extensive computations that are required, even though those computations are conceptually simple.

In the cases where rotations and translations are important, the procedures to do them are simplified by some straightforward matrix calculations. As is true of many elements of matrix arithmetic, the calculations are more tedious than difficult.

Initially, we will consider the problems of manipulating a two-dimensional image. The image must be scaled, translated, and/or rotated before it is drawn. To do so, the X–Y coordinates are represented in world coordinates as a vector, $[X\ Y\ 1]$. After completing the various transformations, we will obtain the normalized device coordinate values $[X'\ Y'\ 1]$. Each result will be obtained from multiplication of a 3 by 3 transformation matrix:

$$[X'\quad Y'\quad 1] = [X\quad Y\quad 1] \begin{vmatrix} a & d & 0 \\ b & e & 0 \\ c & f & 1 \end{vmatrix}$$

The matrix is formed according to the transformation that is to be performed. Each operation will be presented without mathematical explanation. The emphasis of the discussion is to show that the operation is uncomplicated when expressed in terms of matrices.

In the case of translation, the image is moved left or right by the dimension

T_x and up or down by the dimension T_y. This is achieved through multiplication by the following matrix:

$$\begin{vmatrix} 1 & 0 & 0 \\ 0 & 1 & 0 \\ T_x & T_y & 1 \end{vmatrix}$$

When the image is to be rotated, the angle of rotation, ϕ, must be specified; the transformation matrix is

$$\begin{vmatrix} \cos\phi & -\sin\phi & 0 \\ \sin\phi & \cos\phi & 0 \\ 0 & 0 & 1 \end{vmatrix}$$

When only scaling is required, the X and Y coordinates are multiplied by S_x and S_y, respectively. In matrix form,

$$\begin{vmatrix} S_x & 0 & 0 \\ 0 & S_y & 0 \\ 0 & 0 & 1 \end{vmatrix}$$

Movements in two dimensions can be described by those three operations, taking some care to perform them in the order that corresponds to the desired operation.

Three-dimensional graphics require a *four-by-four* matrix. The computations must operate on X–Y–Z data, and a corresponding increase in computation is required.

It is apparent that three-dimensional image manipulation is greatly enhanced by the addition of computational hardware to the system, particularly an array processor. Array processors for microcomputers are poised for entry into the microcomputer market. It is probably now a safe assumption that, within this decade, array processor hardware will become commonplace in the laboratory for graphics, as well as other processing problems.

10.3. GRAPHICS HARDWARE

Graphics communication is a two-way exchange between computer/software and the investigator. However, it is an unnatural communication when compared to the coded world of digits and characters. Specialized equipment has been developed that converts the digital and the analog–graphical domains.

In the following sections, we will outline the equipment. There are devices that can produce images on screen, paper, and for the input of graphical information. The devices that produce images on screen, most often a CRT, offer speed and the capacity for selective erase and animation; the devices that produce images on paper offer higher resolution and hard, reproducible output.

10.3.1. Screen Graphics

Most computer systems now include some form of graphics screen, usually incorporating a CRT. With it, transient images can be drawn and manipulated. An image on a screen has three advantages over an image drawn on paper. (1) it can be drawn *rapidly*, since mechanical movements are not required; (2) it can be *interactively* modified, with only portions of the screen being changed using "electronic" erasure; and (3) since the image is illuminated and often colored, a *powerful visual communication* is easily produced.

Images can be generated on a screen in two ways. The first, a raster display, is more common. It uses the techniques of broadcast television display: The image is built from a matrix of dots that are in fixed positions. In the second method, a vector display, the image is drawn by directing the beam in straight lines (strokes or vectors) from one X–Y position to another on a CRT terminal that is essentially an X–Y oscilloscope.

Raster displays equipped for high-resolution graphics are extensions of VDUs based on the CRTs discussed in Chapter Two. Electron gun deflectors move the beam position in a regular pattern such that, if the beam was always on, the screen would be white; this pattern, illustrated in Figure 10.8, is a **raster** (scanning) pattern. The CRT controller only modulates the intensity of the beam, turning it on or off at appropriate points depending on whether the point is to be light or dark.

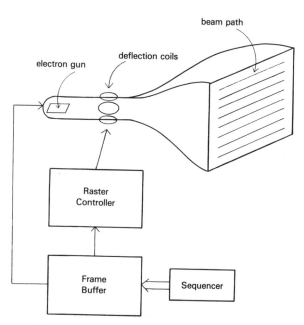

FIGURE 10.8. A raster display. The beam always follows the same path. Data from the frame buffer simply turns the beam on and off.

Storage of the image requires creation of a map in semiconductor memory. For a monochrome display, each location on the screen corresponds to a bit in memory. The CRT controller maps the memory onto the screen by sequencing through that memory, turning the beam on whenever the corresponding bit is a "1". Consequently, the term **bit-map** is often applied.

Images are created by turning ON and OFF the appropriate bits in the bit-map memory. The algorithm for a primitive dot-drawing command must find the absolute memory location for the byte and the bit in that byte which corresponds to the destination point in the display. When a particular point is to be lit, the byte and bit in memory are located, and a logical "1" must be written into that memory bit.

The primitive line-drawing command necessitates another algorithm. In order to draw lines on the screen, the raster graphics system requires an intelligent controller to sort out the points between the endpoints that must be activated. Rapid determination of which bits in an 8--128-K memory array must be reversed in order to draw a straight line is a nontrivial matter. In high-capacity graphics systems, LSI circuits can accomplish this task, but in most small computers, the main CPU must do it.

The resolution limits of a raster-type CRT are constrained by several factors. The amount of memory is fundamental, since there must be 1 bit for each image element. Consequently, to design a screen with a resolution of 512 × 480 dots (**picture elements** or **pixels**), the *minimum* screen memory requirement is

$$\frac{(512 \times 480 \text{ bits})}{(8 \text{ bits/byte})} = 30{,}720 \text{ bytes}$$

Related to the amount of memory is the computational overhead of executing the graphics primitives. Unfortunately, the graphics primitives are executed in most small computers by the same CPU that executes the application program. If the resolution is doubled, the number of dots to be toggled when a line is drawn also is doubled, and the number of dots affected by a fill operation is quadrupled; this greatly affects performance, particularly if the data are to be displayed concurrently with their acquisition.

Beyond the memory limitation, the horizontal and vertical resolutions are separately determined by the CRT. As the beam is moved across the screen horizontally, the video monitor must respond to the input signal for a single dot by turning the beam on and off; consequently, the number of dot positions that can be produced horizontally is determined by the bandwidth of the video monitor, a measure of the speed with which the beam can be turned on and off. A practical limit for good quality monitors is 600–800 resolution elements per line.

Vertically, the resolution is limited by the number of scan lines made by the video monitor in the visible area of the screen. In standard monitors, that number is approximately 240; consequently, no more than 240 vertical elements can be displayed.

The vertical resolution possibly can be doubled by using an **interlaced** CRT scan. Normally, the same data is displayed on each scan. When the scan is interlaced, the horizontal scan lines are shifted slightly lower on alternate scans, appearing in between the scan lines of the previous scan. Separate data are displayed on these two scans, so that half the points are displayed on each scan. Unfortunately, the result in ordinary monitors is a very visible flicker, since each dot is refreshed only half as often as in non-interlaced displays. To combat this problem, longer persistence phosphors are required. Monitors with these phosphors are normally more expensive, and some operators find the slow decay irritating, although less irritating than the flicker.

Further increases in resolution beyond 700 by 480 can only be obtained by using video monitors that do not follow the standard video sequence and/ or scan at higher speed. These are substantially more expensive because they are nonstandard and the beam movements must be more precise.

A raster display typically offers lower resolution than the vector displays described later, and the raster display is somewhat less pleasing for the display of graphs, since the discrete pixels yield diagonal lines that have the stepped appearance of Figure 10.3. However, since the market for graphics VDUs is dominated by business-oriented applications (low-resolution bar graphs and pie charts), the color raster unit receives the greatest mass production benefit and is often the best choice at reasonable cost.

Gray scale, the variation of intensity, and **color** provide two additional methods of either highlighting part of the data or providing further dimensionality to the representation of information that has many variables.

Adding capability to present this information requires more memory. One bit of memory per pixel can define only whether the pixel is on or off. Two bits of memory per pixel can define four levels of intensity, corresponding to the available values 0, 1, 2, and 3. In color raster displays, 3 bits of memory are required for each pixel to define red, green, and blue in a color monitor; using the combinations, eight colors can be obtained. The parallel areas of memory are called **bit planes**.

To obtain more than the basic eight colors, the three primary colors must

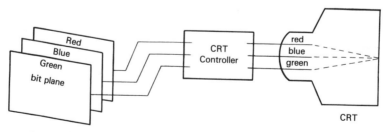

FIGURE 10.9. Three parallel bit planes control the three color guns of a color raster monitor.

be mixed in varying intensities. If the number of bit planes is brought up to 6, each of the primary colors can be off or have one of three intensities; these combinations can produce 64 different colors. Commonly, a display will have 6 bit planes, but allow the four intensity settings to take on any of 16 magnitudes; that capability allows the software designer to work with any 64 colors out of a choice (a palette) of 256^3, about 16 million.

Vector or Stroke display technology is essentially analog. In its simplest form, the vector display positions spots on a screen. When a point corresponding to an X–Y data pair is to be displayed, the pair of digital data values is output to a pair of DACs as represented in Figure 10.10. The voltage outputs then feed the X and Y deflection circuitry of the CRT, and when both DACs are stable, a pulse turns on the beam, generating a spot. An image can be built from a series of spots on the screen, one for each X–Y position addressed.

The image also can be built of **vectors**; if the beam is turned on in one position and left on when it moves to the next, a "streak" or vector results. Some additional circuitry is required to cause the beam to move at a uniform speed in order to draw a straight and even line.

As is true for an oscilloscope, the vector CRT may have either a storage tube or a normal persistence phosphor. In the storage tube, the image of a line drawn on the screen by the electron beam is retained on a fine mesh wire grid adjacent to the phosphor screen. A separate electron gun floods the grid with electrons that only reach the screen where passed by the grid. The image, retained electrostatically in the grid, is not stored in any readable memory; the line can be erased only by blanking the entire surface. The

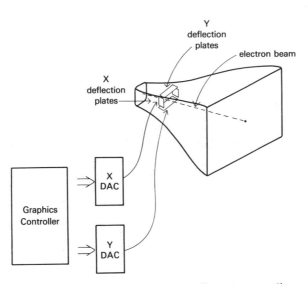

FIGURE 10.10. A vector or stroke display. The controller sequences through a list of coordinates and the beam is deflected to each coordinate.

resolution of this display could be limited by that of the DAC, but since 14- –16-bit DACs can be used, the limitation is more often the resolution of the grid or width of the line. Storage phosphors tend to "bloom"—that is, a line tends to spread somewhat with time on the phosphor screen. Consequently, the practical perceptual resolution is limited to about 500 (horizontal) by 500 (vertical). Not long ago, when memory was costly, this storage technique was the least expensive because no memory was required.

If a nonstorage screen is used, a refresh system is required, since an image persists on a short-persistence screen phosphor for only 10 ms. The starting and stopping locations of each vector are stored in a list in internal memory, and the $X-Y$ locations are sequentially fed to the DACs. The screen is redrawn (refreshed) at a rate of 30 Hz or more, which allows the resulting image to appear constant. However, as the list of vectors becomes longer, the time for a single refresh must increase; eventually the screen appears to flicker. The number of displayable vectors can be maximized either by using a phosphor with a longer decay time or by speeding the sequences. Some commercial systems produce flicker-free display of up to 40,000 vectors, refreshed at 30 times per second. The resolution of such displays depends primarily on the DACs, which, at a typical 12 bits, make possible a 4096 by 4096 display.

Commercial systems in these categories contain a graphics control unit that accepts $X-Y$ data, usually from a conventional serial interface, and formats the data for display. When alphanumeric characters are required, internal look-up tables in ROM are often included.

10.3.2. Plotters

Since information in the form of reports, research papers, and proposals is usually disseminated in hard copy, a method for producing hard copy (a copy on paper as opposed to the image on the CRT) is a necessity. As was the case for CRT graphics devices, both vector and bit-mapped approaches can be used.

Vector Plotters. A digital vector plotter responds to digitally encoded instructions from a computer to move a pen in straight lines with the pen either down or up, according to whether a line is to be drawn or the pen is to be positioned. These plotters can be considered random-positioning devices since they move directly from any position on the bed to any other.

All contemporary plotters can execute line-drawing primitives, drawing a straight line on receipt of the endpoints. Nearly all plotters have some additional primitive capabilities, such as the capability to generate simple alphanumeric characters upon receipt of only the ASCII code; these are drawn in at least four and up to 360 angles and in variable size. Other intelligent plotters can automatically change pens, scale the image, control the drawing speed, and so on.

When writing graphics primitives for plotters, one must consider the tradeoffs between having various intelligent functions performed with the plotter's internal firmware and having these functions carried out under the control of the host computer. For example, alphanumeric characters can be drawn either by sending the ASCII code and asking the plotter to draw the character, or by sending a series of primitives that create the character. The tables of character specifications required to encode the moves and draws for alphanumeric characters are rather cumbersome for small computers, and the use of a plotter's own set is convenient. However, the internal character sets of most plotters are rather crude; use of the character specifications stored in the host computer gives access to better-formed characters and/or a variety of fonts.

Though rather sophisticated commands may be included in the plotter firmware, remember that a graphics software package must cater to the lowest common denominator if it is to support a wide variety of devices. The GKS primitives interface with the plotter's primitives and must provide the capabilities that the plotter's firmware might lack, such as area fill or alphanumeric characters. The most sophisticated commands in firmware might not actually be used.

Plotter pen positioning can be divided into two categories according to the type of motors used to drive the pen. **Servo motors**, much like those used in analog strip-chart and $X-Y$ recorders, have the advantage of being very fast; however, because of their advanced engineering, they are rather expensive. **Stepping motors** are slower and less expensive, but tend to be more robust.

Many digital plotters share a physical form with $X-Y$ flat-bed analog recorders, using similar motors to drive a pen holder in each dimension over a fixed bed (Fig. 10.11). However, plotter manufacturers recently have realized that moving the pen in one dimension and the paper in the other can be much simpler and more efficient, since the paper has much less intertia than the pen holder. The paper is moved by a pair of pinch rollers, and the pen need be moved in only one dimension. Such a plotter is faster, simpler, and cheaper than older plotters.

The resolution of vector pen plotters can be very high due to their mechanical precision. One plotter, which accepts $8\frac{1}{2} \times 11$ in. paper, demonstrates a resolution of $7650 \times 10,300$. An example output is shown in Figure 10.12.

Printer/Plotters. Most dot-matrix printers also can be used as graphic hardcopy devices. When ordinary text is printed, the characters are formed by the internal printer software which determines the dot combinations in response to a single ASCII character. Operated in graphics mode, each pin in the print head can be controlled independently by the host computer. The result is a slow raster, which produces a collection of dots that form the image.

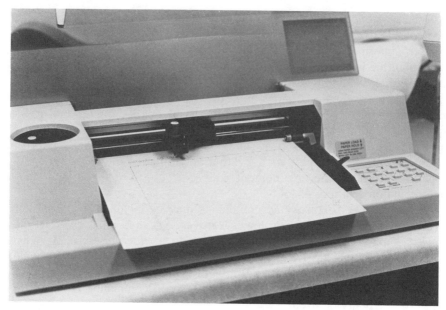

FIGURE 10.11. A typical vector plotter. The pinch wheels which move the paper are visible.

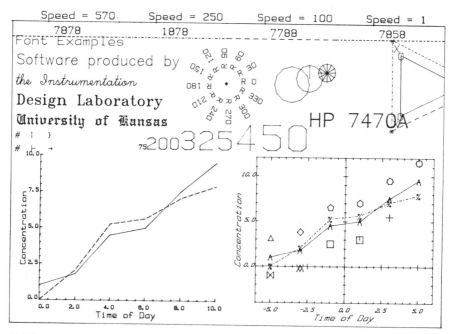

FIGURE 10.12. A demonstration plot from a typical vector plotter. The font examples are drawn with software fonts. Other alphanumeric characters show the limitations in typical plotter firmware fonts.

Because the paper movement in most printers cannot be reversed, a bit map of the image must be formed in memory before the plot is made. The raster video display of many computers conveniently provides that bit map. Therefore, the operator can create the image by interacting with a raster CRT graphics system, after which production of the hard-copy image using a printer/plotter is simpler than the production of a vector plotter image. To dump the image, each row of dots on the screen is transmitted as a row of dots on the printer. An example of a screen dump from a medium-resolution screen is shown in Figure 10.13.

When a screen dump is performed, the resolution of the hard copy is limited to that of the screen, usually much less than is possible on the printer. The alternative is to create a bit map in memory, but that ofter requires too large a memory array (960 dots by 1000 dots would require 960 × 1000/8 = 120 kbytes). Figure 10.14 was dumped from a 44-kbyte memory array to an inexpensive dot-matrix printer.

The future of this practice may well be defined by several current trends: the plummeting cost of memory, improved microprocessors, and the entrance of laser printers into the market. Laser printers can print graphics images with a resolution of 300 dpi; a full-page bit-mapped image requires nearly a megabyte which can be managed by current microprocessors.

10.3.3. Graphics Input Hardware

Our efficiency when entering and manipulating information in graphical form can be much greater than when that information is typed in as lists, just as

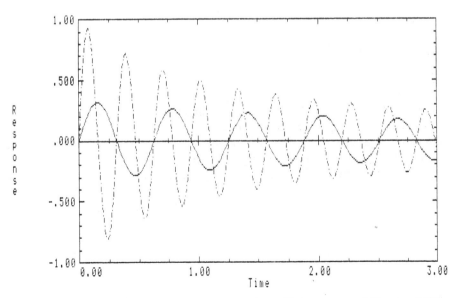

FIGURE 10.13. A screen dump to a dot-matrix printer. The screen resolution was 640 by 225; the resolution of the copy can be no better.

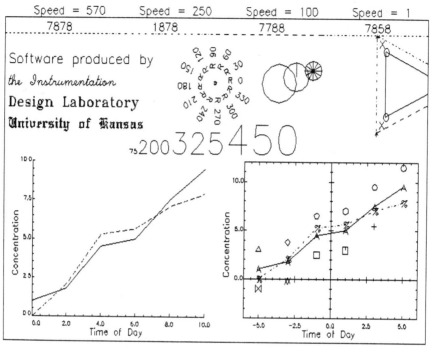

FIGURE 10.14. A high-resolution plot on an inexpensive printer. Printers with three times the resolution are becoming common.

the users of computers are more efficient in the interpretation of images than they are in the interpretation of lists. In today's *ideal* computerized laboratory, entry of data by typing lists is not necessary, since the data either are acquired directly from an experiment in real time or they result from a computation. However, in the real laboratory, some data must be entered manually, and efficient alternatives to typing lists are available.

Consider the problem of a taxonomist who must acquire, store, and manipulate the outlines of fishes from a collection of photographs. There are several possible techniques. In the most ideal, a video camera can acquire the image, each pixel of which can be digitized, so that the image is stored in memory or on disk for subsequent analysis. The least efficient data-entry technique would be to trace the photographs of these fishes on graph paper in order to list the X–Y coordinates of the outline, and subsequently enter the data on the keyboard.

After the data are acquired and stored, they may be compared by the taxonomist visually, using shape parameters. It would be useful to be able to interactively command the computer to scale the images and move the images on a screen for visual comparison. Keyboard entry of all parameters is possible, but the ability to *point* greatly improves interaction.

In this section, we will consider some alternative methods for the acquisition and manipulation of the images that will assist the taxonomist and many others.

Digitizing Tablets are extremely versatile devices for input of graphics information. Although they vary widely in the technology used for digitizing, most have much in common with the one shown in Figure 10.15. A hand-held cursor or pen is moved over a surface; on command from the computer or cursor, the X–Y coordinates are transmitted to the computer. One technology involves measurement of the time for an acoustic or electroacoustic wave to move between the cursor and an edge of the tablet. The uncertainty of the measurement of the position is typically less than 0.05 mm. A standard serial interface (RS-232-C) is common although parallel interfaces may be used and have some advantages in simplicity.

A tablet makes short work of the task of entering the outline of a fish on a photograph, since the fish need only be traced. Other similar applications could include digitization of a strip-chart record of a signal representing a spectrum or chromatogram; if only a few records need be entered, the tablet provides a more efficient method of data entry than direct connection to an instrument.

Pointing Devices in general have seen a great deal of development over the past decade. A pointing device is used in much the same manner as the digitizing tablet but often at lower precision; consequently, the types of problems that it solves are different. The process of *pointing* to a location of the screen in order to select the position or the part of the position located there is intuitively easier than typing estimates of the coordinates into the computer.

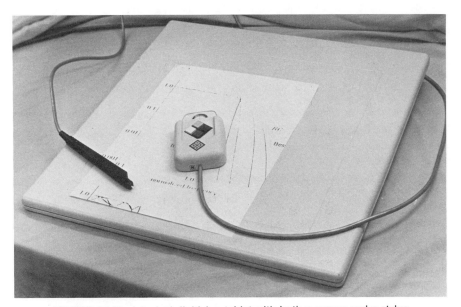

FIGURE 10.15. A typical digitizing tablet with both a cursor and a stylus.

Pointing devices have three types of applications. In graphics generation, the operator can interactively choose the best location to place a particular element in the display—for example, the title of a graph—simply by moving the cursor on the tablet. Pointing devices also can be used to indicate a datum that is to be modified, deleted, or otherwise selected. Finally, if a list of possible procedures (a menu) is displayed on the screen, one of the options can be chosen by "pointing" to it with the hand-held cursor. This application applies equally to graphical and nongraphical information.

The first type of pointing device is one that has already been described, the **digitizing tablet.** The operator can easily move the screen cursor by moving the digitizer cursor if the computer program places the former in a position scaled to the position of the latter. The tablet, although more expensive than the other pointing technologies, has the advantage of being easily interfaced, useful as a pointer with CRT or vector plotter, multifunctional, and operable in a very comfortable position.

The **light pen** is another popular device. The name results from the fact that it may create the illusion of writing on the screen. The light pen, diagramed in Figure 10.16, is actually a photodetector mounted in a pen-shaped probe. When the raster sequence moves the beam past the pen position, the flash of light produces an electrical pulse which can be used by the graphics controller to latch that beam location for acquisition by the computer. If the executing program subsequently reverses the state of the corresponding pixel, the pen appears to write a black mark on the screen. However, a more typical application is the use of a light pen to point to an entry on a menu.

The light pen suffers from several disadvantages: first, it must be interfaced directly to the CRT controller, and currently many controllers cannot accept the pen; second, when compared to using a device on the table surface, it is somewhat more fatiguing.

Touch panels also can be placed over the screen to sense the position of a pointing device. A transparent conductive surface or a matrix of transparent conductors can be mounted directly over the screen; the position of a finger pressed against the surface can be transmitted to the computer with

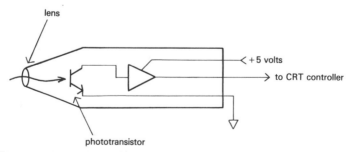

FIGURE 10.16. A light pen. A light pen actually *detects* light from the screen; a pulse to the screen controller latches the coordinates of the pixel that produced the light.

a precision of less than 1 cm. In another design, arrays of infrared LEDs and detectors on opposite sides of the screen in both the X and Y directions can detect interruption of the beam and subsequently transmit the position of the interruptor.

Trackballs, joysticks, and **paddles** also can be used as pointing devices. The latter two tend to be somewhat difficult to use with great precision, but they are quite inexpensive. Video game enthusiasts may bring great skill to their use, however. A joystick is usually a lever that activates one or two of four microswitches when deflected, yielding direction information. Paddles are electrically equivalent to a variable resistor, and a low-resolution interface is easily devised. The trackball, essentially a single, large, ball bearing whose rotation can be transmitted by shaft encoders to the computer,

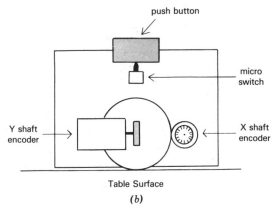

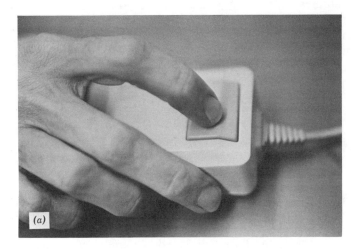

FIGURE 10.17. A mouse. A pair of shaft encoders, turned by a ball that rolls on the table surface, generate pulses used to encode position. A mouse may come with one, two, or three pushbuttons.

can have limitless resolution if many rotations are required to move the cursor across a screen.

The **mouse** follows somewhere between the tablet and the trackball. As a hand-held cursor, it rides on a ball, and moves about on any table surface; the ball movements are transmitted to the computer and control the screen cursor. As illustrated in Figure 10.17, the ball rotates a pair of shaft encoders which separately track movement in the horizontal and vertical directions.

The mouse is employed indentically to the tablet when the latter is used as a pointer except that most require no special surface. Various versions have one to three buttons that can be assigned functions according to the operation in process. For example, movement of the mouse might move a cursor to a datum or a location on the screen, after which pressing the button causes the program to record the datum.

Some studies, particularly at the Xerox Palo Alto Research Center, suggest that a mouse is the most natural pointer; the skill is easily developed while it is not fatiguing to use. The mouse is becoming particularly popular for use with software that is intended for users who are not comfortable with traditional computers. Originally designed for the Xerox Star, it also was an important part of the Apple Lisa, and finally became ubiquitous as an essential component of the Apple Macintosh.

BIBLIOGRAPHY

Bruce A. Artwick. *Applied Concepts in Microcomputer Graphics*, Prentice-Hall, Englewood Cliffs, NJ, 1984.

F. R. A. Hopgood, D. A. Duce, J. R. Gallop, and D. C. Sutcliffe, *Introduction to the Graphical Kernel Systems (GKS)*, Academic Press, New York, 1983.

William M. Newman, and Robert F. Sproull, *Principles of Interactive Computer Graphics*, McGraw-Hill, New York, 1979.

Roy E. Myers, *Microcomputer Graphics*, Addison-Wesley, Reading, MA, 1982.

Theo Pavlidis, *Algorithms for Graphics and Image Processing*, Computer Science Press, Rockville, Md, 1982.

CHAPTER ELEVEN

Computational Techniques for Laboratory Experimentation and Data Processing

In this chapter, the discussion is restricted to techniques which have *special application to data acquired on-line in the laboratory*. For specialized techniques in statistics, calculus, or numerical analysis, the reader is referred to books listed at the end of the chapter. We will consider some general principles for optimizing the acquisition and processing of data after which two important algorithms will be presented.

11.1. HARDWARE–SOFTWARE TRADE-OFFS

Frequently, when an interface between a computer and an instrument or experiment is designed, someone must decide whether to carry out a function in the computer software or in the electronic hardware of that interface; there is a hardware–software trade-off. In this book, there have been cases in which equivalent hardware and software techniques have been presented. In such cases, a general rule is: one should *never do in hardware what can be done in software*.

The reason for this rule is to take full advantage of the *flexibility* of the computer. Changes in the way a signal or datum is handled are usually much easier to make by changing the program than by changing the electronic circuits.

An example is useful. A logarithmic transformation is commonly required in order to linearize raw data from an experiment, and there are two methods of logarithmically transforming raw data. The first is to use a logarithmic amplifier (see Section 5.2.10.) before the ADC. The second is to acquire the datum using an ADC, and then to perform a logarithmic computation. Our rule suggests preference for the latter for which there are two clear benefits: (1) reduced hardware complexity and (2) elimination of the need to maintain calibration of the amplifier.

Another example stresses the need for understanding the second part of the rule: knowing what can be done. If the signal from the experiment lies within a range of 0–100 mv while the ADC accepts a range of 0–10 V, we may have a choice, either to use an external gain amplifier to boost the signal by a factor of 100 or to multiply (in software) each acquired datum by 100. In fact, the use of software does not meet the usual definition of "what can be done", since the goal is to acquire data with the necessary precision, and the conversion of 0- to 100-mV signal to digital values of 0 to 41 does not meet that criterion (there may be exceptions, of course). Consequently, the external gain amplifier is nearly always required.

Returning to the example concerning logarithmic transformations, one can envision a case where a logarithmic amplifier *is* required. The data rate might be very high so that no time exists after the acquisition of the data to compute many logarithmic transformations. In such a case, the hardware log amplifier fits the rule.

Another case exists in which aliasing (discussed in Ch. 4) results when high-frequency noise with a periodic pattern is present; the conclusion was that, even though digital filtering in software is available, the noise must be filtered before digital conversion. In most other cases, however, the digital filtering approach is preferable. Later in this chapter, digital filtering will be presented as a software alternative to the filters discussed in Chapter Five.

11.2. SIMPLEX OPTIMIZATION

Most laboratory experiments elicit responses that are a function of two or more experimental variables which can be controlled by the experimenter. In many studies, the measured response must be maximized or minimized (i.e., optimized). Examples could include the optimization of yield in the production of yeast by controlling both nutrients and temperatue, or the optimization of a chromatographic separation by controlling flow rates, temperature, and mobile phases.

One route to an optimum when a small number of variables is involved, is to perform the experiment with all possible combinations of values for the operational variables. The best response from that list was obtained under the optimum conditions. Obviously, a separate experiment for every combination of the variables is not an efficient route. An orderly procedure for finding a direct route to that optimum is required.

Many optimization approaches have been taken. An operationally and conceptually simple procedure is **simplex optimization**. The rules for the procedure can best be understood with an example, shown in Figure 11.1.

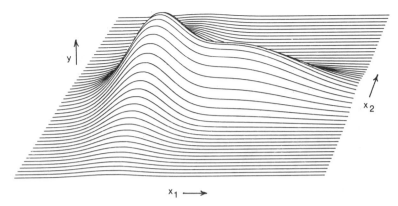

FIGURE 11.1. A response surface for optimization. The response, y, is plotted as a function of the controlled variables, x_1 and x_2.

We envision an experiment in which the response (y) is dependent on only two controlled variables, x_1 and x_2. In this case, x_1 and x_2 are the two horizontal dimensions, and the height at any point is the experimental response. The figure illustrates the graph that would be obtained, were we to make a measurement at every x_1–x_2 combination. The contour map of the surface in Figure 11.1 is shown in Figure 11.2. The objective of the optimization is to find the highest point without making all of those measurements.

Using Figure 11.1 as a model, we could redefine the problem as finding the highest point on the "hill" while blindfolded. A good optimization procedure finds the highest point in the surface with the smallest number of "moves" (experiments) when the shape of the surface is unknown. The simplex method offers an orderly and efficient route to that maximum.

The procedure begins by mapping onto the surface a **simplex**, which is the geometric figure defined by one more measurement than the number of variables. In this two-variable experiment, three beginning measurements are made whose parameters form the triangular simplex, B-M-P. In the example, vertex B yields the best response of the three, P is the poorest, and M is in the middle.

The next experiment requires a good choice for a new set of parameters from which to make a measurement, a position where there will be a good likelihood of obtaining an even better response. A new simplex is built beginning with the best two responses of the old simplex (Fig. 11.3). Vertex P is eliminated by reflecting it across the opposite face through the midpoint of M-B, C_{MB}. The line P-C_{MB} is extended to obtain a new combination of variables at R, and a response is obtained for the values of x_1 and x_2 at the point denoted R. The directions for subsequent moves on the surface can be determined by simply repeating the procedure of the previous paragraph.

In most cases, that procedure will move closer to the maximum; the reflection step is simply repeated, reflecting the worst point of each simplex, until a maximum is found. Unfortunately, it suffers from two drawbacks.

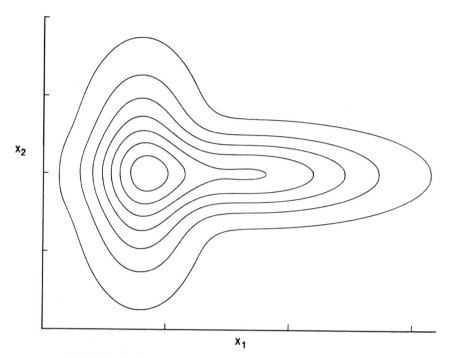

FIGURE 11.2. The data of Figure 11.1 displayed as a contour plot.

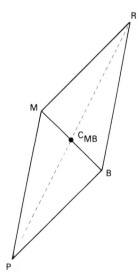

FIGURE 11.3. Generation of a new simplex. When P is the poorest response of the simplex B-M-P, it is reflected to R; C_{MB} is the midpoint of M-B.

First, the simplex can easily be trapped on a ridge or in a valley; if both the discarded point *P* and the new point *R* are the worst responses of their respective simplexes, the procedure will oscillate between them. Second, when all reflections are the same size, no "big jumps" for rapid movement are possible nor are "tiny steps" allowed to facilitate closing in on the optimum.

A preferred procedure involves determining whether the new or old simplex is closer to the optimum, the response at *R* is compared with those of the previous simplex before determining the position of the new vertex. This variation is termed the **Modified** or **Variable-Size Simplex** optimization method.

When the newest response is evaluated, four possibilities exist, as shown in this method (Fig. 11.4):

1. *The response at R is better than that at B:* that is, it is the best overall. In this case, the simplex is probably moving in the best direction, and the optimum is likely to be even farther from *B* than *R*. An *expansion* of the simplex size can be attempted. A new simplex is formed, *M-B-S*, where *S-C_{MB}* is twice *P-C_{MB}*. The search for the following simplex will begin by reflecting *M* across *B-S*. However, if the response at *S* is not better than that at *R*, the expansion fails, and the process is continued with simplex *B-M-R*.

2. *The response at R is between those at B and M.* In this case, *R* is retained as the new vertex, and the simplex *B-M-R*, remains the same

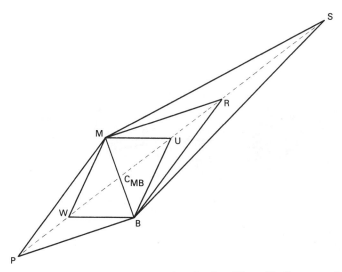

FIGURE 11.4. Generation of a new variable-size simplex. When *P* is the poorest response, *R* is tried, and depending on the response at *R*, the vertex of the new simplex may be *W*, *U*, *R*, or *S*.

size. The search for the following simplex will begin by reflecting M across B-R.

3. *The response at R is poorer than the response at M but better than that at P.* The new vertex is likely to be closer to that of R than that of P and is chosen to be U, the point bisecting RC_{BM}. The resulting simplex, B-M-U, is a *contraction*.

4. *The response at R is even poorer than that at P.* This suggests that the simplex has straddled a ridge and has moved in the wrong direction. Again the simplex is contracted, but the new vertex is at point W, bisecting PC_{BM}.

The process now is repeated until the simplex reduces in size to some value within the limits suitable for the experiment.

Strict adherence to these rules may occasionally lead to a proposed vertex outside the possible boundaries; for example, a negative concentration might be requested when the concentration of a material is one of the variables. In such a case, the unmeasurable response is assumed to be the worst, and the simplex is continued.

A full simplex procedure for Figure 11.1 is shown as the contour plot in Figure 11.5. The optimum is determined by the variable-size simplex method starting with simplex A-B-C. The following steps led to the simplexes shown:

A-B-1:   reflection and expansion of A-B-C

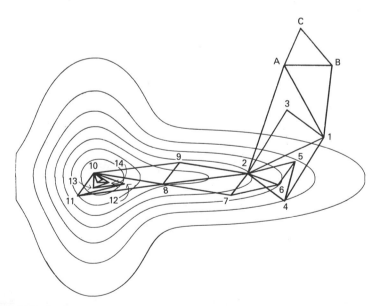

FIGURE 11.5. Application of the variable-size simplex method to the data of Figures 11.1 and 11.2.

A–1–2: reflection and expansion
1–2–3: contraction after a failed reflection
1–2–4: simple reflection
2–4–5: contraction after a failed reflection
2–5–6: contraction after a failed reflection
2–6–7: reflection
2–7–8: reflection and a successful expansion
2–8–9: reflection
8–9–10: reflection
8–10–11: reflection
10–11–12: contraction after a failed reflection
10–12–13 and the next several steps are contractions

While a substantial number of responses were determined during the process of approaching the optimum, the overall number of determinations will generally be much smaller than would be required if an unstructured search were pursued.

The fact that a small number of precise rules define the procedure leads to the possibility of automated optimization of responses to those variables which can be directly computer-controlled. The simplex procedure is readily adaptable to optimizing conditions in experiments where there are many controllable variables that do not interact linearly with each other: optimizing growing conditions, chromatography, analytical atomic spectroscopy, nuclear magnetic resonance, and chemical reaction conditions are examples. Simplex optimization also provides a method for finding a numerical iterative solution to otherwise intractable equations; new estimates are chosen following the same rules, and a minimum is sought.

The application must meet three requirements before the automated simplex procedure is applicable

1. The response surface must have a single optimum; if the surface has multiple local optima (more than one peak on a surface), the simplex may become trapped on an optimum which is not the largest.

2. The response must be amendable to acquisition *and* evaluation by the computer. If the response cannot be measured or if there is no method to determine which response is best, the computer cannot be used.

3. The experimental parameters must be controllable from the computer, preferably without the intervention of the operator. However, the technique still could be used in a manual experimental design where the operator enters responses into the computer and it responds with the display of a message directing the operator to make the parameter changes.

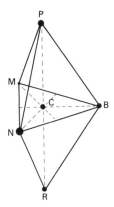

FIGURE 11.6. Generation of a new simplex when there are three controlled variables. Beginning with the three-dimensional simplex *M-B-P-N* at the left, *P* is found to be the poorest. Therefore, *P* is reflected through the centroid of the *M-B-N* surface to *R*, shown on the right.

The illustrations have dealt with a two-dimensional simplex only because the progress can be shown graphically with little difficulty. This should not imply a limit on the number of variables. For example, a three-variable experiment will generate a three-dimensional simplex with four vertices, illustrated in Figure 11.6. The worst response on that tetrahedron must be reflected through the centroid of the opposite surface to produce the next vertex, *R*, by defining the next experimental conditions. When more variables are added, the simplex path can no longer be shown graphically, but the procedure is even more valuable.

11.3. POLYNOMIAL LEAST-SQUARES CONVOLUTION TECHNIQUES

Many computational techniques exist for digitally filtering and transforming data sets. One of the most flexible and elegant techniques is also simple and efficient to apply; while advanced study might lead to better digital filters for particular applications, we will concentrate on the technique of Polynomial Least-Squares Convolution (PLSC).

In this method, the data set is simply convolved with a set of integers. Using only multiplication by integers, these algorithms can filter out noise, differentiate and integrate data sets, and interpolate between data.

Before detailing the filtering technique, an example might be useful. Assume that data from some phenomenon, the acidity of a waterway for example, are acquired at regularly spaced intervals such as ten times/min. A pH electrode is a suitable transducer. Although only the changes that take place on the time scale of hours are important, random minute-to-minute changes will appear as the result of both electrical noise and local, insignificant, variations in the water stream. The objective of the filtering process will be to remove the noise which has a higher frequency. Thereby the S/N can be enhanced without affecting the low-frequency signal. That is, the data are "smoothed" without introducing any distortion of the information.

The procedures in this section will be illustrated with the aid of a typical curve, a Gaussian-shaped peak with 24 data in the interval of the width of the peak at half its height. After adding noise to the peak, the effect of digital filters can be demonstrated. Two parameters can be used to evaluate the process: we want to observe both the noise reduction and the distortion in the shape.

11.3.1. Block and Weighted Averages

A simple smoothing method is a moving **block average**; that is, the data are convolved with a block function. The values inside the block are equal to one while those outside the block are zero. If the block size is nine, the first nine raw data are averaged, and the result becomes the new value for the central (fifth) datum. Then the second through tenth raw data are averaged and the result becomes the new sixth value.

Each new smoothed value is influenced symmetrically by the preceding and following raw points. The amount of smoothing depends on the block size. Smoothing by 23- and 9-point blocks are depicted in Figure 11.7. Clearly, the degree of smoothing increases with the width of the smooth. However, the distortion is best observed in the peaks that do not contain noise, and the example demonstrates that the signal is increasingly distorted as the block size grows; the magnitude of the distortion makes the block averaging technique of questionable merit.

An analog *RC* (first-order low-pass) filter (see Chapter Five) can be simulated with a waveform that is similar to the block average, but the data are not multiplied by 1s. The center datum is multiplied by one, and preceding data are multiplied by exponentially decreasing coefficients; data *after* the center position are multiplied by zero. Like a first-order low-pass filter, the most recent value of the signal is averaged with the earlier values; the weight decreases exponentially with time. Unlike the block average, this filter is asymmetric, having no capacity to "look into the future"; it cannot include contribution from the signal which follows and weight that equally with the signal that precedes. The lack of symmetry leads not only to distortion but to a shift in the positions of peaks in the data.

The results of this filter's application are shown in Figure 11.8. From the case with noisy input data, it appears that this filter does a better job of filtering than the 9-point block, and the second case demonstrates that the distortion in the height and width is less than that observed for the 23-point block. However, the filtered signal *lags* the unfiltered data.

A symmetrical **weighted average** should be an improvement. When the

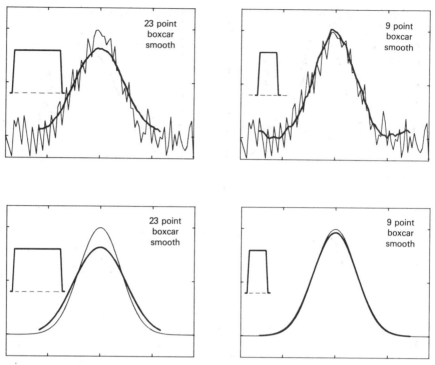

FIGURE 11.7. Application of convolute integers to a Gaussian curve. Top and bottom left, a 23-point boxcar integer set results in good smoothing but severe distortion. Top and bottom right, a narrower integer set results in poorer smoothing but less distortion. The PLSC integers are plotted in the inset.

smoothed value of a point is computed, the points both preceding and following that datum are multiplied by weights less than unity. The central value has the largest effect, and the weights decrease as the position moves farther from the center. This method is an improvement, but a problem in this method lies in finding the optimum weighting factors.

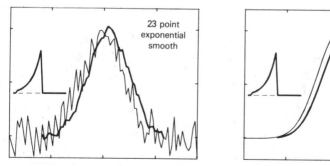

FIGURE 11.8. Application of convolute integers that model a first-order *RC* filter. Both distortion and phase shift (lag) can be observed.

11.3.2. The Polynomial Least-Squares Convolution Method

One solution to the problem of obtaining weighting coefficients comes from the use of linear least-squares (LS) methods. A set of weighting integers are computed which can be used in the weighted average method; in this section we will find that the method is exactly equivalent to performing LS fits of polynomial equations to the data.

Smoothing by Fitting Least-Squares Equations. First, how does LS smoothing operate on the data? If we take any contiguous set of data points, an equation can be found by the method of LS which approximately fits the data. The equation will not usually pass through the data points, but will form a good "smooth" line through those data points. A polynomial equation typically cannot be found that will fit an entire data set unless the data are nearly monotonic. However, an equation can fit short segments.

In Figure 11.9, a portion of the noisy curve in Figure 11.8 is expanded for study. If a subset of the data, consisting of only 23 data points is used, a polynomial equation (in this case, a quadratic) can be used to compute the new "smoothed" estimate for the center point. The dashed line in Figure 11.9 is the best fit of a line to the first subset of 23 data, and a new smoothed datum is found at the center of the subset. The block of 23 points is then shifted one position and the process is repeated; the best fit of a quadratic to subset #2 is the dotted line. Again a new, smoothed point is found. By repeating the procedure in this manner, an entire new set of data points will result with the higher frequency noise suppressed.

Note that the new points are *not* used in later calculations, but the fit of

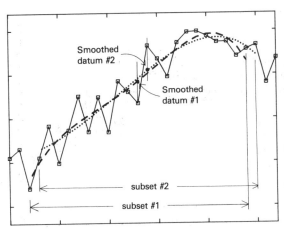

FIGURE 11.9. Least-squares polynomial filter. A fit of a cubic equation to 23 points generates one smoothed datum. The process is shifted by one datum to generate a second smoothed datum.

the equation is always made to the raw data. To use the new data in the filter calculation for later data would be **recursive**. This filter is described as being **nonrecursive**.

The approach requires substantial computational effort, since for each LS computation, the set of raw data, the squares of the data, and several cross products must be summed. Fortunately however, the LS procedure can be reduced to a set of weighting integers so that the same filter can be produced exactly by the procedure of a weighted average. Our ability to use this solution depends on the property that the data are evenly spaced on the axis of the independent variable. This property is typical of data acquired by a computer.

Computing Integers for PLSC. The integers used as coefficients for the PLSC method can be calculated from formulae or looked up in tables. In this section, we will see how those integers are derived.

As a representative case, seven contiguous data points can be taken from a data set in order to find a smoothed value for the central value. The first step with this set is a transformation of the independent variable. For the purposes of the procedure, the independent variables, x_{-3}, x_{-2}, x_{-1}, x_0, x_1, x_2, and x_3, are assigned the values -3, -2, -1, 0, 1, 2, and 3; because the data are evenly spaced on the x axis, this step is simply a linear scaling and translation procedure. The corresponding dependent variables are y_{-3}, y_{-2}, y_{-1}, y_0, y_1, y_2, and y_3.

Using the PLS procedure for a cubic fit, we assume that the best coefficients in a cubic equation are found by minimizing the sums of the squares of the differences between the raw data and the points computed from the equation. That is, χ^2 is minimized where χ^2 is the sum of the squares

$$\chi^2 = \sum \{y_i - (b_0 + b_1 x_i + b_2 x_i^2 + b_3 x_i^3)\}^2 \qquad (11.1)$$

When that equation is expanded, we note that because of the way in which the independent variables were transformed, the sums of the odd powers of the x values are zero, $\sum x_i = 0$ and $\sum x_i^3 = 0$. χ^2 is easily minimized by computing the partial derivatives of χ^2 with respect to b_0, b_1, b_2, and b_3 and setting them to zero. The resulting set of equations is

$$\sum y_i \quad = nb_0 + b_2 \sum x_i^2 \qquad (11.2a)$$

$$\sum x_i y_i = \quad b_2 \sum x_i^2 + b_3 \sum x_i^4 \qquad (11.2b)$$

$$\sum x_i^2 y_i = \quad b_0 \sum x_i^2 + b_2 \sum x_i^4 \qquad (11.2c)$$

$$\sum x_i^3 y_i = \quad b_1 \sum x_i^4 + b_3 \sum x_i^6 \qquad (11.2d)$$

For the seven point fit in question, the summations are carried out over the range $i = -3$ to $i = +3$.

Next we note that the "new," smoothed value of y at x_0, (where $x = 0$) will simply be b_0, and b_0 can be obtained from the simultaneous solution of

(11.2a) and (c). Note that if a quadratic rather than a cubic were used in (11.1), then (11.2a) and (c) still would have remained the same; the results for the fit of a quadratic and a cubic are identical.

The resulting "smoothed" value of y_0 is

$$y_0 = b_0 = \frac{\sum y_i \sum x_i^4 - \sum x_i^2 y_i \sum x_i^2}{n \sum x_i^4 - (\sum x_i^2)^2} \qquad (11.3)$$

In the case of the seven-point smooth introduced, we can readily compute the integers. First, the summations are computed, and for the seven-point smooth considered, $\sum x_i^4 = 196$; $\sum x_i^2 = 28$. Then

$$b_0 = \frac{7 \sum y_i - \sum x_i^2 y_i}{21} \qquad (11.4)$$

so that

$$b_0 = [-2(y_{-3} + y_3) + 3(y_{-2} + y_2) + 6(y_{-1} + y_1) + 7y_0]/21 \qquad (11.5)$$

The resulting computation simply involves multiplication of the raw data by integers, although it is an exact fit of an equation to seven contiguous data. A simple change of the limits of the summations in (11.3) extends the computations to encompass smoothing over more points.

One must be mindful of the limitation that for a smooth of n points, smoothed values of the first and last $(n - 1)/2$ data cannot be included. There are techniques, only slightly more complicated, which will accommodate such a smooth, but these are beyond the scope of this discussion.

Fitting a cubic equation to data is, in many cases, inappropriate since there may be higher resolution detail that is real. Consequently, a higher order equation may be appropriate. Addition of quartic and quintic terms to (11.1) requires that the equations be solved in matrix form. However, the result remains a simple table of integers. The tables of smoothing integers are presented in the Appendix.

Computation of integers for the PLSC technique places two requirements on the experimentally derived data for the procedure to have validity:

1. The data must be equally spaced on the axis of the independent variable (in this case, x).
2. As for any LS procedure, the noise component of the data must be random, averaging to zero.

11.3.3. The Application of PLSC Digital Smoothing

Use of LS convolute integers is, in most cases, inexact since a polynomial usually does not describe the data precisely. Consequently, some distortion is always introduced; this is true when any smoothing procedure is applied to data.

Some of the flavor of the smoothing procedure can be obtained from examples. In Figures 11.10 and 11.11, several demonstrations are presented of the application of PLS smoothing integers to the same data set as was treated in earlier figures. Again, a visual impression of the decrease in the noise can be obtained from viewing the results of smoothing a noisy peak, but the distortion is best observed from application to a noise-free peak.

First a 23-point quadratic/cubic smooth is applied to noisy data; visual inspection suggests that the noise has been nearly removed. In the case of the 23-point quartic smooth, the removal of the noise was less effective, and the 9-point quadratic smooth was even less effective.

Consideration of Figure 11.11 shows observable distortion in the technique only for the 23-point quadratic smooth, which was the most effective filter. A 23-point set of quartic/quintic integers yielded reduced error because of the capacity of a higher order equation to fit a tighter curve.

There are three variables to be considered in the application of this procedure:

1. The order of the LS fit: quadratic/cubic, quartic/quintic, or possibly higher.

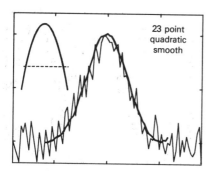
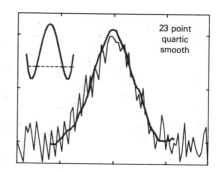

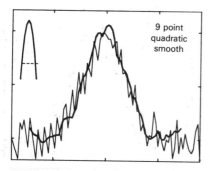

FIGURE 11.10. Application of PLSC integers filtering to a noisy Gaussian curve. The widest quadratic gives the most smoothing.

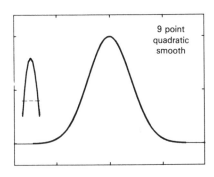

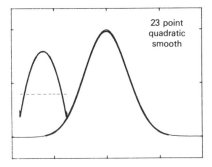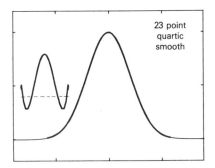

FIGURE 11.11. Application of PLSC integer filtering to a noise-free Gaussian curve. The PLSC integers are plotted in the inset. Only the 23-point quadratic shows visible distortion (a slight loss of peak height).

2. The number of points, always odd, over which the smooth is applied.
3. The number of times the same smooth is applied to the data set.

Clearly, when a lower order polynomial equation is graphed, it generates a smoother line than does a higher order equation. Similarly, the fit of a lower order equation will have a greater smoothing effect (a lower frequency cutoff) than a higher order polynomial.

The choice of the number of points to include has received attention from Enke and Nieman (1976). A general recommendation can be made—when distortion is to be minimized, the **smoothing ratio** should not exceed two; this is the ratio of the number of points in the smooth to the full-width-at-half-maximum of the peak or similar structure in the data (see Fig. 11.12). Examination of the tables of integers shows that if the integers are plotted, they take a peak-like appearance, and as a convolution technique, the convolute integers should take the shape of the structure of the data to be convoluted if distortion is to be minimized.

On the other hand, the data in Figure 11.13 suggests that the relative S/N enhancement is obtained with a smoothing ratio of two.

One possible way to increase S/N enhancement without the distortion of

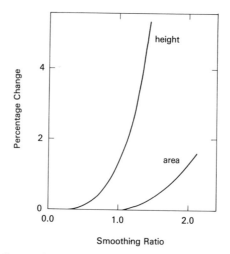

FIGURE 11.12. Distortion resulting from application of PLSC filtering. The smoothing ratio is the ratio of smoothing width to the peak width of the data. The error is shown for quadratic/cubic smoothing. (Reprinted with permission from C. G. Enke and T. A. Niemen, *Anal. Chem.,* **48,** 705A, 1976. Copyright 1976, American Chemical Society.)

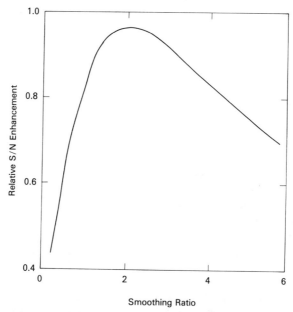

FIGURE 11.13. The fraction of the maximum possible S/N enhancement shown as a function of Smoothing Ratio. (Reprinted with permission from C. G. Enke and T. A. Niemen, *Anal. Chem.,* **48,** 705A, 1976. Copyright 1976, American Chemical Society.)

the large smoothing ratio is to repeat the same smoothing operation several times using the smoothing ratio which minimizes distortion. However, the operation of the smooth procedure occurring more than once on a single data set carries a penalty in that the beginning and ending $(n - 1)/2$ data are lost on each application. This can be avoided by either retaining un-smoothed values or setting the first $(n - 1)/2$ values to that of the smoothed n^{th} value, and doing the same with the last $(n - 1)/2$ data. The improvement in S/N was found to vary approximately with the inverse eighth root of the number of smooths. Overall, a single pass of optimum size is to be preferred to multiple passes of smaller size.

11.3.4. Computing Derivatives with Convolute Integers

A modification of this procedure generates the derivative of the data set with respect to the x axis simultaneously with smoothing. We note that at $x = 0$, the value of the first derivative of the equation

$$y = b_0 + b_1 x + b_2 x^2 + b_3 x^3$$

is simply b_1. Therefore, this case differs from smoothing only because we must solve for a different constant. The integers corresponding to b_1 are again computed from (11.2b) and

$$b_1 = \frac{\Sigma x_i y_i \Sigma x_i^6 - \Sigma x_i^3 y_i \Sigma x_i^4}{(\Sigma x_i^4)^2 - \Sigma x_i^6 \Sigma x_i^2} \tag{11.6}$$

In the case of the first derivative using 7-point convolute integers for a cubic (and quartic) fit, the solution of (11.6) yields

$$b_1 = \frac{-22(y_3 - y_{-3}) + 67(y_2 - y_{-2}) + 58(y_1 - y_{-1}) + 0y_0}{252}$$

Similarly, the smoothed value of the second derivative of the fit of a cubic equation to the data at $x = 0$ is $2b_2$, computed from the solutions of (11.2a) and (c). The smoothed value of the third derivative is $3b_3$. Higher order derivatives require a higher order model equation but also can be obtained.

Examples of the application of 23-point cubic-quartic sets of first- and second-derivative integers to both clean and noisy Gaussian curve are shown in Figure 11.14.

This technique, using convolute integers to obtain derivatives, is a very powerful alternative to analog methods described in Chapter Five since

1. The latter are very sensitive to high-frequency noise while this digital technique provides simultaneous filtering.

2. The derivative can be calculated while the integrity of the original data is maintained in memory or on disk.

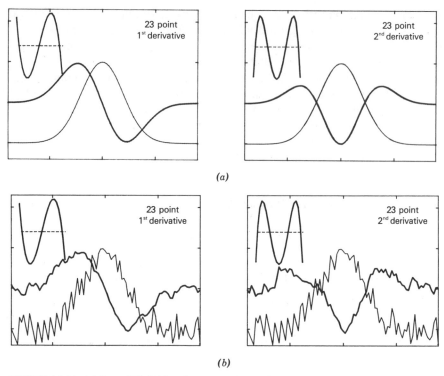

(a)

(b)

FIGURE 11.14. (a) Use of PLSC integers to compute the derivative of the Gaussian curve. The PLSC integers are plotted in the inset. Cubic/quartic integers are used in each case. (b) Use of PLSC integers to compute the derivatives of noisy curves, as in Figure 11.4.

11.3.5. Interpolation with Convolute Integers

Interpolation can be useful when, for example, doubling the data density leads to a more pleasing visual image. An adaptation of the same type of integers and techniques is readily made by computing a new value for y_0 when the values on the x axis are assigned only to odd numbers. For example, if six data are transformed on the x axis to have values of x of -5, -3, -1, 1, 3, and 5, the resultant value of b_0 will correspond to $x = 0$, halfway between $x = -1$ and $x = 1$.

Taking (11.3) but computing the summations in x over only the odd values from -5 to $+5$, the smoothed and interpolated value b_0 is

$$b_0 = [-3(y_{-5} + y_5) + 7(y_{-3} + y_3) + 12(y_{-1} + y_1)]/32 \quad (11.7)$$

Several sets of interpolating integers are presented in the Appendix.

11.3.6. Integration with Convolute Integers

If the fit of a polynomial to the data is appropriate, that polynomial can be used to determine the integrated value of the data over the desired range.

If the cubic equation is used again, the integral of y is determined by deriving the integral of the polynomial:

$$\int y = b_0 x + b_1 x^2/2 + b_2 x^3/3 + b_3 x^4/4 \qquad (11.8)$$

Carrying through the integration over the limits of $x = -1$ to $x = 1$, we obtain

$$\int y = 2b_0 + 2b_2/3 \qquad (11.9)$$

A full peak in a data set can be integrated by summing the results of the area between each pair of adjacent data using the formula in (11.9).

The integer tables for integration can be obtained readily by computing tables for b_0 and b_2 and appropriately adding them. Although this computation of the convolute integers is slightly more complicated than for other procedures, the application is not as involved.

11.3.7. Two-Dimensional Techniques with Convolute Integers

The problem of smoothing data with two independent variables cannot properly be solved by the filtering techniques discussed so far. However, the simple cubic equation in (11.1) can be expanded to obtain a smoothing function in two dimensions by simply expanding the equation:

$$y = b_0 + b_1 x + b_2 x^2 + b_3 x^3 + b_4 z + b_5 z^2 + b_6 z^3 \qquad (11.10)$$

Assuming that the smoothing function is applied equally in the x and z dimension, the 7×7 matrix of data has eightfold symmetry as shown in Figure 11.15. A set of 10 integers is required.

$$b_0 = [2(y_{x-3z0} + y_{x3z0} + y_{z-3x0} + y_{z3x0})$$
$$+ 7(y_{x-2z0} + y_{x2z0} + y_{z-2x0} + y_{z2x0})$$
$$+ 10(y_{x-1z0} + y_{x1z0} + y_{z-1x0} + y_{z1x0})$$
$$- 7(y_{x-3z3} + y_{x3z3} + y_{z-3x3} + y_{z-3x-3})$$
$$+ 3(y_{x-2z2} + y_{x2z2} + y_{z-2x2} + y_{z-2x-2})$$
$$+ 9(y_{x-1z1} + y_{x1z1} + y_{z-1x1} + y_{z-1x-1})$$
$$- 2(y_{x-3z2} + y_{x3z2} + y_{z-3x2} + y_{z3x2} + y_{x-2z3} + y_{x-2z-3} + y_{z-2x3} + y_{z-2x-3})$$
$$+ 1(y_{x-3z1} + y_{x3z1} + y_{z-3x1} + y_{z3x1} + y_{x-1z3} + y_{x-1z-3} + y_{z-1x3} + y_{z-1x-3})$$
$$+ 6(y_{x-1z2} + y_{x1z2} + y_{z-1x2} + y_{z1x2} + y_{x-2z1} + y_{x-2z-1} + y_{z-2x1} + y_{z-2x-1})$$
$$+ 11y_{x0z0}]/147 \qquad (11.11)$$

Several tables of integers are found in the Appendix. It may or may not

o	o	o	o	o	o	o	o	o	o
o	-3.3	-2.3	***-1.3***	*0.3*	***1.3***	2.3	3.3	o	o
o	-3.2	**-2.2**	-1.2	**0.2**	1.2	**2.2**	3.2	o	o
o	***-3.1***	-2.1	*-1.1*	**0.1**	*1.1*	2.1	***3.1***	o	o
o	*-3.0*	**-2.0**	**-1.0**	0.0	**1.0**	**2.0**	*3.0*	o	o
o	***-3.-1***	-2.-1	*-1.-1*	**0.-1**	*1.-1*	2.-1	***3.-1***	o	o
o	-3.-2	**-2.-2**	-1.-2	**0.-2**	1.-2	**2.-2**	3.-2	o	o
o	-3.-3	-2.-3	***-1.-3***	*0.-3*	***1.-3***	2.-3	3.-3	o	o
o	o	o	o	o	o	o	o	o	o
o	o	o	o	o	o	o	o	o	o

FIGURE 11.15. Data matrix for application of two-dimensional smoothing. The members that are to be multiplied by the same coefficient are printed in the same typeface.

be appropriate to use the same number of data in both the x and z directions, and tables for these cases are also included.

This procedure also has been applied effectively in a derivative mode (Edwards, 1982) to enhance photographic images; the process enhances gradients in the image, and thereby is useful for detecting the edges of objects.

11.4. ASPECTS OF LINEAR LEAST-SQUARES CURVE FITTING

Fitting an equation to data is a common task in scientific investigation. This task may have two purposes: The first may be to summarize a set of data for calibration curves, interpolation, or visual interpretation. The second may be to test a theoretical relationship or determine coefficients in a theoretical model that applies to an experiment.

The concept of using a LS approach to curve fitting was introduced earlier in the context of developing sets of convolute integers for smoothing and transforming data. Most scientists are familiar with LS fitting in a more general sense; the technique is used frequently for fitting a model to a set of data. However, there are several caveats regarding use of this procedure with experimental data which are discussed here.

In this section, the approach to fitting straight lines to data is reviewed after which the importance of proper weighting procedures is covered. Finally the techniques for implementing multiparameter procedures in small computers are discussed.

11.4.1. Fitting a Straight Line to Data

Most experimentalists are quite familiar with the necessity of fitting a straight line to a set of data. In real measurements, the data will exhibit scatter around the ideal relationship between the controlled parameter (independent variable) and the measured response (dependent variable). The LS approach provides a way to *estimate* the *most likely* coefficients of an equation which would describe the relationship if there were either no scatter or an infinite number of data to average.

The straight line normally follows the equation

$$y = \beta_0 + \beta_1 x \tag{11.13}$$

where x represents the independent variable and y represents the dependent variable; the objective is to estimate coefficients β_0 and β_1, given a set of experimental values x_i and y_i.

According to the LS method, we will compute values of the coefficients in the equation that minimize the sum of the squared deviations of the observed values from those that the equation predicts. For the straight line and other polynomial equations, the coefficients can be calculated exactly.

Use of the LS approach is dependent on three assumptions:

1. *The appropriate form of the equation must be chosen*; in the present case, the coefficients have no meaning if a straight line model is not reasonable.
2. *The data must be representative of the phenomenon under study*; this criterion is the most difficult to test, requiring accurate knowledge of the phenomenon under study.
3. *The data must be statistically uncorrelated*; that is, the random error for any point must be truly random, not correlated with the error for any other point.

In the case of the straight line, estimates of β_0 and β_1 are b_0 and b_1 and are determined from the n values of x_i and y_i by use of the formulae

$$b_0 = \Delta^{-1}(\Sigma w_i x_i^2 \, \Sigma w_i y_i - \Sigma w_i x_i \, \Sigma w_i x_i y_i) \tag{11.14a}$$

and

$$b_1 = \Delta^{-1}(\Sigma w_i \, \Sigma w_i x_i y_i - \Sigma w_i x_i \, \Sigma w_i y_i) \tag{11.14b}$$

where

$$\Delta = \Sigma w_i \, \Sigma w_i x_i^2 - (\Sigma w_i x_i)^2 \tag{11.14c}$$

All sums are carried out from $i = 1$ to n data points, and w_i denotes the weighting factor (the reciprocal of the variance) for each point. (See the references for the derivation.)

One of the first problems to be expected is that **data may be unequally reliable**. Consequently, a scheme that favors the more reliable data is desirable. In fact, the LS procedure calls for *weighting* the data, and each value is weighted by the reciprocal of the variance. (The variance is the square of the standard deviation.) Details on the use of that weighting factor are reviewed in Section 11.4.2. If there is no basis on which to provide weights, all weights must be set to one. However, one should be careful to verify that all data are equally reliable.

A second problem arises from the transformation of a **nonlinear model** to a polynomial model. An example is the common case of a relaxation or decay curve, applicable to the discharge of a capacitor, the progress of certain chemical reactions, and a great many other phenomena. The model might follow the form

$$R = Ge^{-kt} \tag{11.15}$$

where R is the measured response, t is time, and G and k are the desired constants.

Computation of G and k can be enabled if a transformation can be made which makes the equation linear in its coefficients. In this case, linearization is accomplished by taking the log of both sides of (11.15):

$$y = \ln(R) = \ln(G) - kt = \beta_0 + \beta_1 x \tag{11.16}$$

After calculating β_0 and β_1, the reverse transformation will provide k.

We might consider for comparison the case of

$$R = Ge^{-k_1 t} + Fe^{-k_2 t} \tag{11.17}$$

In this case, no transformation will make the equation linear in its coefficients, so nonlinear iterative methods are required which are described in the references.

To complicate a simple matter, we must remember that the variances also are transformed. It becomes important to observe weighting considerations *even if all observations are equally reliable* as will be shown later.

One special case of fitting a straight line to data is the case where the line should be constrained to pass through zero. If all measurements are made as differences between the response when the independent variable is zero and the response at another independent variable, a nonzero value of β_0 has no meaning. This is the case when the instrument is "zeroed" between each measurement. In this case, the model equation should be

$$y = \beta_1 x \tag{11.18}$$

and the estimate b_1 is obtained from

$$b_1 = \Sigma w_i x_i y_i / \Sigma w_i y_i^2 \tag{11.19}$$

11.4.2. Weighting the Data in LS Fitting

After pointing out the need to weight the data, the question of how to determine the weights arises. Since the weighting factor is the reciprocal variance, one also must ask how the weight is affected by a transformation.

There are fundamentally two methods of determining the weights: measurement of the variance and derivation of the variance.

In the former case, a single data point stored in memory or on disk may be the average of many observations. The computation of the variance of that measurement as it is being acquired is probably a trivial additional task for the computer.

In some cases, the relationship of the variance to the measured response may be well known. For example, when scintillation or photon-counting measurements are made, the resultant Poisson distribution predicts that the variance in the measurement is equal to the value of the measurement. In other cases, we may know or have determined that the variances of all observations are equal or that variance in the observation is proportional to the value of the observation. In these cases, it may be possible to write the LS equations in terms of the known formula for the weights.

An additional problem arises when the measured values are transformed. Consider for example the decay case presented in the previous section. Even if all measurements of R have the same uncertainty, the corresponding uncertainties in the values of $\ln(R)$ would not be equal.

In this case, we must make use of the mathematics of the propagation of errors. If $Y = \ln(R)$, the variance in Y, σ_Y^2, is related to the variance in R, σ_R^2, by the relationship

$$\sigma_Y^2 = \sigma_R^2 \left(\frac{\partial Y}{\partial R}\right)^2 \tag{11.20}$$

In our simple example, if the uncertainty in all measurements of R was constant, a weighting factor of R^{-1} still would be required in the LS equations.

The effect of weighting can be seen in an example in Figure 11.16. The

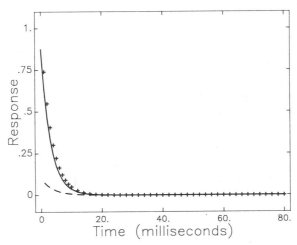

FIGURE 11.16. Illustration of errors in unweighted linear least-squares analysis. The data (+) result from an exponential decay including a small amount of noise (not easily visible in the figure). The solid line is a weighted fit of the exponential function to the data; the dashed line results from an unweighted fit.

decay curve, described in (11.15), was generated with added noise, and is fit by the method of LS using both the weighted and unweighted scheme. Constant variance in the observations was assumed. (The noise is present in the data, but is difficult to see in the figure.) Because data were collected well into the tail, a small amount of error in that region can cause an inordinate effect on the exponential decay parameters unless proper weighting is used.

The dashed line illustrates a linear LS fit of an exponential decay equation to the data without weighting, whereas the solid curve is the result of a weighted fit. Because of the log transformation that was used to linearize the equation, the uncertainty in the data that lie in the "tail" of the curve is magnified with respect to the uncertainty in the data at the beginning of the curve.

11.4.3. Computing Multiple Linear Regression Parameters on Small Computers

If more complicated models are required, a simple formula for computing the coefficients cannot be written. In this case we resort to a matrix formulation of the calculation that presents some problems in a computer of limited precision and memory.

The most common model used in fitting equations to data is

$$y = \beta_0 + \beta_1 x + \beta_2 x^2 + \beta_3 x^3 + \cdots \tag{11.21}$$

That the equation follows that progression of terms is not required, however, and any relationship for which a function of x can be computed for each x will work.

The values of the coefficients can be computed from the matrix equation

$$B = (X'WX)^{-1} (X'WY) \tag{11.22}$$

where B is the matrix of m coefficients, Y is the matrix of the n measured observations, X is an m by n matrix of independent variables that contribute to each member of Y, and W is the matrix of weights.

Examination of the algebra involved suggests the need to store several matrices in memory, the largest of which will be a square matrix containing the square of the number of values of y. In many small computers, the fitting of an equation to as few as 100 data points would be prohibited.

The memory requirement is greatly reduced by accumulating several sums where the m by n matrix $G = X'WX$ is composed of

$$G_{ij} = \sum_{k=1}^{n} w_k x_{ki} x_{kj} \tag{11.23a}$$

and the m by 1 matrix $R = X'WY$ is composed of

$$R_i = \sum_{j=1}^{n} w_j x_{ji} y_j \tag{11.23b}$$

Then

$$B = G^{-1}R \tag{11.23c}$$

If very large data sets are to be analyzed, the data can be stored on disk, and these sums can be accumulated as the data is read from the disk. Thereby, a very minimal memory area is required.

As part of the fit, the determination of the validity of each coefficient may be necessary. Details of the computation of the uncertainties in the members of B can be obtained from references.

One final caveat: The computation of B still requires taking differences between extremely large numbers (such as products of elements of G). Consequently, the use of double-precision variables for matrices R and G may be prudent.

BIBLIOGRAPHY

Polynomial Least-Squares Smoothing

T. R. Edwards, *Anal. Chem.,* **54,** 1519–1524, 2638, 1982.

C. G. Enke, T. A. Nieman, *Anal. Chem.,* **48,** 705A, 1976.

R. L. LaFara, *Computer Methods for Science and Engineering*, Hayden, Rochelle Park, NJ, 1973.

H. H. Madden, *Anal. Chem.,* **50,** 1383, 1978, and references therein.

A. Savitzky and M. J. E. Golay, *Anal. Chem.,* **36,** 1627, 1964.

J. Steinier, Termonia, T., and Deltour, *J. Anal. Chem.,* **44,** 1906, 1972.

Simplex Optimization

S. N. Deming, L. R. Parker, *CRC Crit. Rev. Anal. Chem.,* 187, 1978.

J. A. Nelder, R. Mead, *Computer J.,* **7,** 308, 1965.

Linear Least-Squares

P. R. Bevington, *Data Reduction and Error Analysis for the Physical Sciences*, McGraw-Hill, New York, 1969.

C. L. Lawson, R. J. Hanson, *Solving Least-Squares Problems*, Prentice-Hall, Englewood Cliffs, NJ, 1974.

C. Daniel, and F. S. Wood, *Fitting Equations to Data*, Wiley-Interscience, New York, 1971.

CHAPTER TWELVE

The Overall Task

We have covered many subjects in the previous chapters. The study began with the independent small computer and continued through the components of the interface. In this chapter, we step back and look at the overall process of solving an experimental problem through the development of a laboratory computer application.

The solution begins with an analysis of the problem, namely, the data types, the time scale of the measurements and the time available for real-time processing, the memory requirements, the required precision of the result, and the human interaction factors that will determine the approach to the solution.

Deciding which computer is right for the job and design of the application follows. Consideration also must be given to hardware–software trade-offs, anticipation of future needs, and planning of the user interface.

The application hardware and software must be constructed and rid of errors (debugged). In most cases, full debugging requires a second period of testing by persons not involved thus far.

Documentation must be written if the application is to have a significant lifetime. The importance of fully documented hardware, software, and procedures for their use would be very difficult to overstress.

Finally, we will consider how the entire system can be maintained.

12.1. ANALYSIS OF THE PROBLEM

The analysis of the problem requires answers to two questions.

1. *Can it be done?*

That question should probably be answered both under ideal conditions and in view of available support.

2. *What constraints must be satisfied in order to solve it?*

Taken under ideal conditions, the answer to the first question is invariably "yes," but the answer to the second determines whether it will be worth doing. Collection of the information for that question starts by getting "the big picture" and working down.

12.1.1. The Overall System

The first step is to sketch a block diagram of the technical parts of the entire system. It shows the sources of data, the devices that need to be controlled, and the ways in which the investigator will interact with the system during the experiment?

The designer's block diagram is an analog of an author's outline. It can be carried to varying degrees of detail, but at the first level, the main components of the system are shown: computer, ADC subsystem, transducers, and the experimental system under study. Where single analog or digital signals are shown, single lines are drawn. Where signals are taken in parallel, such as the lines required to transmit the output of a panel meter, a broad arrow or a line with a slash is drawn.

A moderately complex automation example, involving a treadmill-based exercise-physiology experiment, is shown in Figure 12.1. The experimental objective is to measure many physiological parameters while requiring the exercising subject to produce effort at a level that will produce a predetermined increase in heart rate; the effort, assessed by the heart rate, can be controlled by varying the speed of the treadmill. The experiments may test the effect of temperature on physiological performance; consequently the temperature, both of the environment and of the subject, must be continuously monitored.

The instrument in the figure determines the subject's effort (physical work) from the treadmill rate and the degree of incline. Since a multiplexed ADC will be available within the experiment, the incline can be monitored by an analog transducer, a potentiometer (Fig. 8.8). A shaft encoder, even of very low resolution, can precisely monitor the speed, since rate can be determined by precisely timing the *period* between pulses; using an even simpler method, a magnet mounted to the wheel can activate a reed relay (Fig. 8.4) on each revolution. In order to make precision period measurements, the pulses must trigger interrupts; the following completion routine reads a hardware clock.

During the experiment, exhaled air is monitored to determine breath volume, breathing rate, and changes in both O_2 and CO_2 concentrations. The O_2 and CO_2 concentrations are determined by existing electrochemical and infrared spectroscopic instruments. The instruments' recorder outputs can

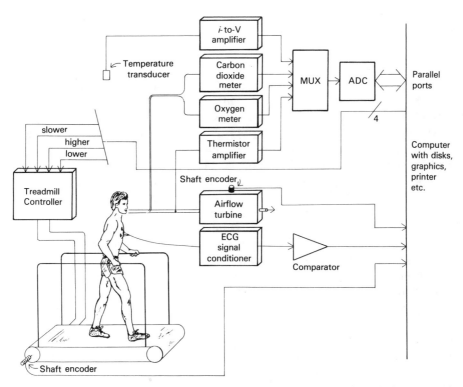

FIGURE 12.1. Block diagram of a laboratory automation project. Signals from several analog and digital sources are acquired, and the treadmill is under the control of the computer.

be monitored using an ADC. The acquisition rate for each must be at least 5–12 Hz, since one group of measurements involves averaging the gas concentration over the period of a breath.

The breathing rate and volume are measured with a turbine flowmeter coupled to a shaft encoder. Real-time processing is required to measure both breathing rate and volume. Therefore, this encoder also triggers interrupts, and during the interrupt completion routine, the processor must determine whether a single breath has ended by timing the period between pulses. A precision timer, with a resolution of 0.01 or less, must be part of the system.

The treadmill speed and incline can be controlled and measured using either mechanical or solid-state relays in parallel with the switches on the treadmill controller's console. It must be made failsafe; if the computer fails while the speed is being increased, the increase should stop.

12.1.2. Specifications for the Interface Subsystem

Having a diagram that shows the signal, the requirements of those signals must be specified. The overall design of the interface is determined by the answers to questions of "How fast?" and "How much?" We have previously shown examples of how the computer both accepts and receives information and how that information may be analog or digital; this discussion will focus on the acquisition of analog information.

Acquisition Rate. The typical experiment entails acquisition of analog data using an ADC and/or acquisition of digital data *via* a parallel interface. The data rate, cost, and precision of the measurement are interrelated factors.

The frequency of data acquisition for a single signal must be matched to that signal under two constraints. We must prevent meaningful high-frequency signals from being aliased by sampling at too low a rate to yield an incorrect signal frequency. In addition, we must prevent the aliasing of higher frequency noise signals from creating an incorrect data component. These problems were illustrated graphically in Chapter Four.

The ADC conversion rate also depends on the need for a multiplexer. Many experiments call for the acquisition of more than one analog signal requiring a multiplexer. The minimum acquisition rate for the ADC subsystem must then be the product of the acquisition rate for a single signal and the number of channels to be sampled by the ADC.

The treadmill experiment to be automated in the example may have previously used a polygraph to amplify and record an array of physiological signals (heart rate, respiration rate, muscle activity, etc.) A typical polygraph has a bandwidth of over 100 Hz, so that relatively short "spikes" (< 10 ms) may be accurately amplified. These spikes probably have no importance in the experiment, and in fact the typical strip-chart recorder filters them out. However, peaks of 200 ms in duration may be important.

Using the rough guidelines for the exercise physiology experiment, each signal must be sampled at about 12 Hz; to acquire five channels of information, our ADC must be capable of a minimum conversion rate of 60 Hz. A filter with a time constant of about 5–30 ms applied to the input signal will aid in reducing noise effects. The 60-Hz conversion rate is compatible with some integrating ADCs although a successive approximation converter might be required.

Rates above 100 Hz require successive approximation ADCs. If the experimental acquisition rate exceeded 20–30 kHz, direct memory access would be required, and if the rate exceeds a megahertz, a flash ADC in a transient digitizer would probably be necessary.

The Acquisition Time Base. Digital data are acquired at particular instants in time, and temporal precision is often as important as the precision of the data. The timing of some experiments is **event-driven**; that is, a datum is acquired after some event has taken place (the next sample is ready, an animal has moved past a detector, a heat pulse has been delivered, etc.). The timing of other experiments may be **time-driven**; the data is acquired at evenly spaced intervals (for example, in the acquisition of a chromatographic peak). In many cases, *both* types of timing is necessary: For example, although the beginning of the acquisition of data from an experiment is event-driven, the acquisition of an array of data is time-driven. In the case of the treadmill, data are acquired in an event-driven sequence, but precise timing between events is required.

When the acquisition is time-driven, the rate places constraints on the timing technique as well as on the ADC. As described in Chapter 4, the ADC can be triggered either by external pulses or by program-generated pulses. A clock or programmable timer in the interface can either send pulses to the ADC or to the interrupt circuitry of the computer; in the latter case, the ADC is triggered as part of the interrupt service routine.

Alternatively, the program can monitor a clock, and when the proper time is detected, the acquisition is triggered. Most small computers have a time-of-day clock which can be read with a resolution of either $\frac{1}{100}$ or $\frac{1}{60}$ s. This clock can be read *via* a program in a wait loop until the proper time has elapsed; the program then generates a trigger pulse. The need for programmable timer hardware is removed, but the method has two disadvantages. First, the fastest rate that can be controlled is between 0.01 and 1 s. Second, there will be both a delay (latency) and an uncertainty in the delay between the recognition of the correct time and the generation of a trigger pulse. Consequently, if the conversion rate is much faster than 1 Hz and the time base must be accurate, a programmable clock must be used to trigger the ADC.

It may be possible to use the time-of-day clock hardware to measure time periods, however. The timer that is part of the computer system can be interrogated after each event to yield precise timing information even though it is not reprogrammed by the user. This would be a suitable technique for measuring rates of encoder movement in the treadmill experiment.

Resolution. Increases in ADC resolution bring increases in price and usually in conversion time. The conversion time of a successive approximation ADC is directly proportional to the number of bits, and the cost increases similarly. Furthermore, if a high-resolution ADC is used, more attention to the signal conditioning stage may be necessitated.

The minimum resolution is determined from the knowledge of two factors: the precision with which the magnitude of the signal must be known and the noise level in the data. If 0.5% precision easily suffices for all foreseen measurements, an 8-bit ADC, good to 1 part in 256, can be specified. This, of course, assumes that the full-scale range of the signal matches the full-scale input of the ADC.

If the signal contains too large a noise component, a moderate resolution (12-bit) ADC will not provide advantage over low resolution, unless the measurement can be made repeatedly and the results averaged. If the noise in an otherwise static signal is *random*, is *low in frequency* relative to the acquisition rate, and *exceeds one quantization level*, the precision of the average will improve with the square root of the number of measurements.

The technique of **dithering**, for increasing the resolution of the measurement of a stationary signal, was described in Chapter Four. It involves addition of noise to the signal and signal averaging; the improvement in the resolution can equal the square root of the number of conversions that are averaged.

Speed–Storage Trade-Offs. In the typical experiment, acquired data must be stored on disk. However, the mechanics of the storage make it a slow step, somewhat difficult to synchronize with the experiment.

When an instruction in the program calls for writing data to the disk, the OS actually places the information in a buffer in the computer's memory until that buffer fills; only then does the physical writing of the data actually take place. The occurrences of disk activity are relatively random, and in many computers this activity cannot take place concurrently with high-speed data acquisition.

Consequently if data acquisition rates are very high, one cannot write to disk *during* the experiment; the data must all be stored in memory which might be a limited commodity.

In analyzing data storage needs, one needs to determine whether a short delay can be accommodated. That delay will be longest with floppy disks, much shorter with hard disks, and even shorter with semiconductor "pseudo-disks" (RAM drives). If the rates are too high, the data must be stored in program memory, each datum requiring 2, 4, or 8 bytes for integers, floating-point numbers, and double-precision numbers, respectively.

Real-Time Data Processing. In a laboratory experimental system, some data processing during the experiment will be necessary. In some cases, a variable will need to be computed which determines the feedback to the experiment; possibly the ratio of two measured values must be computed to determine whether a motor or valve is to be actuated. In other cases, some preliminary data reduction allows several measurements to reduce to a single stored datum thereby reducing storage requirements.

Computations during the acquisition of a burst of data may take computer resources from the acquisition process. This can be true, though less likely, even if an interrupt-driven system and/or a DMA interface are employed making the acquisition and processing tasks appear concurrent. Some idea of the burden on the processor will help to determine maximum data acquisition rates.

The treadmill experiment requires concurrent processing using interrupt completion routines. Timing of breaths and averaging of analog data are performed in the completion routine, while data storage and higher level processing of speeds and parameters to compute work and efficiency take place in the main program. Each of the processes require resources from the computer.

12.1.3. The Needs of the Operator/User

All too often when custom systems are designed and built, operation requires that the user enter what appears to be cryptic messages and set switches and knobs according to a mysterious sequence. If the *only* person to ever use the system is someone who revels in this sort of anachronistic computer operation, savings in development time that accrues from that style are justified. However, the "if" part of that statement can seldom be guaranteed. The system must present a humane user-interface if it is to be used with enough frequency and confidence to justify the development.

The extent to which the operator of such a system must be made to feel comfortable with the computer may determine the style of interaction between program and operator. The profile of the user(s) of the system can determine the attention that must be paid to providing on-line help, error checking, easy error correction, and automatic setup. If the typical operator is an occasional user, help in the form of on-line instruction and automatic double-checking of entries is useful. If the clientele is a small group that uses the system intensively, the help is useful, but provision for bypassing extensive verbiage is mandatory.

Unfortunately, many systems after automating actually take control away from the operator by not placing interactive control in the software. For example, an analog spectrophotometer allows interaction *during* the recording of absorbance as a function of wavelength; the scale or the scan rate may be changed, or the scan may be stopped. Some automated instruments completely lock out the operator until the operation is complete, since that interaction in real time is a difficult programming task.

12.2. CHOOSING A COMPUTER

The illusion of being able to simply lay out all of the facts and make rational decisions regarding computers should be enough to cause us to pursue the possibility. This approach is valuable even in the "real world." However the potentially objective decision maker typically is beset with both emotional and bureaucratic baggage which weight the decision, even in those rare cases when the merits of the options are clearly understood. Persons giving advice are often emotionally attached to a particular type of machine; complicated bureaucratic channels may dictate a suboptimum choice.

The planning information in the previous section provides a basis for generating the bare minimum specifications: The computer must be capable of acquiring data at requisite rates, of performing the real-time processing necessary in the experiment, and of storing the acquired data. As noted in Chapter Two, that computer will usually have an OS, disk storage, a bus, and graphics; these specifications are satisfied by most current personal computers.

After satisfying those requirements, there are many other questions to be answered. The tendency is to look first at the CPU's capabilities: cycle time, memory capacity, instruction set, and word length. However, there are other factors that should play even greater importance.

Software development should be considered the most critical. If available software packages solve the problems to which the computer is to be applied, a rule of thumb is to choose the software and buy the computer that will execute it. For example, if a primary goal is to execute a program for the control of a Fourier-Transform Infrared Spectrophotometer, and sophisticated data analysis is available, the choice of computer should be based on the execution of that software.

Similarly, if one particular member of a research group or a service facility is to prepare the application software, better software will usually be written quicker if the development can build on previous experience. The availability of sound advice for a particular computer is also important. Unfortunately, this also can be a trap; a person having expertise with an outmoded or poorly suited computer can present a significant energy barrier to the acquisition of a more suitable computer.

If these criteria still leave choices, one can look at other factors.

What is the quality of secondary applications software (word processing, graphics and graphing, information management, etc.)? Quite often the computer that was purchased for experimental work will actually be used more with these productivity tools.

What is the availability and quality of the programming languages that the user of the computer or other software writers might use?

What is the quality of the graphics? Compare resolution, color, and ease of use.

What is the selection of ancillary hardware and peripherals? Popular bus-oriented computers stimulate the development of a wide array of special-purpose interface boards, storage devices, graphics tools, etc.; if the computer becomes popular, third-party vendors will make up for many deficiencies in the original machine.

The factors just listed can easily make or break the development of a successful system.

The CPU plays a large part in determining the performance of any computer; if we were to design a computer, the CPU is where we would start. Presently, there are three generations of microprocessors:

1. 8-bit microprocessors, such as Intel 8080/8085s and Zilog Z-80s (computers executing CP/M such as KayPro and CompuPro), and MOS Tech 6502s (Apple II and Commodore-64).

2. 16-bit microprocessors, such as Intel 8088/8086/80286s (IBM PC and

PC/AT, Zenith Z-100/150 and clones), and Motorola 68000s (Macintosh, Masscomp, Apollo, Sun, Charles River).

3. 32-bit microprocessors, such as MicroVax II, Intel 80386, Motorola 68020, National 32332, and Zilog Z80,000.

Computers based on first-generation microprocessors are no longer less expensive to manufacture than later machines; the argument for choosing one would have to be based on compatibility with other available computers or desired software. Limited memory space and instruction sets restrict the range and size of applications, and concurrent operations are difficult to accommodate.

Computers based on second-generation CPUs have mature software, particularly those computers using the Intel 8086 family/MS-DOS OS combinations. Computers using the 68000 are attractive since few argue that the 8086 and relatives are more powerful; it has a richer instruction set, a faster clock speed, and larger memory space. Unfortunately, most computers using the 68000 are ill-suited to data acquisition and experiment control because of large OS overhead.

We now await the third-generation computers. A computer with the most powerful available architecture and CPU is always attractive, but a risk must be assessed—a computer can become an "orphan" if the manufacturer can no longer support it. A single orphan computer is a particular problem if it should fail (see Section 12.4); having two available diminishes the risk.

A good choice of computer for the treadmill experiment might be an IBM PC/AT with an Intel 80286 CPU, although it is not the most powerful small computer available. It has concurrent processing capability necessary for the handling of interrupts from the experiment, a large memory space, and a hefty amount of software.

12.3. THE DEVELOPMENT PROCESS

With the specifications of the project, one is prepared to develop the design. Both hardware and software must be designed; the hardware and software must both be constructed according to the design; the errors must be removed; the design should be tested after which some redesign is typically required; finally the subsystem must be documented.

12.3.1. Designing the System

The design process must take into account two general criteria. The first is obvious: The final system must work to specification; if the specifications cannot be met, the limitations must be known before construction proceeds. The second is less obvious and more difficult, the development process should be efficient.

The experimental systems dealt with in this book are similar to *prototypes*, and functionality supersedes appearance to a greater extent than it does in a commercial product. Usually no plans need to be made for mass production. Consequently, the temptation continually arises to program without design and to construct without documentation. Such approaches are inevitably short-sighted; frequently the development process will be slowed, and few projects will not require later changes due to differences in experimental objectives.

There are, however, important differences between the development of a preproduction prototype and a one-of-a-kind custom system. The engineer developing a commercial product is rewarded for reducing the size of a circuit board or the amount of software stored in the system's ROM. The time invested is ideally amortized over cost-savings in a large production run. For the custom system, there is no similar way to recoup the costs of development, and when possible, one should take advantage of time-savings which can result from additional or more expensive hardware.

Usually the design is carried out in two closely related components: hardware and software. Often in custom laboratory systems, the hardware will be designed first, taking careful account of the specifications. The software may then be considered the component that "makes the hardware go."

Given the importance of assessing hardware/software tradeoffs, it may be just as important to see the hardware as the component of the interface that "makes the software go." If the project involves a design team, they must be able to coordinate their designs. In any case, we will apply again the tradeoff axiom introduced earlier, "never do in hardware what can be done in software."

Hardware Design. The hardware component of the interface should have three initial characteristics. It should be **minimal, flexible,** and **modular.** It should, in many occasions, be made **failsafe.**

That the hardware is minimal and flexible is guaranteed by the tradeoff axiom repeated earlier. Circuitry for such functions as linearization of nonlinear responses should be forsaken unless the function cannot be properly handled in software. The consequence is that flexibility is gained; in this example, the linearization model can be easily revised.

Another part of designing for flexibility is to add as many functions as can be conveniently included. For example, if a function in the instrument can be actuated with a single TTL signal, but the output port in the interface has several bits available, make them all accessible even though no immediate application is envisioned. Later, the system may be enclosed or otherwise becomes less accessible, and, should the software designer wish to use such a function, the effort of modifying the hardware will not inhibit that proposal.

Modularity begets flexibility. On the one hand, the cost of placing the entire interface on a single circuit board is lower than the cost of separate boards and connectors. On the other hand, if the bus interface and possibly the ADC/DAC subsystems are on one board while the application-oriented signal modifying circuitry are separate, adaptation to a new or additional application will be simplified.

A second benefit of modularity is improved debugging. In this case, the modularity need not necessarily be physical. Even when all subsystems are on a single board, if they can be tested as units with discrete inputs and outputs, errors can be quickly isolated, and design errors are less probable.

The failsafe functionality should be checked in all interfaces. One must consider whether there exists the possibility that a programming error or computer malfunction could enable devices in a way that will cause catastrophic failure. For example, both positive and negative current might be accidentally switched to a dc motor causing a direct low-resistance current path and catastrophic failure of switching transistors, power supply, and more. Driving a lamp with high current when the fan failed could destroy the lamp and maybe the housing. Another example might be a motor that drives a leadscrew; if a software failure caused the motor to drive to the physical limit of the movement, there could be damage to the motor or apparatus. Failure to govern the speed control circuitry, even in case of computer failure, could expose the subject to dangerously fast speeds. Interlocks, thermostats, and limit switches should be incorporated which would disable power if the software fails.

Software Design. Most software projects encompassed by the discussions in this book are relatively small, certainly when compared with large commercial application packages (word processors, compilers, NMR instrumentation, etc.). Often they are undertaken by single programmers. Consequently we are tempted to write software to support scientific experimentation without design.

However, some form of written advance design is valuable for all lab automation projects. Planning helps bring order and efficiency to the development and execution of the program.

At this point, we need not specify how the software design should be committed to paper. The standard IBM flow chart is no longer in great favor. An alternative is to show program loops on a printed page by indenting each step within a loop several spaces from the lines which limit the loop (see Fig. 3.3); loops within loops are then easily identified.

One concrete suggestion can be made, and that is to make the software *modular*. By software modularity, we mean that each function in the program should be a separate subroutine with distinct point of entry and exit.

Modular software has at least three advantages. First, modular programs are more easily debugged. Second, there will be fewer errors in the first place since the programmer will be writing a module at a time, only worrying about the information entering and leaving that module. Finally, the modules will then be portable; the program can be built from functions written for a previous program.

12.3.2. Programming and Debugging

Once the design is complete, the preparation of the program can begin. Computer trade journals frequently question whether programming is an art, skill, or science. Only part of the answer is at all clear; as we will see, debugging can be treated as a science.

The software project will go through several phases. The first is to write and debug a program that is *workable*. The second is to make it *tolerant*. The third might be to make it *humane*.

Preparing a Program that Works. Having the software design in hand, it is only necessary to write the code, and one absolute rule should be restated from Chapter Three: **Comments should be included as though someone else would have to understand the program in one reading.** It is easy for a programmer to assume that the code is understandable and that he or she will remember why a particular algorithm was used, but unwritten explanation is surely lost after a few days. Second, wherever possible **use variable names that indicate the function of the variable.**

Ridding the program of "bugs" can be treated as a science rather than a trial-and-error travail. This is the stage at which the programmer is grateful that the program is modular. The program can now be checked by testing small modules one at a time.

When a malfunction in a module occurs, the science truly takes over. We attempt to isolate the problem, continually breaking it into smaller pieces, by using the the scientific method. Perturbations in the execution of the program must be made which will test a hypothesis; by changing either part of the code or the input information, it should be possible to determine whether the hypothesis is true. Without the technique of testing hypotheses, the average programmer will grope around, blindly making changes until the problem is found, a very inefficient path. The scientist's training, on the other hand, probably creates demand for computer programmers with laboratory science background.

Making the Program Tolerant. If the program is to be tolerant, it must not fail no matter what information it receives, whether from the keyboard or from the experiment. A programming error or unanticipated problem can generate misleading results or loss of data from a crucial experiment. One example might be a program that would "crash" if there is insufficient space on a disk to store the data that was laboriously collected from an experiment. Another would be a program that would accept experimental parameters which, taken in combination, would be nonsensical.

This step almost always requires testing by another person. Typically, the programmer will not think to provide a "nonsense" answer to a prompt on the screen that he or she wrote, and second-party (*beta* site) testing is necessary. This person must be ready and willing to report all problems and failures. (Successful development requires a programmer who graciously accepts all these reports as well.)

Making the Program Humane. In the personal computer industry, this feature is usually called "user-friendliness." The extent to which the program must be humane depends on the system's users. If the program is to become commercial, it is likely that as much as half of the effort will go into this task.

Some of the following features might be considered:

Should the computer assist the user whenever help is desired?

Does the user need to type, or can the user control the experiment by "pointing"?

Must the user memorize any of the operating commands?

The nature of the "user interface" is the subject of a great deal of debate in the software industry. Seldom has the debate considered the needs of the laboratory user, however, who is not only sitting at the computer console but spends much of the time interacting with other components of the experiment; when are "function keys" or a mouse required, and are they as much a help to an operator on his or her feet as to a sitting operator? A greater emphasis on speech I/O and on graphics may be necessary.

12.3.3. Documentation

Complete documentation of the system hardware and software must be written for three reasons:

1. A complete record is needed for the time when the developer wishes to make changes or improvements to the system.
2. A complete description is needed for additional users of the system.
3. This record makes portability to other laboratories possible.

The document will normally include an operating description for users, the operating manual that explains how to set up the system, and how to use it. A prose technical description of the hardware should accompany schematics; it is useful to include documentation on unusual integrated circuits. A prose description of the software also should accompany the listings of the software. The necessity for including comments in the software listing should be reemphasized; some professional software development houses set up rules requiring as much as a comment for every line of code.

12.4. MAINTAINING THE SYSTEM

The automation project is now "complete"; it appears to operate in every desired respect. The next objective is to keep it working, even through changes in operators and experimental designs.

Much of the maintenance work has already been done. The design of the hardware and software was based on modular organization. Both hardware and software were fully documented, with copies in more than one place.

Maintenance of the hardware includes two parts, updates and repairs. Updates differ from the development process described earlier only in that documenting the update can be more difficult. Software updates can be identified by the time–date stamp in the disk directory, but must contain the name of the programmer, the date of the update, and the nature of the modification in the initial comments; persons who fail to set time and date on their computer also should accept this comment as yet another warning. Hardware updates must be recorded in a master file. These procedures should not be bypassed *even if the same person is the designer, programmer, and user.*

When does a small computer fail? Fortunately, it does not happen often. In older computers, most failures could be traced to three sources: power supplies, connectors and cables, and storage devices. If the computer is totally dead, check the various power supplies (typically 5, 12, −12 V). Whether dead or showing intermittent failure, reseat the boards, and, if the ICs are in sockets, reseat them by pushing them in again. Check the interface cable and the connectors on each end for wear or damage. Storage devices wear out quicker than solid-state devices because they have moving parts.

Protection against damaging voltage spikes and surges, mostly caused by electrical storms and nearby heavy electrical equipment (such as elevator motors), is available from most vendors of computer accessories; the need depends on the environment. The computer is no more expensive than other essential instrumentation in the laboratory; if the instrumentation does not require protection, the computer system probably does not either. Most computer power supplies already offer some protection. However, a disaster need strike only once to prove one's judgment wrong.

Most modern computers are reasonably well protected against electromagnetic interference (EMI) in the power lines; if not, a line conditioner or isolation transformer may help.

If blackouts of more than a few milliseconds are common, and the experience of having the computer halt or rest is unacceptable, an Uninterruptible Power Supply (UPS) may be required; unfortunately, they are quite expensive.

Finally, after checking those components, there may still be a problem. A few additional hypotheses might be tested. If the error is intermittent, a connection problem most likely exists, even in a trace on the printed circuit board. If the problem occurs only after the computer warms up, there may be thermal failure in a connector or IC; electronics shops sell aerosol cans of coolant which can be directed at single ICs to find the one that "gets well" upon cooling.

After that process, identification of the problem depends upon having an identical system available. Nearly all computer repair persons pinpoint the board by swapping boards until the problem board is identified; they will then exchange the problem board and send the faulty one back to the factory. Another possibility is to exchange the ICs, possibly a half-dozen at a time from the problem board into a known good board, until the problem appears in the good board. If the problem board is not repaired by a transfusion of chips, the failure may be in a passive component (a resistor burned open or a shorted capacitor) or in the printed circuit board (and many of these have several layers of interconnections, sandwiched together); an electronics technician or a board swap may be necessary.

12.5. FINAL CONSIDERATIONS

The project of automating an experiment through an interface to a computer is never as simple as connecting a few wires. Software development, testing, *beta*-site testing, and documentation usually consume more time than the construction of the interface.

Fortunately, the automation tools are improving and becoming more readily available: better computers, interface hardware, software, and word processors.

Woody Allen may not have been talking about laboratory automation, but his statement still may apply:

> Summing up, it is clear that the future holds great opportunities. It also holds pitfalls. The trick will be to avoid the pitfalls, seize the opportunities, and get back home by six o'clock.

APPENDIX ONE

ASCII Code

ASCII (American Standard Code for Information Exchange)[a]

Dec	Hex	Char		Description	Dec	Hex	Char	Dec	Hex	Char	Dec	Hex	Char
0	0	^@	NUL	Null	32	20		64	40	@	96	60	`
1	1	^A	SOH	Start of heading	33	21	!	65	41	A	97	61	a
2	2	^B	STX	Start of text	34	22	"	66	42	B	98	62	b
3	3	^C	ETX	End of text	35	23	#	67	43	C	99	63	c
4	4	^D	EOT	End of transm'n	36	24	$	68	44	D	100	64	d
5	5	^E	ENQ	Enquiry	37	25	%	69	45	E	101	65	e
6	6	^F	ACK	Acknowledge	38	26	&	70	46	F	102	66	f
7	7	^G	BEL	Bell	39	27	'	71	47	G	103	67	g
8	8	^H	BS	Backspace	40	28	(	72	48	H	104	68	h
9	9	^I	HT	Tab	41	29	)	73	49	I	105	69	i
10	A	^J	LF	Linefeed	42	2A	*	74	4A	J	106	6A	j
11	B	^K	VT	Vertical tab	43	2B	+	75	4B	K	107	6B	k
12	C	^L	FF	Form feed	44	2C	,	76	4C	L	108	6C	l
13	D	^M	CR	Carriage return	45	2D	-	77	4D	M	109	6D	m
14	E	^N	SO	Shift out	46	2E	.	78	4E	N	110	6E	n
15	F	^O	SI	Shift in	47	2F	/	79	4F	O	111	6F	o
16	10	^P	DLE	Data link escape	48	30	0	80	50	P	112	70	p
17	11	^Q	DC1	Device control 1	49	31	1	81	51	Q	113	71	q
18	12	^R	DC2	Device control 2	50	32	2	82	52	R	114	72	r
19	13	^S	DC3	Device control 3	51	33	3	83	53	S	115	73	s
20	14	^T	DC4	Device control 4	52	34	4	84	54	T	116	74	t
21	15	^U	NAK	Negative ack	53	35	5	85	55	U	117	75	u
22	16	^V	SYN	Synchr. idle	54	36	6	86	56	V	118	76	v
23	17	^W	ETB	Block	55	37	7	87	57	W	119	77	w
24	18	^X	CAN	Cancel	56	38	8	88	58	X	120	78	x
25	19	^Y	EM	End of medium	57	39	9	89	59	Y	121	79	y
26	1A	^Z	SUB	Substitute	58	3A	:	90	5A	Z	122	7A	z
27	1B		ESC	Escape	59	3B	;	91	5B	[	123	7B	{
28	1C		FS	File separator	60	3C	<	92	5C	\	124	7C	|
29	1D		GS	Group separator	61	3D	=	93	5D	]	125	7D	}
30	1E		RS	Record separator	62	3E	>	94	5E	^	126	7E	~
31	1F		US	Unit separator	63	3F	?	95	5F	_	127	7F	DEL

[a] Dec = decimal value; Hex = hexadecimal value; Char = printed character if the decimal value is greater than 31, the control character if less than 27. For nonprinting characters, the next column lists the ASCII name and a description; however, many of the descriptions do not describe functions common for small computers.

Polynomial Convolution Integers

Smoothing Convolution Integers

Quadratic/cubic function

Convolution Width:	29	25	21	17	13	9	7
Point Number							
0	629	467	329	43	25	59	17
1	624	462	324	42	24	54	12
2	609	447	309	39	21	39	-3
3	584	422	284	34	16	14	
4	549	387	249	27	9	-21	
5	504	342	204	18	0		
6	449	287	149	7	-11		
7	384	222	84	-6			
8	309	147	9	-21			
9	224	62	-76				
10	129	-33	-171				
11	24	-138					
12	-91	-253					
13	-216						
14	-351						
Norm:	8091	5175	3059	323	143	231	35

Quartic/quintic function

Convolution Width:	29	25	21	17	13	9
Point Number						
0	54251	4253	44003	883	677	179
1	53040	4125	42120	825	600	135
2	49470	3750	36660	660	390	30
3	43730	3155	28190	415	110	-55
4	36135	2385	17655	135	-135	15
5	27126	1503	6378	-117	-198	
6	17270	590	-3940	-260	110	
7	7260	-255	-11220	-195		
8	-2085	-915	-13005	195		
9	-9820	-1255	-6460			
10	-14874	-1122	11628			
11	-16050	-345				
12	-12025	1265				
13	-1350					
14	17550					
Norm:	445005	30015	260015	4199	2431	429

Note that these convolution integers are symmetrical; points numbered -1 to -14 have the same values as those numbered 1 to 14.

First Derivative Convolution Integers

Quadratic Function

Convolution Width: Point Number	29	25	21	17	13	9	7
0	0	0	0	0	0	0	0
1	1	1	1	1	1	1	1
2	2	2	2	2	2	2	2
3	3	3	3	3	3	3	
4	4	4	4	4	4	4	
5	5	5	5	5	5		
6	6	6	6	6	6		
7	7	7	7	7			
8	8	8	8	8			
9	9	9	9				
10	*10	10	10				
11	11	11					
12	12	12					
13	13						
14	14						
Norm:	2030	1300	770	408	182	60	10

Cubic/quartic Function

Convolution Width: Point Number	29	25	21	17	13	9
0	0	0	0	0	0	0
1	108992	8558	29592	358	832	126
2	213581	16649	56881	673	1489	193
3	309364	23806	79564	902	1796	142
4	391938	29562	95338	1002	1578	-86
5	456900	33450	101900	930	660	
6	499847	35003	96947	643	-1133	
7	516376	33754	78176	98		
8	502084	29236	43284	-748		
9	452568	20982	-10032			
10	363425	8525	-84075			
11	230252	-8602				
12	48646	-30866				
13	-185796					
14	-477477					
Norm:	35341488	1776060	3634092	23256	24024	1188

Note that these convolution integers have an inverted symmetry; points numbered -1 to -14 are the negative of those numbered from 1 to 14.

Second Derivative Convolution Integers

Quadratic/cubic function

Convolution Width:	29	25	21	17	13	9	7
Point Number							
0	-70	-52	-110	-24	-14	-20	-2
1	-69	-51	-107	-23	-13	-17	-1
2	-66	-48	-98	-20	-10	-8	2
3	-61	-43	-83	-15	-5	7	
4	-54	-36	-62	-8	2	28	
5	-45	-27	-35	1	11		
6	-34	-16	-2	12	22		
7	-21	-3	37	25			
8	-6	12	82	40			
9	11	29	133				
10	30	48	190				
11	51	69					
12	74	92					
13	99						
14	126						
Norm:	56637	26910	33649	3876	1001	462	7

Quartic/quintic function

Convolution Width:	29	25	21	17	13	9
Point Number						
0	-101192	-147290	-14322	-2820	-3780	-370
1	-97136	-139337	-13224	-2489	-3016	-211
2	-85219	-116143	-10061	-1557	-971	151
3	-66194	-79703	-5226	-207	1614	371
4	-41316	-33342	626	1256	3504	-126
5	-12342	18285	6578	2405	2970	
6	18469	69193	11451	2691	-2211	
7	48356	112067	13804	1443		
8	74056	138262	11934	-2132		
9	91804	137803	3876			
10	97333	99385	-12597			
11	85874	10373				
12	52156	-143198				
13	-9594					
14	-105651					
Norm:	18512208	17168580	980628	100776	58344	1716

Note that the convolution integers are symmetrical; points numbered -1 to -14 have the same values as those numbered 1 to 14.

Interpolation Convolution Integers

Quadratic/cubic function

Convolution Width:	28	24	20	16	12	8	4
Point Number							
1	117	429	297	189	21	9	9
3	115	419	287	179	19	7	-1
5	111	399	267	159	15	3	
7	105	369	237	129	9	-3	
9	97	329	197	89	1		
11	87	279	147	39	-9		
13	75	219	87	-21			
15	61	149	17	-91			
17	45	69	-63				
19	27	-21	-153				
21	7	-121					
23	-15	-231					
25	-39						
27	-65						
Norm:	1456	4576	2640	1344	112	32	16

Quartic/quintic function

Convolution Width:	28	24	20	16	12	8
Point Number						
1	23400	10725	5940	675	300	225
3	22280	10025	5380	575	220	85
5	20103	8679	4323	393	87	-69
7	16995	6795	2895	165	-45	15
9	13145	4535	1285	-55	-95	
11	8805	2115	-255	-195	45	
13	4290	-195	-1410	-165		
15	-22	-2071	-1802	143		
17	-3690	-3135	-990			
19	-6210	-2955	1530			
21	-7015	-1045				
23	-5475	3135				
25	-897					
27	7475					
Norm:	186368	73216	33792	3072	1024	512

Note that these convolution integers are symmetrical; points numbered -1 to -27 have the same values as those numbered 1 to 27.

Integration Convolution Integers

Quadratic/cubic function

Convolution Width:	28	24	20	16	12	8	4
Point Number							
1	45773	73931	59169	4809	4487	281	13
3	44991	72209	57179	4555	4061	219	-1
5	43427	68765	53199	4047	3209	95	
7	41081	63599	47229	3285	1931	-91	
9	37953	56711	39269	2269	227		
11	34043	48101	29319	999	-1903		
13	29351	37769	17379	-525			
15	23877	25715	3449	-2303			
17	17621	11939	-12471				
19	10583	-3559	-30381				
21	2763	-20779					
23	-5839	-39721					
25	-15223						
27	-25389						
Norm:	570024	789360	526680	34272	24024	1008	24

Quartic/quintic function

Convolution Width:	28	24	20	16	12	8
Point Number						
1	1771422952	20756593	4847964	5281539	238628	13559
3	1686772680	19404813	4392364	4502999	175572	5307
5	1522230435	16805435	3532349	3085753	70657	-3843
7	1287312815	13166823	2370289	1309469	-33723	817
9	996294717	8801523	1059739	-406351	-73977	
11	668209337	4126263	-194561	-1502371	34683	
13	326848170	-338047	-1136686	-1279421		
15	761010	-3966315	-1459526	1101503		
17	-276744050	-6029267	-804786			
19	-467600618	-5693447	1237014			
21	-528984003	-2021217				
23	-413311215	6029243				
25	-68240965					
27	563326335					
Norm:	14136595200	142084800	27688320	24186240	823680	31680

Note that these convolution integers are symmetrical; points numbered -1 to -27 have the same values as those numbered 1 to 27. Also note that, in order to produce an integrated value around zero, consecutive data take odd Point Numbers.

Two-Dimensional Smoothing Convolution Integers

<u>Quadratic/cubic function</u>

Convolution area: 3 by 3
x Point Number:	0	1
z Point Number		
0	5	2
1	2	-1

Norm: 9

Convolution area: 3 by 7
x Point Number:	0	1	2	3
z Point Number				
0	13	12	9	4
1	4	3	0	-5

Norm: 63

Convolution area: 3 by 11
x Point Number:	0	1	2	3	4	5
z Point Number						
0	167	162	147	122	87	42
1	50	45	30	5	-30	-75

Norm: 1287

Convolution area: 3 by 15
x Point Number:	0	1	2	3	4	5	6	7
z Point Number								
0	943	928	883	808	703	568	403	208
1	280	265	220	145	40	-95	-260	-455

Norm: 9945

Convolution area: 7 by 7
x Point Number:	0	1	2	3
z Point Number				
0	11	10	7	2
1	10	9	6	1
2	7	6	3	-2
3	2	1	-2	-7

Norm: 147

Convolution area: 7 by 11
x Point Number:	0	1	2	3	4	5
z Point Number						
0	141	136	121	96	61	16
1	128	123	108	83	48	3
2	89	84	69	44	9	-36
3	24	19	4	-21	-56	-101

Norm: 3003

Convolution area: 7 by 15

x Point Number:	0	1	2	3	4	5	.6	7
z Point Number								
0	2387	2342	2207	1982	1667	1262	767	182
1	2166	2121	1986	1761	1446	1041	546	-39
2	1503	1458	1323	1098	783	378	-117	-702
3	398	353	218	-7	-322	-727	-1222	-1807

Norm: 69615

Convolution area: 11 by 11

x Point Number:	0	1	2	3	4	5
z Point Number						
0	139	134	119	94	59	14
1	134	129	114	89	54	9
2	119	114	99	74	39	-6
3	94	89	74	49	14	-31
4	59	54	39	14	-21	-66
5	14	9	-6	-31	-66	-111

Norm: 4719

Convolution area: 11 by 15

x Point Number:	0	1	2	3	4	5	6	7
z Point Number								
0	2353	2308	2173	1948	1633	1228	733	148
1	2268	2223	2088	1863	1548	1143	648	63
2	2013	1968	1833	1608	1293	888	393	-192
3	1588	1543	1408	1183	868	463	-32	-617
4	993	948	813	588	273	-132	-627	-1212
5	228	183	48	-177	-492	-897	-1392	-1977

Norm: 109395

Convolution area: 15 by 15

x Point Number:	0	1	2	3	4	5	6	7
z Point Number								
0	781	766	721	646	541	406	241	46
1	766	751	706	631	526	391	226	31
2	721	706	661	586	481	346	181	-14
3	646	631	586	511	406	271	106	-89
4	541	526	481	406	301	166	1	-194
5	406	391	346	271	166	31	-134	-329
6	241	226	181	106	1	-134	-299	-494
7	46	31	-14	-89	-194	-329	-494	-689

Norm: 49725

Note that these convolution integers are four-fold symmetrical; points numbered -1 to -7 have the same values as those numbered 1 to 7 in both the x and z directions.

Index